Spring Boot（2.3.x）开发技术指南

吴佳华　编著

中国原子能出版社

图书在版编目（CIP）数据

Spring Boot（2.3.x）开发技术指南 / 吴佳华编著.
-- 北京：中国原子能出版社，2021.10（2023.4重印）
ISBN 978-7-5221-1644-0

Ⅰ.①S… Ⅱ.①吴… Ⅲ.①JAVA 语言－程序设计－指南 Ⅳ.①TP312.8-62

中国版本图书馆 CIP 数据核字(2021)第 210985 号

Spring Boot（2.3.x）开发技术指南

出版发行	中国原子能出版社（北京市海淀区阜成路 43 号　100048）
责任编辑	张书玉
责任印制	赵　明
印　　刷	河北文盛印刷有限公司
经　　销	全国新华书店
开　　本	889 mm×1194 mm　1/16
印　　张	22　　　　字　数　549 千字
版　　次	2021 年 10 月第 1 版　2023 年 4 月第 2 次印刷
书　　号	ISBN 978-7-5221-1644-0　　　　　定　价　98.00 元

网址：http://www.aep.com.cn　　　　E-mail：atomep123@126.com
发行电话：010-68452845　　　　　　版权所有　侵权必究

前　言

本书以 Spring Boot（2.3.1.RELEASE）应用开发为主线，讲解其基础与核心，以及各相关技术知识点和实际应用。本书共分 4 部分：初级篇主要介绍 Spring Boot 入门知识、简单示例和 IDEA 插件；中级篇主要讲解 Spring MVC、WebFlux、各类数据库、Spring 安全、定时任务和 Excel 文件处理等相关开发技术；高级篇主要介绍 Kafka、应用监控、Spider 和搜索引擎等相关应用开发；附录主要包括参考和推荐资料。

本书涵盖的技术知识比较多，故不可能做到面面俱到，因此大都给出一个简单开始的起点，以期起到抛砖引玉的作用。关于每项技术及其各个版本在网络上都有相应的官方文档，以及与技术相关的各式各样的博客和视频资源。希望读者将这些学习资源与本书中的知识相结合，对技术有更深入更全面的理解和掌握，并最终能有所应用。

本书作为 Spring Boot 开发入门和进阶书籍，希望为读者拓宽知识面，也为从事 Spring Boot 开发的专业技术人员提供借鉴和参考。由于时间仓促，书中难免有所疏漏与错误，恳请读者朋友批评指正。

若读者看完本书能有所增进，就是对作者最大的奖励。摘录了几句话，大家共勉：
（1）软件在能够复用前必须先能用。
（2）优秀的代码是它自己最好的文档。当你考虑要添加一个注释时，问问自己"如何能改进这段代码，以让它不需要注释"。
（3）往往会有一些隐含的需求没有明确提出来。如果软件只满足那些精确定义了的需求而没有满足这些隐含的需求，软件质量也得不到保证。

Spring Boot 项目的编程环境主要采用的开源软件和插件，如下表所示。

Spring Boot 编程环境配置表

软件名称	版本
JDK	jdk-8u***-windows-x64
IDE 开发工具	IntelliJ IDEA：ideaIC-2020.1.3.win
IntelliJ IDEA 插件	SonarLint、阿里 Java 规范插件和 Lombok
Maven	apache-maven-3.6.3
数据库	mariadb-10.5.4-winx64
服务器操作系统	CentOS-8.2.2004-x86_64-minimal
Spring Boot	2.3.1.RELEASE

目 录

初级篇 1
 第一章 Spring Boot 介绍 2
 第二章 简单示例 3
  2.1 静态 HTML 示例 6
  2.2 动态 HTML 示例 7
  2.3 字符串示例 8
  2.4 JSON 示例 9
  2.5 JAR 示例 10
  2.6 本章小结 13
 第三章 Spring Boot 基础 14
  3.1 Spring 注解 14
  3.2 Spring Boot 注解 18
  3.3 Starters 23
  3.4 构建代码 24
  3.5 配置类 25
  3.6 自动配置 25
  3.7 Spring Bean 和依赖注入 26
  3.8 本章小结 27
 第四章 SpringApplication 类 28
  4.1 延迟初始化（懒加载） 28
  4.2 定制横幅 29
  4.3 应用程序可用性 30
  4.4 应用程序事件和监听器 31
  4.5 Web 环境 31
  4.6 访问应用程序参数 32
  4.7 ApplicationRunner 和 CommandLineRunner 32
  4.8 应用程序退出 33
  4.9 管理员功能 34
  4.10 本章小结 34
 第五章 外部化配置 35
  5.1 配置随机值 35
  5.2 应用程序属性文件 36
  5.3 YAML 文件 37
  5.4 Profiles 38
  5.5 国际化 38
  5.6 本章小结 40

第六章		公共模块	41
	6.1	公共类	41
	6.2	热加载	45
	6.3	IDEA 插件	48
	6.4	本章小结	54

中级篇 55

第七章		Spring MVC	56
	7.1	DispatcherServlet	57
	7.2	上下文层次结构	58
	7.3	拦截器	58
	7.4	文件上传	61
	7.5	静态资源	63
	7.6	全局异常处理	65
	7.7	跨域资源共享	71
	7.8	过滤器	74
	7.9	函数式端点	77
	7.10	异步请求	80
	7.11	HTTP/2	80
	7.12	验证	81
	7.13	本章小结	85
第八章		Spring WebFlux	86
	8.1	为什么创建 Spring WebFlux	86
	8.2	选择 Spring MVC 还是 WebFlux	86
	8.3	简单示例	87
	8.4	WebClient	89
	8.5	RSocket	92
	8.6	反应式库	93
	8.7	本章小结	94
第九章		单元测试和集成测试	95
	9.1	单元测试	95
	9.2	测试 Spring 应用程序	97
	9.3	测试 Spring Boot 应用程序	98
	9.4	测试工具	100
	9.5	本章小结	103
第十章		日志	104
	10.1	日志格式	104
	10.2	文件输出	104
	10.3	日志组	105
	10.4	Logback	105

	10.5	本章小结	108
第十一章		分布式 ID	109
	11.1	UUID	109
	11.2	数据库自增 ID	109
	11.3	Leaf 号段方案	110
	11.4	Leaf 雪花方案	111
	11.5	雪花算法	111
	11.6	其他方案	112
	11.7	本章小结	112
第十二章		数据库 MariaDB	113
	12.1	JDBC	113
	12.2	连接池	119
	12.3	JPA	121
	12.4	事务	125
	12.5	分布式事务	126
	12.6	本章小结	127
第十三章		Redis	128
	13.1	搭建	128
	13.2	进一步配置	129
	13.3	搭建集群	132
	13.4	Redis 示例	135
	13.5	UI 客户端	138
	13.6	Jedis	139
	13.7	Redisson	140
	13.8	本章小结	142
第十四章		MongoDB	143
	14.1	在线安装	143
	14.2	启动 MongoDB	145
	14.3	Shell 客户端	145
	14.4	本地测试用局域网访问	146
	14.5	离线安装	147
	14.6	Compass 客户端	149
	14.7	MongoDB 示例	152
	14.8	本章小结	154
第十五章		Cassandra	155
	15.1	下载	155
	15.2	安装	156
	15.3	集群搭建	159
	15.4	Cassandra 示例	160

| 15.5 | 本章小结 | 162 |

第十六章 MyBatis Plus ... 163
- 16.1 功能特性 ... 163
- 16.2 简单示例 ... 164
- 16.3 分页 ... 166
- 16.4 单表操作 ... 170
- 16.5 多表操作 ... 172
- 16.6 代码自动生成 ... 173
- 16.7 本章小结 ... 181

第十七章 Spring 安全 ... 182
- 17.1 简单示例 ... 183
- 17.2 @EnableGlobalMethodSecurity ... 186
- 17.3 数据库支持 ... 189
- 17.4 "记住我"身份验证 ... 191
- 17.5 OAuth 2 ... 197
- 17.6 Apache Shiro ... 206
- 17.7 本章小结 ... 215

第十八章 缓存 ... 216
- 18.1 @Cacheable ... 216
- 18.2 Redis 缓存 ... 220
- 18.3 本章小结 ... 221

第十九章 作业调度 ... 222
- 19.1 @Scheduled ... 222
- 19.2 Quartz ... 223
- 19.3 Quartz 示例 ... 223
- 19.4 其他 ... 225
- 19.5 本章小结 ... 226

第二十章 其他技术 ... 227
- 20.1 验证码 ... 227
- 20.2 二维码 ... 230
- 20.3 Swagger ... 232
- 20.4 Excel 处理 ... 237
- 20.5 本章小结 ... 246

高级篇 ... 247

第二十一章 Apache Kafka ... 248
- 21.1 下载 ... 248
- 21.2 单节点安装 ... 250
- 21.3 集群搭建 ... 252
- 21.4 Kafka 安全 ... 254

v

- 21.5 简单示例 ... 255
- 21.6 本章小结 ... 258

第二十二章 应用监控：Actuator ... 259
- 22.1 端点 ... 259
- 22.2 自定义端点 ... 260
- 22.3 应用程序信息 ... 262
- 22.4 度量 ... 264
- 22.5 HTTP 跟踪 ... 269
- 22.6 本章小结 ... 270

第二十三章 网络爬虫 ... 271
- 23.1 Apache Solr ... 271
- 23.2 Apache Nutch ... 273
- 23.3 Spider Flow ... 278
- 23.4 本章小结 ... 283

第二十四章 Elasticsearch ... 284
- 24.1 简介 ... 284
- 24.2 发展历程 ... 284
- 24.3 基本概念 ... 284
- 24.4 下载页面 ... 292
- 24.5 Win 10：Elasticsearch 安装 ... 294
- 24.6 Win 10：Kibana 安装 ... 296
- 24.7 Win 10：Elasticsearch 集群 ... 301
- 24.8 CentOS 8：Java 11 安装 ... 302
- 24.9 CentOS 8：Elasticsearch 集群 ... 304
- 24.10 CentOS 8：Kibana 安装 ... 309
- 24.11 CentOS 8：Elasticsearch 安全 ... 312
- 24.12 CentOS 8：Kibana 安全 ... 315
- 24.13 中文分词 ... 317
- 24.14 简单示例 ... 319
- 24.15 本章小结 ... 323

第二十五章 Apache ZooKeeper ... 324
- 25.1 下载 ... 324
- 25.2 单节点安装 ... 325
- 25.3 集群搭建 ... 328
- 25.4 ZooKeeper UI ... 330
- 25.5 配置中心 ... 335
- 25.6 本章小结 ... 338

附录 ... 339
- 附录 A 参考 ... 339

	A.1	Spring Boot 官方	339
	A.2	开发环境中的软件	339
附录 B		推荐	341
	B.1	Spring	341
	B.2	其他	341

第一部分：初级篇

本篇主要讲解 Spring Boot 的入门知识、注解、简单示例和 IDEA 插件等知识，以期开发者能够快速入门 Spring Boot。第一部分要达到的目标：让开发者尽快上手操作 Spring Boot，学习一些基础技术，以便快速形成有关 Spring Boot 的良好基础。

第一章 Spring Boot 介绍

Pivotal 团队在 2014 年 4 月 1 日发布了 Spring Boot 的 v1.0.0.RELEASE 版本，它是全新开源的轻量级框架，基于 Spring Framework 设计的。它不仅继承了 Spring Framework 的优秀特性，而且还通过简化配置来进一步简化了 Spring 应用的整个搭建和开发过程。另外 Spring Boot 通过集成大量的框架和技术 jar 使得依赖项的版本冲突和引用的不稳定性等问题得到了很好的解决。以下是 Spring 官方给出的 Spring Boot 介绍及功能特性说明。

Spring Boot 使创建独立的、生产级的基于 Spring 的应用程序变得很容易，可以"直接运行"这些应用程序。可以从最少的麻烦开始编写业务代码，大多数 Spring Boot 应用程序只需要很少的 Spring 配置。

功能特性
1) 创建独立的 Spring 应用程序；
2) 直接在应用程序中嵌入 Tomcat、Jetty 和 Undertow（不需要部署 WAR 文件）；
3) 提供现成的"starter"依赖项来简化构建配置；
4) 尽可能自动配置 Spring 和第三方库；
5) 提供生产就绪功能，如度量、运行状况检查和外部化配置；
6) 完全没有代码生成，也不需要 XML 配置。

Spring Boot 所有的历史文档列表，请访问以下网址：
https://docs.spring.io/spring-boot/docs

Spring Boot 有两个非常重要的策略：开箱即用和约定优于配置。开箱即用是指在开发过程中，通过在 MAVEN（或 Gradle）项目的 pom 文件中添加相关依赖项，然后使用注解来代替繁琐的 XML 配置文件以管理对象的生命周期。这个特点使得开发者摆脱了复杂的配置工作和依赖项的管理工作，更加专注于业务逻辑。约定优于配置是一种由 Spring Boot 本身来配置目标结构，由开发者在结构中添加信息的软件设计范式。这一特点虽降低了部分灵活性，增加了 BUG 定位的复杂性，但减少了开发者需要做出决定的数量，同时减少了大量的 XML 配置，并且可以更方便地进行代码编译、测试和打包等工作自动化。

从下一章开始，本书会给出一些简单的入门示例，以此来开启 Spring Boot 之旅的第一步。

第二章 简单示例

本章一共给出了 5 个简单示例，分别展现了 Spring Boot 的不同方面。Spring Boot 官方为我们提供了构建初始化项目的便捷的网页服务，我们可以利用该服务进行简单配置以快速构建自己的项目。在页面中按需进行配置，如图 2-1 所示。

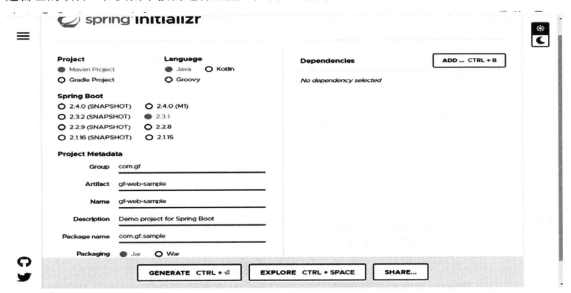

图 2-1 Spring Boot 配置图

配置好后，单击按钮[EXPLORE CTRL + SPACE] 可以在线预览生成的项目结构及其文件内容，如图 2-2 所示。

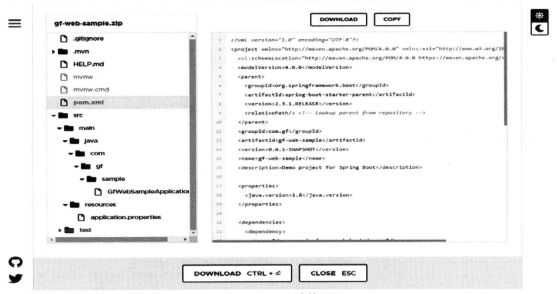

图 2-2 项目结构图

确认无误后，单击按钮[DOWNLOAD CTRL + ↵] 以下载配置好的项目到本地，如图 2-3 所示。

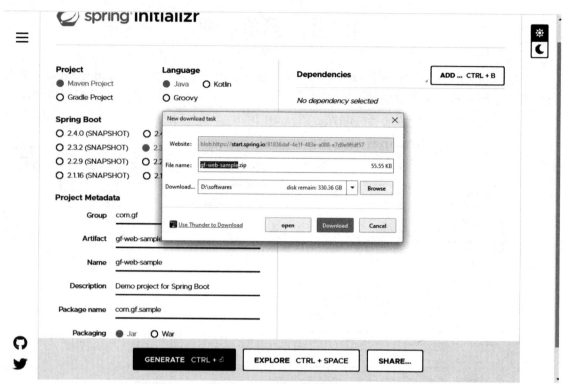

图 2-3 下载项目图

在上面的配置中,并未配置依赖项,但 pom.xml 配置文件中会默认添加一些依赖项。将下载的初始化项目导入到 IntelliJ IDEA 中,并执行 clean 和 compile 等命令,如图 2-4 所示。

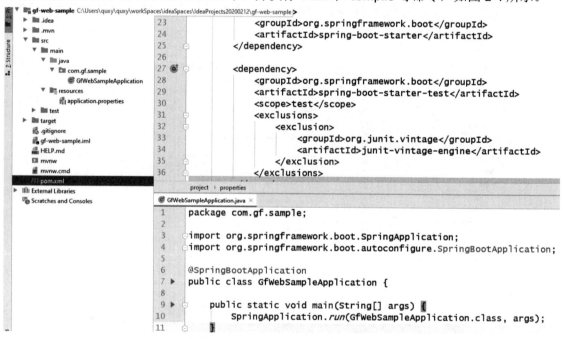

图 2-4 pom.xml 配置文件内容

运行 GfWebSampleApplication 类的主方法，控制台的输出，如图 2-5 所示。

```
"C:\Program Files\Java\jdk1.8.0_171\bin\java.exe" ...
Connected to the target VM, address: '127.0.0.1:54548', transport: 'socket'

  .   ____          _            __ _ _
 /\\ / ___'_ __ _ _(_)_ __  __ _ \ \ \ \
( ( )\___ | '_ | '_| | '_ \/ _` | \ \ \ \
 \\/  ___)| |_)| | | | | || (_| |  ) ) ) )
  '  |____| .__|_| |_|_| |_\__, | / / / /
 =========|_|==============|___/=/_/_/_/
 :: Spring Boot ::        (v2.3.1.RELEASE)

2020-07-07 09:07:18.930  INFO 12416 --- [  restartedMain] com.gf.sample.GfWebSampleApplication     :
2020-07-07 09:07:18.945  INFO 12416 --- [  restartedMain] com.gf.sample.GfWebSampleApplication     :
2020-07-07 09:07:19.133  INFO 12416 --- [  restartedMain] .e.DevToolsPropertyDefaultsPostProcessor :
2020-07-07 09:07:21.165  INFO 12416 --- [  restartedMain] o.s.b.d.a.OptionalLiveReloadServer       :
2020-07-07 09:07:21.196  INFO 12416 --- [  restartedMain] com.gf.sample.GfWebSampleApplication     :
```

图 2-5 控制台

接下来，我们将项目改造为 web 项目，只需要更换 pom.xml 配置文件中的依赖项即可，如下所示：

```xml
<dependency>
    <groupId>org.springframework.boot</groupId>
    <artifactId>spring-boot-starter</artifactId>
</dependency>
```

将上面的依赖项替换为下面的：

```xml
<dependency>
    <groupId>org.springframework.boot</groupId>
    <artifactId>spring-boot-starter-web</artifactId>
</dependency>
```

在终端中执行命令：mvn dependency:tree，可以查看到 spring-boot-starter-web 的依赖关系，如图 2-6 所示。

```
[INFO] --- maven-dependency-plugin:3.1.2:tree (default-cli) @ gf-web-sample ---
[INFO] com.gf:gf-web-sample:jar:0.0.1-SNAPSHOT
[INFO] +- org.springframework.boot:spring-boot-starter-web:jar:2.3.1.RELEASE:compile
[INFO] |  +- org.springframework.boot:spring-boot-starter:jar:2.3.1.RELEASE:compile
[INFO] |  |  +- org.springframework.boot:spring-boot:jar:2.3.1.RELEASE:compile
[INFO] |  |  +- org.springframework.boot:spring-boot-autoconfigure:jar:2.3.1.RELEASE:compile
[INFO] |  |  +- org.springframework.boot:spring-boot-starter-logging:jar:2.3.1.RELEASE:compile
[INFO] |  |  |  +- ch.qos.logback:logback-classic:jar:1.2.3:compile
[INFO] |  |  |  |  \- ch.qos.logback:logback-core:jar:1.2.3:compile
[INFO] |  |  |  +- org.apache.logging.log4j:log4j-to-slf4j:jar:2.13.3:compile
[INFO] |  |  |  |  \- org.apache.logging.log4j:log4j-api:jar:2.13.3:compile
[INFO] |  |  |  \- org.slf4j:jul-to-slf4j:jar:1.7.30:compile
[INFO] |  |  +- jakarta.annotation:jakarta.annotation-api:jar:1.3.5:compile
[INFO] |  |  \- org.yaml:snakeyaml:jar:1.26:compile
[INFO] |  +- org.springframework.boot:spring-boot-starter-json:jar:2.3.1.RELEASE:compile
[INFO] |  |  +- com.fasterxml.jackson.core:jackson-databind:jar:2.11.0:compile
[INFO] |  |  |  +- com.fasterxml.jackson.core:jackson-annotations:jar:2.11.0:compile
[INFO] |  |  |  \- com.fasterxml.jackson.core:jackson-core:jar:2.11.0:compile
[INFO] |  |  +- com.fasterxml.jackson.datatype:jackson-datatype-jdk8:jar:2.11.0:compile
[INFO] |  |  +- com.fasterxml.jackson.datatype:jackson-datatype-jsr310:jar:2.11.0:compile
[INFO] |  |  \- com.fasterxml.jackson.module:jackson-module-parameter-names:jar:2.11.0:compile
[INFO] |  +- org.springframework.boot:spring-boot-starter-tomcat:jar:2.3.1.RELEASE:compile
[INFO] |  |  +- org.apache.tomcat.embed:tomcat-embed-core:jar:9.0.36:compile
[INFO] |  |  +- org.glassfish:jakarta.el:jar:3.0.3:compile
[INFO] |  |  \- org.apache.tomcat.embed:tomcat-embed-websocket:jar:9.0.36:compile
[INFO] |  +- org.springframework:spring-web:jar:5.2.7.RELEASE:compile
```

图 2-6 spring-boot-starter-web 的依赖关系

再次运行 GfWebSampleApplication 类的主方法，控制台的输出，如图 2-7 所示。

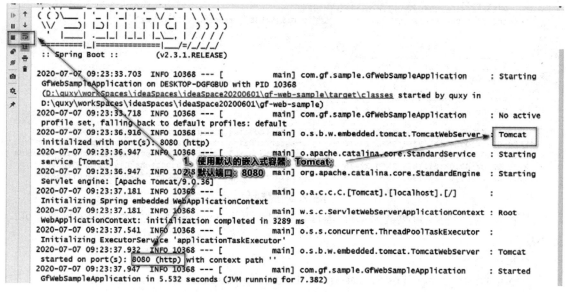

图 2-7 控制台输出

这次，程序运行后没有立即终止，而是一直处于运行状态，监听端口为：8080。到目前为止，我们已经成功启动了一个空的 web 应用程序。接下来，我们以各种方式体验一下这个 web 应用程序。

2.1 静态 HTML 示例

在 src\main\resources\static 目录中新建 index.html 文件，并添加内容，如下所示：

```html
<!DOCTYPE html>
<html lang="en">
<head>
    <meta charset="UTF-8">
    <title>Gf</title>
</head><body>
<h1>欢迎你！</h1>
</body></html>
```

重新启动应用后，在浏览器中访问 URL，如图 2-8 所示。
http://localhost:8080

图 2-8 静态示例

2.2 动态 HTML 示例

上面的例子中,直接返回了静态页面。我们还可以基于 FreeMarker 模板技术响应动态生成的页面。Spring Boot 已经为该技术提供了 starter,将其添加到 pom.xml 配置文件中即可,如下所示:

```xml
<dependency>
    <groupId>org.springframework.boot</groupId>
    <artifactId>spring-boot-starter-freemarker</artifactId>
</dependency>
```

从 Spring Boot 2.2.0 开始,FreeMarker 模板文件的默认后缀是 ".ftlh"。如果仍然希望使用 ".ftl" 作为后缀,请在 application.yml 配置文件中添加配置项,如下所示:

spring.freemarker.suffix=.ftl

在 src\main\resources\templates 目录中,定义 hello.ftlh 模板文件,如下所示:

```html
<!DOCTYPE html>
<html lang="en">
<head>
    <meta charset="UTF-8">
    <title>Gf</title>
</head>
<body>
<h1>模板文件名称:hello.ftlh</h1>
<h3>大家好,我是${person.name},${person.msg}</h3>
</body>
</html>
```

定义 controller 层代码,如下所示:

```java
package com.gf.sample.controller;
import org.springframework.stereotype.Controller;
import org.springframework.ui.Model;
import org.springframework.web.bind.annotation.GetMapping;
import org.springframework.web.bind.annotation.RequestMapping;
import org.springframework.web.bind.annotation.ResponseBody;
import org.springframework.web.servlet.ModelAndView;
import javax.servlet.http.HttpServletResponse;
import java.util.HashMap;
import java.util.Map;
@Controller
@RequestMapping("/hello2")
public class HelloHtmlController {
```

```java
    @GetMapping("/html")
    public ModelAndView html(String name) {
        ModelAndView mv = new ModelAndView("hello");
        Map<String, String> map = new HashMap<>();
        map.put("name", name);
        map.put("msg", "Hello World!");
        mv.addObject("person", map);
        return mv;
    }
    @GetMapping("/html2")
    public String html2(Model model, String name) {
        Map<String, String> map = new HashMap<>();
        map.put("name", name);
        map.put("msg", "你好,世界!");
        model.addAttribute("person", map);
        return "hello";
    }
}
```

启动应用后,要想查看效果,请在浏览器中分别访问 URL,如下所示:

http://localhost:8080/hello2/html?name=北京

http://localhost:8080/hello2/html2?name=北京2

2.3 字符串示例

这里只是简单举个例子,在实际开发中,直接响应纯粹的字符串的情况是比较少的。

```java
package com.gf.sample.controller;
import org.springframework.http.ResponseEntity;
import org.springframework.web.bind.annotation.GetMapping;
import org.springframework.web.bind.annotation.RequestMapping;
import org.springframework.web.bind.annotation.RestController;
import org.springframework.web.servlet.ModelAndView;
import java.util.Collections;
import java.util.HashMap;
import java.util.Map;
@RestController
@RequestMapping("/hello")
public class HelloController {
    /**
```

```java
     * 响应一个字符串
     */
    @GetMapping("/str")
    public String str(String name) {
        return "Hello, " + name;
    }
}
```

2.4 JSON 示例

响应 JSON 数据的方式一共有以下两种，在 RESTful API 风格的编程中，我们大都使用前者。

```java
@RestController
@RequestMapping("/hello")
public class HelloController {
    /**
     * 响应一个 JSON 类型的数据对象
     */
    @GetMapping("/json")
    public ResponseEntity json(String name) {
        return ResponseEntity.ok(Collections.singletonMap(name, "Say Hello!"));
    }
}
```

@ResponseBody 注解用于将 Java 对象转化为 JSON 格式的数据，并响应到客户端。从 Spring Framework 4.0 开始，它可以作为类级别的注解使用了，如下所示：

```java
@Controller
@RequestMapping("/hello2")
public class HelloHtmlController {
    /**
     * 添加 @ResponseBody 后，不再进行视图解析直接响应返回值。但返回值为 ModelAndView 类型的除外。
     */
    @GetMapping("/json")
    @ResponseBody
    public Map<String, String> json(HttpServletResponse response, String name) {
        //response.setContentType(MediaType.APPLICATION_JSON_VALUE);
        Map<String, String> map = new HashMap<>();
```

```
        map.put("name", name);
        map.put("msg", "Hello World!");
        return map;
    }
}
```

2.5 JAR 示例

对于 Spring Boot 应用项目，在大多数时候，其产出物是 jar 文件。要想将项目打包为可执行的 jar 文件，请在 pom.xml 配置文件中添加插件，如下所示：

```xml
<build>
    <plugins>
        <plugin>
            <groupId>org.springframework.boot</groupId>
            <artifactId>spring-boot-maven-plugin</artifactId>
        </plugin>
    </plugins>
</build>
```

另外，还需要注意 pom.xml 配置文件中的这个位置，如图 2-9 所示。

```xml
<?xml version="1.0" encoding="UTF-8"?>
<project xmlns="http://maven.apache.org/POM/4.0.0" xmlns:xsi="http://www.w3
    .org/2001/XMLSchema-instance"
    xsi:schemaLocation="http://maven.apache.org/POM/4.0.0 https://maven.apache.org/xsd/maven-4.0
    .0.xsd">
    <modelVersion>4.0.0</modelVersion>
    <parent>
        <groupId>org.springframework.boot</groupId>
        <artifactId>spring-boot-starter-parent</artifactId>
        <version>2.3.1.RELEASE</version>
        <relativePath/> <!-- lookup parent from repository -->
    </parent>
    <groupId>com.gf</groupId>
    <artifactId>gf-web-sample</artifactId>
    <version>0.0.1-SNAPSHOT</version>
    <name>gf-web-sample</name>
    <description>Demo project for Spring Boot</description>
    <packaging>jar</packaging>

    <properties>
        <java.version>1.8</java.version>
    </properties>

    <dependencies>
        <dependency>
            <groupId>org.springframework.boot</groupId>
            <artifactId>spring-boot-starter-web</artifactId>
        </dependency>
```

记得添加 packaging 元素。

图 2-9 添加 packaging 元素

在终端（例如：CMD 窗口）中，执行 "mvn clean install" 命令来生成可执行的 jar 文件，如图 2-10 所示。

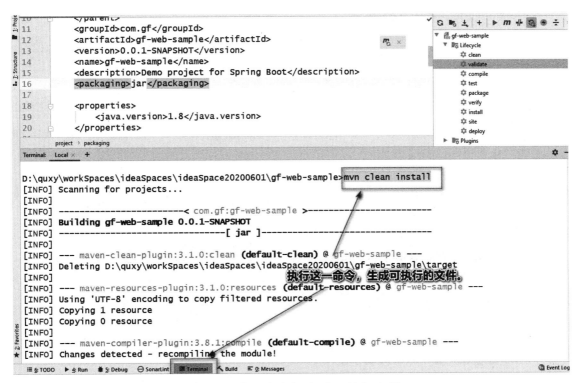

图 2-10 使用命令行生成可执行文件

还可以使用 IntelliJ IDEA 的 Maven 面板中的相应命令,以执行相同的操作,如图 2-11 所示。

图 2-11 使用 Maven 面板命令生成可执行文件

生成的 jar 文件，如图 2-12 所示。

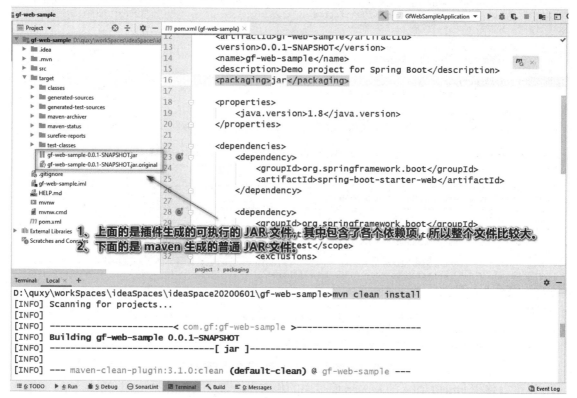

图 2-12 可执行 jar 文件

可执行的 jar 文件生成了，让我们进入到该文件所在目录执行命令来运行它，如下所示：
java -jar gf-web-sample-0.0.1-SNAPSHOT.jar

执行命令后，控制台产生的输出，如图 2-13 所示。

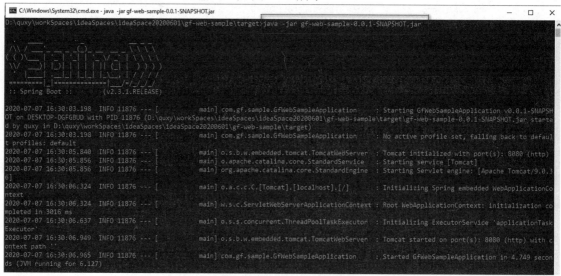

图 2-13 运行 jar 文件

成功启动后，就可以在浏览器或 Postman 之类的客户端中访问定义的 URL 了。

2.6 本章小结

本章讲解了 Spring Boot 的 5 个入门示例，它们分别以不同的形式来展现 URL 的访问或 API 的调用。希望通过本章的学习，读者可以感受到 Spring Boot 的便捷之处，并对 Spring Boot 抱有更多一份的期待。在下一章中，我们将一起探讨 Spring Boot 基础和核心的知识点。

第三章 Spring Boot 基础

Spring Boot 遵循约定大于配置的原则，其中的各个注解正是这一原则的具体表示。Spring Boot 中不同的注解标志了不同的：类和方法的类型、用途或条件等。要想真正地熟练掌握 Spring Boot 的核心技能，只简单了解这些注解是不够的，还需要懂得其具体功用，其定义细节，其相互关系。正所谓，知其然，还要知其所以然。

3.1 Spring 注解

本节讲解 Spring 注解。Spring Boot 是基于 Spring 框架构建的，它的有些注解需要与 Spring 的注解配合才能使用。下面列举了一些常用的 Spring 注解，及其示例。

3.1.1 @Configuration

本注解是从 Spring 3.0 版本开始提供的，它只能用在类型（类、接口、注解和枚举）上。它用于自定义配置类 bean，并且由 Spring 容器管理。它还可以用于嵌套类上。为 Spring Boot 的 web 应用程序添加过滤器的示例代码，如下所示：

```
package com.gf.mvc.config;
import org.slf4j.Logger;
import org.slf4j.LoggerFactory;
import org.springframework.boot.web.servlet.FilterRegistrationBean;
import org.springframework.context.annotation.Bean;
import org.springframework.context.annotation.Configuration;
import javax.servlet.*;
import java.io.IOException;
/**
 * 定义 web 过滤器
 */
public class GfFilter implements Filter {
    private Logger log = LoggerFactory.getLogger(this.getClass());
    @Override
    public void doFilter(ServletRequest servletRequest,
ServletResponse servletResponse, FilterChain filterChain) throws
IOException, ServletException {
        log.info("执行前：逻辑处理代码");
        filterChain.doFilter(servletRequest, servletResponse);
        log.info("执行后：逻辑处理代码");
```

        }
}

```java
package com.gf.mvc.config;
import org.springframework.boot.web.servlet.FilterRegistrationBean;
import org.springframework.context.annotation.Bean;
import org.springframework.context.annotation.Configuration;
@Configuration
public class WebConfiguration {
    /**
     * 注册 web 过滤器
     */
    @Bean
    public FilterRegistrationBean filterRegistrationBean() {
        FilterRegistrationBean registrationBean = new FilterRegistrationBean(new GfFilter());
        registrationBean.addUrlPatterns("/*");
        registrationBean.setName("gfFilter");
        return registrationBean;
    }
}
```

@Configuration 注解的类可以用 @Profile 注解标记，以表示只有在给定的 profile 处于活动状态时才应处理该类中定义的 bean。示例代码，如下所示：

```java
import org.springframework.context.annotation.Bean;
import org.springframework.context.annotation.Configuration;
import org.springframework.context.annotation.Profile;
import javax.activation.DataSource;
@Profile("dev")
@Configuration
public class DevDatabaseConfig {
    @Bean
    public DataSource dataSource() {
        // TODO: 数据库初始化代码
        return null;
    }
}
```

```
@Profile("prod")
@Configuration
public class ProdDatabaseConfig {
    @Bean
    public DataSource dataSource() {
        // TODO: 数据库初始化代码
        return null;
    }
}
```

3.1.2 @Component

本注解是从 Spring 2.5 版本开始提供的，它只能用在类型（类、接口、注解和枚举）上。添加本注解的类被认为是一个组件，在 Spring Boot 启动时，会被扫描成为一个 bean。本注解包含一个 value 字段，用于指定组件的名称。在 @Configuration 注解的定义中使用到了本注解。

3.1.3 @ComponentScan

本注解是从 Spring 3.1 版本开始提供的，它只能用在类型（类、接口、注解和枚举）上。本注解用于配置与 @Configuration 注解一起使用的组件扫描指令。本注解提供与 Spring XML 的 <context:component-scan> 元素并行的支持。

可以指定 basepackageclass 或 basePackages（或其别名值）来定义要扫描的特定包。如果未定义特定的包，将从声明此注解的类所在的包进行扫描。

请注意，<context:component-scan> 元素具有 annotation-config 属性；但是，本注解没有。这是因为在几乎所有情况下，当使用 @ComponentScan 注解时，默认的注解配置处理（例如处理 @Autowired）被假定。此外，当使用 AnnotationConfigApplicationContext 时，注解配置处理器总是被注册，这意味着在 @ComponentScan 级别禁用它们的任何尝试都将被忽略。

3.1.4 @Conditional

本注解是从 Spring 4.0 版本开始提供的，它只能用在类型（类、接口、注解和枚举）和方法上。只有当本注解指定的所有条件都匹配时，组件才有资格进行注册。条件是在 bean 定义被注册之前以编程方式确定的任何状态。

可以以下列任何一种方式使用 @Conditional 注解：
1) 作为直接或间接用 @Component 注解的任何类的类型级注解，包括 @Configuration 类；
2) 作为元注解，用于编写自定义构造型注解；
3) 作为任何 @Bean 方法的方法级注解。

如果 @Configuration 的类标记为 @Conditional，则与该类关联的所有 @Bean 方法、@Import 注解和 @ComponentScan 注解都将受这些条件的约束。

注意：不支持 @Conditional 注解的继承；不考虑来自超类或重写方法的任何条件。为了强制执行这些语义，@Conditional 本身没有声明为 @Inherited；此外，任何使用 @Conditional 进行元注解的自定义组合注解都不能声明为 @Inherited。

3.1.5 @Import

本注解是从 Spring 3.0 版本开始提供的，它只能用在类型（类、接口、注解和枚举）上。本注解表明要导入的一个或多个组件类通常是 @Configuration 类。本注解提供与 Spring XML 中的 <import/> 元素等效的功能。允许导入用 @Configuration 注解的类、ImportSelector 接口和 ImportBeanDefinitionRegistrar 接口的实现，以及常规组件类（从 Spring 4.2 开始，类似于 AnnotationConfigApplicationContext.register）。

在导入的 @Configuration 类中声明的 @Bean 定义应该通过使用 @Autowired 注入来访问。可以自动注册 bean 本身，也可以自动注册声明 bean 的配置类实例。后一种方法允许在 @Configuration 类方法之间进行显式的、对 IDE 友好的导航。

如果需要导入 XML 或其他非 @Configuration bean 定义的资源，请改用 @ImportResource 注解。

3.1.6 @ImportResource

本注解是从 Spring 3.0 版本开始提供的，它只能用在类型（类、接口、注解和枚举）上。本注解表明包含要导入的 bean 定义的一个或多个资源。与 @Import 一样，此注解提供了类似于 Spring XML 中的 <import/> 元素的功能。本注解通常在设计 @Configuration 类时使用，这些类由 AnnotationConfigApplicationContext 引导，但在某些情况下仍然需要一些 XML 功能，如名称空间。

默认情况下，如果以".groovy"结尾，将使用 GroovyBeanDefinitionReader 处理 value 属性的参数；否则，将使用 XmlBeanDefinitionReader 解析 Spring <beans/> XML 文件。可选地，可以声明 reader 属性，允许用户选择自定义的 BeanDefinitionReader 实现。

3.1.7 @Value

本注解是从 Spring 3.0 版本开始提供的，它只能用在字段、方法、参数和注解上。本注解是字段或方法（构造方法）参数级别的注解，它指定受影响参数的默认值表达式。本注解通常用于表达式驱动的依赖注入，也支持处理器方法参数的动态解析，例如在 Spring MVC 中。

一个常见的用例是使用 #{systemProperties.myProp} 样式表达式设定默认字段值。

请注意，@Value 注解的实际处理是由 BeanPostProcessor 执行的，这反过来意味着不能在

BeanPostProcessor 接口或 BeanFactoryPostProcessor 接口及其实现类中使用 @Value。有关 AutowiredAnnotationBeanPostProcessor 类，请查看 javadoc 文档（默认情况下，该类用于检查此注解是否存在）。本注解的使用示例，如图 3-1 所示。

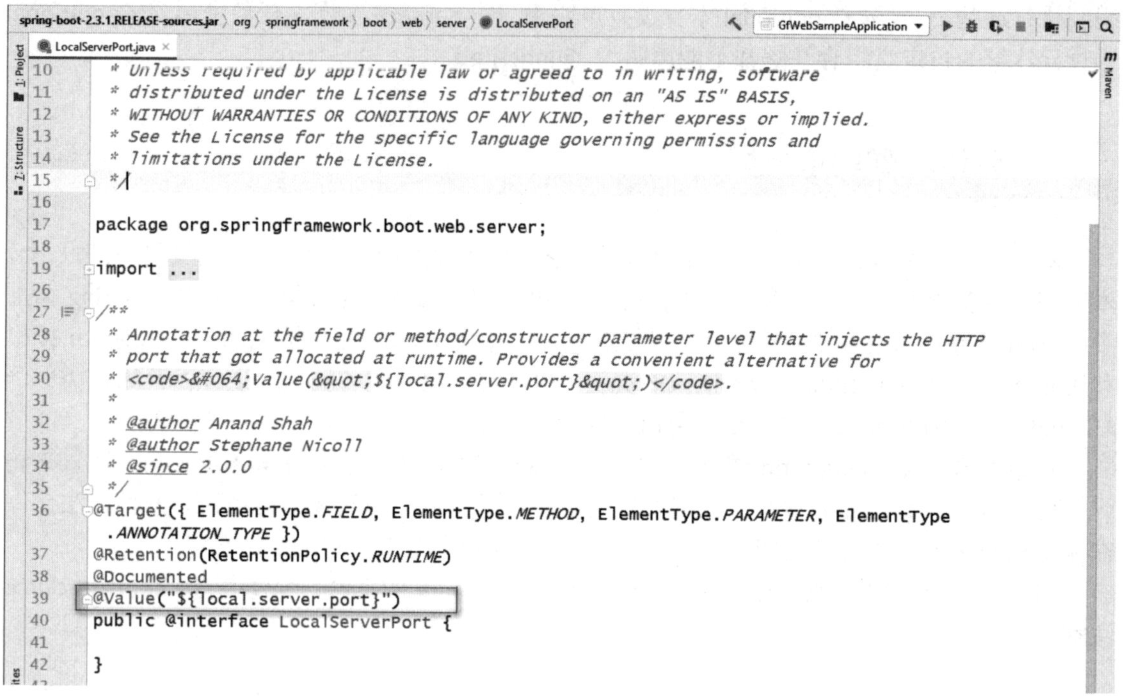

图 3-1 @Value 注解

以上只是列举了在 Spring 项目中几个常用的注解，还有更多的注解，有待大家在实际开发中边学边用。

3.2 Spring Boot 注解

本节讲解 Spring Boot 注解，恰当地使用这些注解可以减少代码量，并提升代码开发的效率。下面列举了一些常用的 Spring Boot 注解，及其示例。

3.2.1 @SpringBootApplication

本注解是从 Spring Boot 1.2.0 版本开始提供的，它只能用在类型（类、接口、注解和枚举）上。本注解用于配置类上，该配置类声明一个或多个 @Bean 方法并触发自动配置和组件扫描。其中，应用程序的 main 方法一般也放入到该配置类中，也就是说该配置类会作为应用程序的主类。这是一个方便的注解，相当于同时声明了 @Configuration、@EnableAutoConfiguration 和 @ComponentScan。本注解的最简单示例代码，如下所示：

```
package com.gf.sample;
import org.springframework.boot.SpringApplication;
```

```
import org.springframework.boot.autoconfigure.SpringBootApplication;
@SpringBootApplication
public class GfWebSampleApplication {
    public static void main(String[] args) {
        SpringApplication.run(GfWebSampleApplication.class, args);
    }
}
```

3.2.2 @SpringBootConfiguration

本注解是从 Spring Boot 1.4.0 版本开始提供的,它只能用在类型(类、接口、注解和枚举)上。使用本注解的类提供 Spring Boot 应用程序的 @Configuration。它可以用作 Spring 的标准 @Configuration 注解的替代,以便可以自动找到配置(例如在测试中)。

应用程序应该只包含一个 @SpringBootConfiguration,大多数 Spring Boot 应用程序一般情况下从 @SpringBootApplication 继承它。

3.2.3 @EnableAutoConfiguration

本注解是从 Spring Boot 1.0.0 版本开始提供的,它只能用在类型(类、接口、注解和枚举)上。本注解用来启用 Spring 应用程序上下文的自动配置,尝试猜测和配置可能需要的 bean。自动配置类通常基于类路径和定义的 bean 来应用。例如,如果 tomcat-embedded.jar 在类路径上,则可能需要为环境提供一个 TomcatServletWebServerFactory bean(除非你定义了自己的 ServletWebServerFactory bean)。

在使用 @SpringBootApplication 注解时,会自动启用上下文的自动配置,因此添加本注解不会产生任何附加效果。

自动配置会尽可能智能,并且会随着自定义更多自己的配置而后退。你始终可以手动 exclude() 任何不想应用的配置(如果没有访问权限,请使用 excludeName())。你也可以通过 spring.autoconfigure.Include 属性排除它们。自动配置始终在注册用户定义的 bean 后应用。

用 @EnableAutoConfiguration(通常通过 @SpringBootApplication)注解的类的包具有特定的意义,通常用作"默认值"。例如,在扫描 @Entity 类时将使用它。通常建议将 @EnableAutoConfiguration(如果不使用 @SpringBootApplication)放在根包中,以便可以搜索所有子包和类。

自动配置类是常规的 Spring @Configuration bean。它们使用 SpringFactoriesLoader 机制(针对该类设置键)进行定位。一般来说,自动配置 bean 是 @Conditional bean(通常使用 @ConditionalOnClass 和 @ConditionalOnMissingBean 注解)。

3.2.4 @ConfigurationProperties

本注解是从 Spring Boot 1.0.0 版本开始提供的，它只能用在类型（类、接口、注解和枚举）和方法上。本注解是外部化配置的注解。如果要绑定和验证某些外部属性（例如：从 .properties 配置文件中），请将其添加到类定义或 @Configuration 类中的 @Bean 方法上。

绑定可以通过对带注解的类调用 setter 来执行，如果使用 @ConstructorBinding，则可以通过绑定到构造方法参数来执行。请注意：与 @Value 相反，本注解不支持 SpEL 表达式，因为属性值是外部化的。本注解的使用示例，如图 3-2 所示。

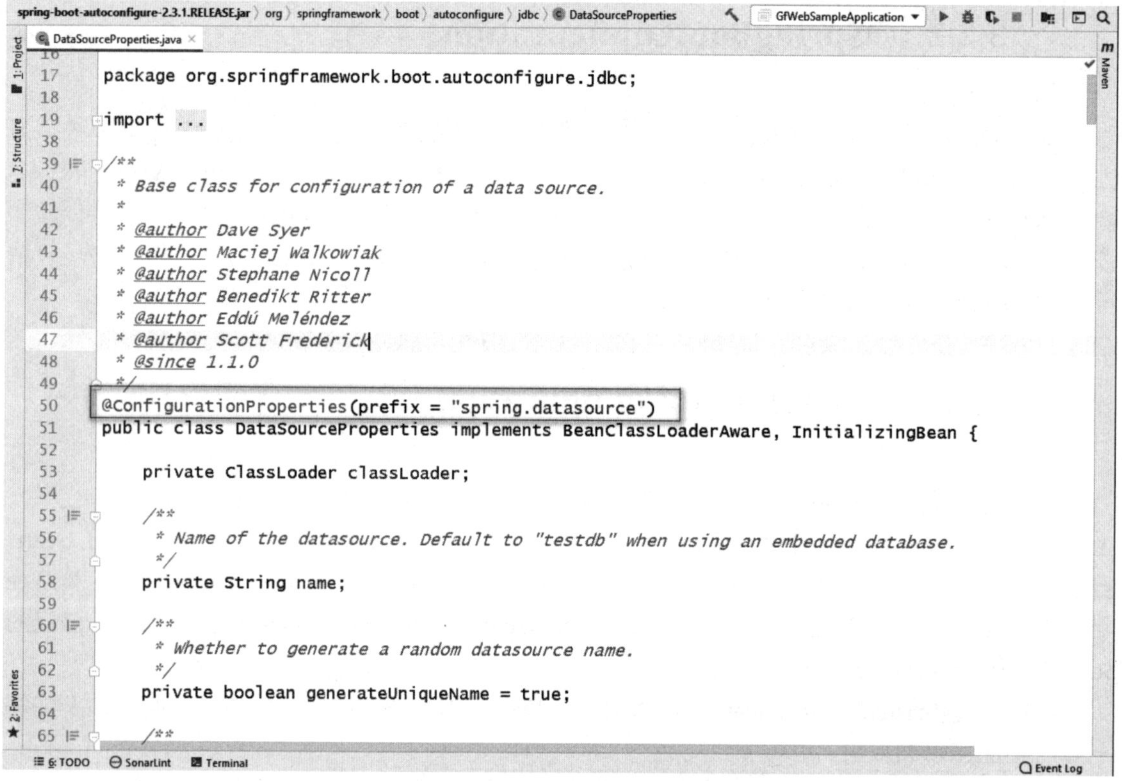

图 3-2 @ConfigurationProperties 注解

3.2.5 @ConditionalOnBean

本注解是从 Spring Boot 1.0.0 版本开始提供的，它只能用在类型（类、接口、注解和枚举）和方法上。本注解是只有当 BeanFactory 中已经包含满足所有指定要求的 bean 时才匹配的 @Conditional。要使条件匹配，必须满足所有要求，但不必由同一个 bean 满足这些要求。当放置在 @Bean 方法上时，bean 类默认为 factory 方法的返回类型：

```
@Configuration
public class MyAutoConfiguration {
```

```
@ConditionalOnBean
@Bean
public MyService myService() {
    // TODO: 具体代码
}
}
```

在上面的示例中，如果 BeanFactory 中已经包含 MyService 类型的 bean，那么条件将匹配。该条件只能匹配到目前为止已由应用程序上下文处理的 bean 定义，因此，强烈建议仅在自动配置类上使用该条件。如果候选 bean 可能由另一个自动配置创建，则还请确保使用此条件的 bean 在之后运行。

3.2.6 @ConditionalOnMissingBean

本注解是从 Spring Boot 1.0.0 版本开始提供的，它只能用在类型（类、接口、注解和枚举）和方法上。本注解是只有当 BeanFactory 中没有包含满足指定要求的 bean 时才匹配的 @Conditional。要使条件匹配，必须不满足任何要求。当放置在 @Bean 方法上时，bean 类默认为 factory 方法的返回类型：

```
@Configuration
public class MyAutoConfiguration {
    @ConditionalOnMissingBean
    @Bean
    public MyService myService() {
        // TODO: 具体代码
    }
}
```

在上面的示例中，如果 BeanFactory 中没有包含 MyService 类型的 bean，那么条件将匹配。该条件只能匹配到目前为止已由应用程序上下文处理的 bean 定义，因此，强烈建议仅在自动配置类上使用该条件。如果候选 bean 可能由另一个自动配置创建，还请确保使用此条件的 bean 在之后运行。

3.2.7 @ConditionalOnClass

本注解是从 Spring Boot 1.0.0 版本开始提供的，它只能用在类型（类、接口、注解和枚举）和方法上。本注解是只有当指定的类在类路径上时才匹配的 @Conditional。

value() 可以安全地在 @Configuration 类上指定，因为在加载类之前使用 ASM 解析注解元数据。在放置 @Bean 方法时需要格外小心，请考虑在单独的 Configuration 类中的隔离条件，特别是在方法的返回类型与条件的目标相匹配时。

3.2.8 @ConditionalOnMissingClass

本注解是从 Spring Boot 1.0.0 版本开始提供的，它只能用在类型（类、接口、注解和枚举）和方法上。本注解是只有当指定的类不在类路径上时才匹配的 @Conditional。这正好与 @ConditionalOnClass 注解的意思相反。

3.2.9 @ConditionalOnWebApplication

本注解是从 Spring Boot 1.0.0 版本开始提供的，它只能用在类型（类、接口、注解和枚举）和方法上。本注解是当应用程序是 web 应用程序时才匹配的 @Conditional。默认情况下，任何 web 应用程序都将匹配，但可以使用 type() 属性将其缩小。

3.2.10 @ConditionalOnNotWebApplication

本注解是从 Spring Boot 1.0.0 版本开始提供的，它只能用在类型（类、接口、注解和枚举）和方法上。本注解是只有当应用程序上下文不是 web 应用程序上下文时才匹配的 @Conditional。

3.2.11 @ConditionalOnJava

本注解是从 Spring Boot 1.1.0 版本开始提供的，它只能用在类型（类、接口、注解和枚举）和方法上。本注解是基于应用程序运行的 JVM 版本匹配的 @Conditional。本注解的示例代码，如下所示：

```
/**
 * 当 Java 的版本小于 1.8 时，可以使用以下 factory 方法获取 bean。
 */
@Bean
@ConditionalOnJava(range = ConditionalOnJava.Range.OLDER_THAN,
        value = JavaVersion.EIGHT)
public MyObject getMyObject() {}
```

3.2.12 @ConditionalOnResource

本注解是从 Spring Boot 1.0.0 版本开始提供的，它只能用在类型（类、接口、注解和枚举）和方法上。本注解是只有当指定的资源在类路径上时才匹配的 @Conditional。本注解的示例代码，如图 3-3 所示。

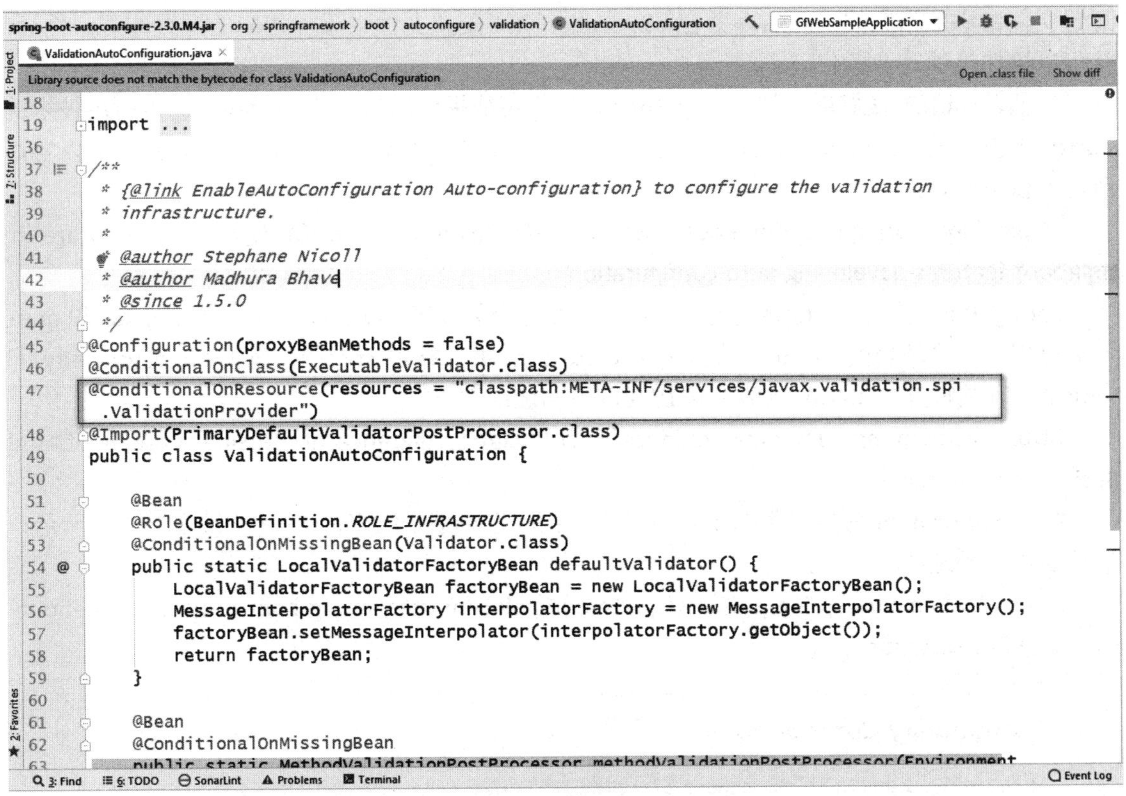

图 3-3 @ConditionalOnResource 注解

3.2.13 @ConditionalOnCloudPlatform

本注解是从 Spring Boot 1.5.0 版本开始提供的，它只能用在类型（类、接口、注解和枚举）和方法上。本注解是当指定的云平台处于活动状态时才匹配的 @Conditional。

以上几节中列出的注解也只是作了简短的说明，要想真正地学会用会，还需自己动起手来在自己的项目中用起来。

3.3 Starters

Spring Boot 的每个版本都提供了其所支持的依赖项列表，以提供统一的版本管理，在项目中无需要指定依赖项的版本。这样方便所有依赖项跟随 Spring Boot 一起升级，也避免了依赖项之间不兼容的问题。Spring Boot 的每个版本都与 Spring Framework 的基本版本相关联，强烈建议不要单独指定 Spring Framework 版本。

starter 是一组方便的可以在应用程序中包含的依赖项描述符，它封装了繁琐，把简单留给开发者。它把向 Spring Boot 中引入某项技术所需的所有依赖项（包括 Spring Boot 风格的基础代码和配置获取方式）打包在一起。若想在 Spring Boot 项目中引入该技术只需在 pom.xml 配置文件中添加相应的 starter，并在 application.yml 配置文件中添加相应的配置即可。然后，就可以愉快地编写业务代码了。

虽然，starter 很好用，但 Spring Boot 官方不可能提供所有技术的 starter。当所用技术的 starter 未被提供时，Spring Boot 允许我们自己定义一个 starter。有关自定义 starter 的内容，请访问以下网址：

https://docs.spring.io/spring-boot/docs/2.3.1.RELEASE/reference/html/spring-boot-features.html#boot-features-developing-auto-configuration

Spring Boot 官方把 starter 分为 3 类。一类是应用程序 starter，例如：mongodb 和 aop；一类是用于生产环境的 starter，例如：actuator；一类是技术 starter，例如：reactor-netty 和 tomcat。有关官方的 starter 列表，请访问以下网址：

https://docs.spring.io/spring-boot/docs/2.3.1.RELEASE/reference/html/using-spring-boot.html#using-boot-starter

除了 Spring Boot 官方提供的 starter 之外，社区也贡献了很多，详细列表，如图 3-4 所示，请访问以下网址：

https://github.com/spring-projects/spring-boot/blob/master/spring-boot-project/spring-boot-starters/README.adoc

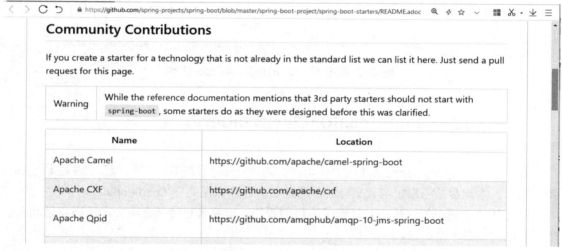

图 3-4 github 社区

3.4 构建代码

Spring Boot 不强制要求任何特定的代码布局，但遵守一些最佳实践，可以事半功倍。强烈建议不要把类放到默认包中，也就是类不在任何包内，这样会带来一些不必要的问题。例如，会造成 @ComponentScan 无法以最佳状态工作。

通常情况下，我们把主类放在其他类之上的根包中，并把 @SpringBootApplication 注解放在主类之上。这隐式为某些项定义了一个基本搜索包，例如：组件扫描默认会使用此根包。

项目结构及其主类代码,如图 3-5 所示。

图 3-5 项目结构图

3.5 配置类

Spring Boot 支持基于 Java 的配置。不需要将所有 @Configuration 放入一个类中,@Import 注解可用于导入其他配置类。或者,可以使用 @ComponentScan 注解自动获取所有 Spring 组件,包括 @Configuration 类。

如果绝对必须使用基于 XML 的配置,我们建议仍然从 @Configuration 类开始。然后,可以使用 @ImportResource 注解加载 XML 配置文件。

3.6 自动配置

Spring Boot 自动配置尝试根据添加的依赖项 jar 自动配置 Spring Boot 应用程序。例如,如果 HSQLDB 位于类路径上,并且尚未手动配置任何数据库连接 bean,则 Spring Boot 会自动配置内存中的数据库。

逐步取代自动配置

自动配置是非侵入性的。在任何时候,都可以开始定义自己的配置来替换自动配置的特定部分。例如,如果添加自己的数据源 bean,则默认的嵌入式数据库支持会后退。

如果需要找出当前正在应用的自动配置以及原因,请使用 --debug 开关启动应用程序。这样做可以为选择的核心日志器启用调试日志,并将条件报告输出到控制台。

禁用特定的自动配置类

如果发现不希望应用的特定自动配置类,可以使用 @SpringBootApplication 的 exclude 属性禁用它们,如下所示:

```
package com.gf.sample;
import org.springframework.boot.SpringApplication;
```

```
import org.springframework.boot.autoconfigure.SpringBootApplication;
import org.springframework.boot.autoconfigure.jdbc.DataSourceAutoConfiguration;
@SpringBootApplication(exclude = DataSourceAutoConfiguration.class)
public class GfWebSampleApplication {
    public static void main(String[] args) {
        SpringApplication.run(GfWebSampleApplication.class, args);
    }
}
```

如果类不在类路径上，则可以使用该注解的 excludeName 属性并指定完全限定名。最后，还可以在 application.yml 配置文件中使用 spring.autoconfigure.exclude 属性。

3.7 Spring Bean 和依赖注入

按照"构建代码"的一般约定去构建代码，@Component、@Service、@Repository 和 @Controller 等注解的类都会自动注册为 Spring bean。在 @RestController bean 中，获取所需的 MenuSevice3 bean 有 3 种方式。第一种是使用 @Autowired 注入，如下所示：

```
@RestController
@RequestMapping("/mvc/menu3")
public class MenuController3 {
    @Autowired
    private MenuService3 service;
    ……
}
```

第二种方式与第三种基本一致，区别是：当只有一个构造方法时，可以忽略 @Autowired，也就是第三种方式。第二、三种方式分别如下所示：

```
@RestController
@RequestMapping("/mvc/menu3")
public class MenuController3 {
    private final MenuService3 service;
    private boolean flag;
    @Autowired
    public MenuController3(MenuService3 service) {
        this.service = service;
    }
    public MenuController3(MenuService3 service,boolean flag){
```

```
        this.service = service;
        this.flag=flag;
    }
    ……
}


@RestController
@RequestMapping("/mvc/menu3")
public class MenuController3 {
    private final MenuService3 service;

    public MenuController3(MenuService3 service) {
        this.service = service;
    }
    ……
}
```

3.8 本章小结

本章首先讲解了 Spring 和 Spring Boot 的重要的注解，接着讲解了 Starters 和构建代码的最佳实践。然后，讲解了配置类的应用和 Spring Boot 自动配置的运行机制。最后，讲了哪些类会被加载为 Spring bean，以及如何在需要的地方注入 bean。

在下一章中，我们讲解有关 SpringApplication 类的具体内容，它是 Spring Boot 中非常重要的一个类。

第四章 SpringApplication 类

SpringApplication 类提供了一种方便的方法来引导从 main() 方法启动的 Spring 应用程序。在大多数情况下，可以委托给其静态 run 方法，如下所示：

```
package com.gf.sample;
import org.springframework.boot.SpringApplication;
import org.springframework.boot.autoconfigure.SpringBootApplication;
@SpringBootApplication
public class GfWebSampleApplication {
    public static void main(String[] args) {
        SpringApplication.run(GfWebSampleApplication.class, args);
    }
}
```

4.1 延迟初始化（懒加载）

SpringApplication 允许应用程序被延迟初始化。启用延迟初始化后，将根据需要而不是在应用程序启动期间创建 bean。因此，启用延迟初始化可以减少应用程序启动所需的时间。在 web 应用程序中，启用延迟初始化将导致许多与 web 相关的 bean 直到收到 HTTP 请求才被初始化。

延迟初始化会导致有些问题在启动应用程序时不能被发现，只有在因使用需要初始化相应的 bean 时才能发现问题，例如：配置错误、依赖的 bean 缺失和 JVM 内存不足等等。由于这些问题，默认情况下是禁用延迟初始化的。延迟初始化可以通过代码方式或配置方式被启用。其中，代码示例，如下所示：

```
import org.springframework.boot.SpringApplication;
import org.springframework.boot.autoconfigure.SpringBootApplication;
import org.springframework.boot.builder.SpringApplicationBuilder;
@SpringBootApplication
public class GfWebSampleApplication {
    public static void main(String[] args) {
//        // 方式一
//        SpringApplicationBuilder builder = new SpringApplicationBuilder(GfWebSampleApplication.class);
//        // 启用延迟初始化
//        builder.lazyInitialization(true);
//        builder.run(args);
```

```
    // 方式二
    SpringApplication springApplication = new
SpringApplication(GfWebSampleApplication.class);
    // 启用延迟初始化
//springApplication.setMainApplicationClass(GfWebSampleApplication.
class);
    springApplication.setLazyInitialization(true);
    springApplication.run(args);
  }
}
```

另外，还可以通过在 application.properties 配置文件中添加配置项来启用延迟初始化，如下所示：

spring.main.lazy-initialization=true

4.2 定制横幅

横幅在应用程序启动时会打印到控制台，默认情况下，打印"Spring"字样及其版本。要想定制横幅，请在 src\main\resources 目录的 banner.txt 文件中添加相应的文本，如图 4-1 所示。

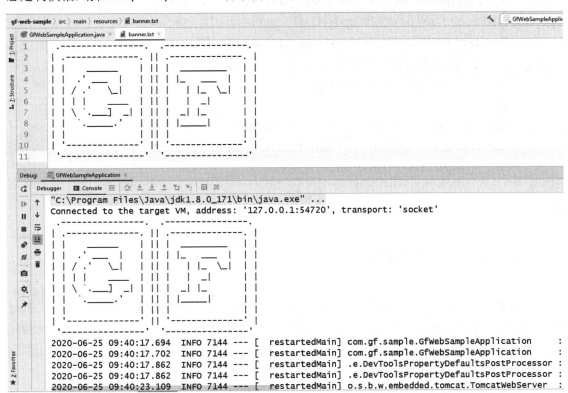

图 4-1 横幅定制

上面的自定义横幅是通过在线制作工具完成的，网址如下所示：
http://patorjk.com/software/taag/#p=display&h=0&f=Blocks&t=GF

想要更改 banner.txt 文件的位置及其文本编码，请在 application.properties 配置文件中指定相应的属性，如下所示：

spring.banner.location=classpath:/banner.txt
spring.banner.charset=UTF-8

除了可以添加文本横幅外，还可以添加图片横幅，支持的格式有：git、jpg 和 png。还有，在 banner.txt 文件中可以添加版本等占位符。有关以上这些，请访问如下网址：

https://docs.spring.io/spring-boot/docs/2.3.1.RELEASE/reference/html/spring-boot-features.html#boot-features-banner

还可以先把图片转化为相应的文本，然后将其放入 banner.txt 文件中。想要把图片转换为文本，请访问如下网址：

https://www.degraeve.com/img2txt.php

除了修改 banner.txt 文件中的内容之外，还可以添加有关 Banner 的代码来设置横幅，如下所示：

```java
@SpringBootApplication
public class GfWebSampleApplication {
    public static void main(String[] args) {
        SpringApplication springApplication = new SpringApplication(GfWebSampleApplication.class);
        Banner banner = (environment, sourceClass, out) -> {
            //TODO: 有关横幅的具体代码。
            // 可以参考 org.springframework.boot.SpringBootBanner 类
        };
        springApplication.setBanner(banner);
        springApplication.setBannerMode(Banner.Mode.CONSOLE);
        springApplication.run(args);
    }
}
```

4.3 应用程序可用性

在部署到平台上时，应用程序可以使用 Kubernetes 探针等基础设施向平台提供有关其可用性的信息。Spring Boot 包括对常用的"活性"和"就绪"可用性状态的开箱即用支持。如果使用的是 Spring Boot 的 Actuator 支持，那么这些状态将作为健康端点组公开。此外，还可以通过将 ApplicationAvailability 接口注入自己的 bean 来获得可用性状态。

一般来说，"活性"状态不应该基于外部检查，例如：运行状态检查。如果是这样，发生故障的外部系统（数据库、Web API 和外部缓存）将触发跨平台的大规模重新启动和级联故障。

Spring Boot 应用程序的内部状态主要由 Spring ApplicationContext 表示。如果应用程序上下文已成功启动，则 Spring Boot 假定应用程序处于有效状态。一旦上下文被刷新，应用程序就被认为是活动的。

应用程序的"就绪"状态表示其是否准备好处理请求。若应用程序未处于的"就绪"状态，则告诉平台现在不应该将请求路由到该应用程序。这通常发生在启动过程中，同时正在处理 CommandLineRunner 和 ApplicationRunner 组件，或者在应用程序确定其太忙而无法进行其他通信的任何时候。一旦调用了应用程序和命令行运行程序，就认为 Spring Boot 已准备就绪。

4.4 应用程序事件和监听器

有些事件实际上是在创建 ApplicationContext 之前触发的，因此不能将监听器注册为 @Bean。但是，可以使用 SpringApplication 的 addListener 方法或 SpringApplicationBuilder 的 listeners 方法注册它们。

如果希望这些监听器自动注册，而不考虑应用程序的创建方式，可以添加 META-INF/spring.factories 文件到项目中，并使用 org.springframework.context.ApplicationListener 键引用监听器，如下所示：

org.springframework.context.ApplicationListener=com.example.project.MyListener

通常不需要使用应用程序事件，但是知道它们的存在会很方便。在内部，Spring Boot 使用事件来处理各种任务。

应用程序事件通过使用 Spring Framework 的事件发布机制发送。此机制的一部分确保在子上下文中发布给监听器的事件也在任何祖先上下文中发布给监听器。因此，如果应用程序使用 SpringApplication 实例的层次结构，则监听器可能会接收同一类型应用程序事件的多个实例。

为了让监听器区分其上下文的事件和子上下文的事件，它应该请求注入其应用程序上下文，然后将注入的上下文与事件的上下文进行比较。Spring Boot 的上下文可以通过实现 ApplicationContextAware 注入，如果监听器是 bean，则可以使用 @Autowired 注入。

4.5 Web 环境

SpringApplication 试图创建正确类型的 ApplicationContext。用于确定 WebApplicationType 的算法相当简单：
1) 如果存在 Spring MVC，则使用如下注解：
AnnotationConfigServletWebServerApplicationContext
2) 如果 Spring MVC 不存在并且 Spring WebFlux 存在，则使用如下注解：
AnnotationConfigReactiveWebServerApplicationContext
3) 否则，将使用 AnnotationConfigApplicationContext

这意味着，如果在同一个应用程序中使用 Spring MVC 和 Spring WebFlux 中的新 WebClient，那么默认情况下将使用 Spring MVC。可以通过调用以下方法覆盖它：

setWebApplicationType(WebApplicationType)

还可以完全控制调用 setApplicationContextClass(...) 所使用的 ApplicationContext 类型。在 JUnit 测试中使用 SpringApplication 时，通常需要调用以下方法：

setWebApplicationType(WebApplicationType.NONE)

4.6 访问应用程序参数

如果需要在应用程序中访问传递给 SpringApplication 的 run 方法的应用程序参数，则可以注入 ApplicationArguments bean。ApplicationArguments 接口提供对原始 String[] 参数以及已分析的 option 和 non-option 参数的访问，如下所示：

```
import org.springframework.boot.ApplicationArguments;
import org.springframework.stereotype.Component;
import java.util.List;
@Component
public class ArgsConfig {
    public ArgsConfig(ApplicationArguments args) {
        boolean debug = args.containsOption("debug");
        List<String> names = args.getNonOptionArgs();
        //TODO: other code
    }
}
```

Spring Boot 还向 Spring Environment 注册 CommandLinePropertySource，这样就可以使用 @Value 注解访问单个应用程序参数了。

4.7 ApplicationRunner 和 CommandLineRunner

如果在 SpringApplication 启动后需要运行某些特定代码，可以实现 ApplicationRunner 或 CommandLineRunner 接口。这两个接口以相同的方式工作，并提供一个单独的 run 方法，该方法在 SpringApplication.run(...) 完成之前被调用。

CommandLineRunner 接口以简单的字符串数组形式提供对应用程序参数的访问，而 ApplicationRunner 接口的实现类使用前面讨论的 ApplicationArguments 接口。带有 run 方法的 CommandLineRunner 的实现类，如下所示：

```
import org.springframework.boot.CommandLineRunner;
import org.springframework.stereotype.Component;
@Component
public class MyCommandLineRunner implements CommandLineRunner {
    @Override
    public void run(String... args) throws Exception {
```

> *//TODO：加载初始化数据、缓存数据。*
    }
}

如果定义了几个必须按特定顺序调用的 CommandLineRunner 或 ApplicationRunner bean，则可以另外实现 org.springframework.core.Ordered 接口或使用以下注解：
org.springframework.core.annotation.Order

4.8 应用程序退出

每个 SpringApplication 都向 JVM 注册一个关闭钩子，以确保 ApplicationContext 在退出时优雅地关闭。可以使用所有标准的 Spring 生命周期回调（例如：DisposableBean 接口或 @PreDestroy 注解）。

此外，如果 bean 希望在调用 SpringApplication.exit() 时返回特定的退出代码，则可以实现 org.springframework.boot.ExitCodeGenerator 接口。然后，可以将此退出代码传递给 System.exit()，以将其作为状态代码返回，如下所示：

```java
import org.springframework.boot.ExitCodeGenerator;
import org.springframework.boot.SpringApplication;
import org.springframework.boot.autoconfigure.SpringBootApplication;
import org.springframework.context.ConfigurableApplicationContext;
import org.springframework.context.annotation.Bean;
@SpringBootApplication
public class GfWebSampleApplication {
    @Bean
    public ExitCodeGenerator exitCodeGenerator() {
        return () -> 42;
    }
    public static void main(String[] args) {
        ConfigurableApplicationContext applicationContext =
                SpringApplication.run(GfWebSampleApplication.class, args);
        int status = SpringApplication.exit(applicationContext);
        System.exit(status);
    }
}
```

此外，ExitCodeGenerator 接口可以通过异常实现。当遇到此类异常时，Spring Boot 返回由实现的 getExitCode() 方法提供的退出代码。要启用优雅地关闭，请在 application.properties 配置文件中添加配置项，如下所示：

server.shutdown=graceful

要想给这个优雅关闭加个超时的话，如下所示：

spring.lifecycle.timeout-per-shutdown-phase=20s

4.9 管理员功能

通过在 application.properties 配置文件中指定 spring.application.admin.enabled 属性，可以为 Spring Boot 应用程序启用与管理相关的功能。这公开了平台 MBeanServer 上的 SpringApplicationAdminMXBean。可以使用此功能远程管理 Spring Boot 应用程序，此功能对于任何服务包装器实现也很有用。

如果需要知道应用程序在哪个 HTTP 端口上运行，请使用键 local.server.port。

4.10 本章小结

本章主要围绕着 SpringApplication 类而展开来讲解的，依次讲解了：延迟初始化、定制横幅、可用性、事件/监听器和 Web 环境等等。还讲解了如何访问应用程序参数，如何在应用启动时添加自己的业务代码，如何优雅地退出应用程序。

在下一章中，我们将一起探讨如何让同一个 Spring Boot 应用程序在不同的环境中使用。

第五章　外部化配置

Spring Boot 允许将配置外部化，这样就可以在不同的环境中使用相同的应用程序代码。可以使用属性文件、YAML 文件、环境变量和命令行参数来外部化配置。属性值可以使用 @Value 注解直接注入 bean，可以通过 Spring 的 Environment 抽象对其进行访问，也可以通过 @ConfigurationProperties 绑定到结构化对象。

Spring Boot 按照一定的顺序处理不同的属性源，其优先级从上至下依次递减，如下所示：

1) devtools 全局设置属性：它是在 $HOME/.config/spring-boot 目录中的，并且当 devtools 处于活动状态时才生效；
2) 测试上的 @TestPropertySource 注解；
3) 测试的 properties 属性：它在 @SpringBootTest 和测试注解中可用，用于测试应用程序的特定部分。
4) 命令行参数；
5) 来自 SPRING_APPLICATION_JSON 的属性：它是嵌入在环境变量或系统属性中的内联 JSON；
6) ServletConfig 初始化参数；
7) ServletContext 初始化参数；
8) 来自 java:comp/env 的 JNDI 属性；
9) Java 系统属性（System.getProperties()）；
10) 操作系统环境变量；
11) RandomValuePropertySource：它的属性仅在 random.* 中；
12) 在打包的 jar 之外的特定 profile 的应用程序属性：它们在 application-{profile}.properties 和 YAML 变体中；
13) 打包在 jar 中的特定 profile 的应用程序属性：它们在 application-{profile}.properties 和 YAML 变体中；
14) 在打包的 jar 之外的应用程序属性：它们在 application.properties 和 YAML 变体中；
15) 打包在 jar 中的应用程序属性：它们在 application.properties 和 YAML 变体中；
16) @Configuration 类上的 @PropertySource 注解。请注意，在刷新应用程序上下文之后，才会将这样的属性源添加到 Environment 中，这时配置某些属性（例如，在刷新开始之前读取的 logging.* 和 spring.main.*）为时已晚。
17) 默认属性：它是通过设置 SpringApplication.setDefaultProperties 方法指定的。

5.1 配置随机值

RandomValuePropertySource 用于注入随机值（例如，注入测试用例中）。它可以生成整数、长整型数、UUID 或字符串，如下所示：

1) my.secret=${random.value}
2) my.number=${random.int}
3) my.bignumber=${random.long}

4) my.uuid=${random.uuid}
5) my.number.less.than.ten=${random.int(10)}
6) my.number.in.range=${random.int[1024,65536]}
可以利用这一功能生成随机端口号，如下所示：
server.port=${random.int[1000,1999]}

5.2 应用程序属性文件

SpringApplication 从以下位置的 application.properties 配置文件中加载属性，并将它们添加到 Spring Environment 中：
1) 当前目录的 /config 子目录
2) 当前目录
3) 类路径 /config 包
4) 类路径根

这些位置的优先级从上至下依次递减，其中的当前目录指的是可执行的 jar 所在的目录。也可以使用 YAML 文件（.yml）代替".properties"。

可以通过 spring.config.name 环境属性改变 application.properties 配置文件的名称，可以通过 spring.config.location 环境属性指定配置文件的位置。通过 spring.config.additional-location 环境属性在默认位置的基础上追加配置文件的位置。有关详细信息，请访问以下网址：

https://docs.spring.io/spring-boot/docs/2.3.1.RELEASE/reference/html/spring-boot-features.html#boot-features-external-config-application-property-files

Spring Boot 配置文件的优先级，也可以通过逻辑图的形式来展现，上面的大于下面的，如图 5-1 所示。

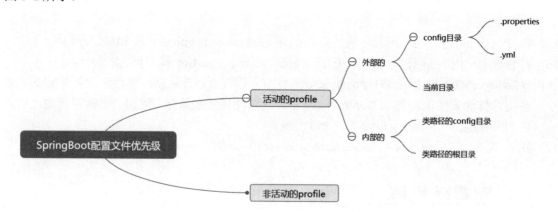

图 5-1 Spring Boot 配置文件优先级逻辑图

在 application.properties 配置文件中，后定义的配置项可以引用之前定义的配置项，如下所示：

gf.web-name=中国北京
gf.msg=${gf.web-name}是首都。

5.3 YAML 文件

YAML 是 JSON 的超集，它是用于指定分层配置数据的便捷格式。当类路径上有 SnakeYAML 库时，SpringApplication 类会自动支持 YAML 作为属性文件的替代者。在 Spring Boot 的 web 应用中，YAML 相关的依赖项是自动引入的，开发者无需关心这一点。

Spring Framework 提供了两个方便的类，可用于加载 YAML 文档。YamlPropertiesFactoryBean 将 YAML 加载为 Properties，而 YamlMapFactoryBean 将 YAML 加载为 Map。

可以通过自定义 @ConfigurationProperties 类，来把 YAML 文件的配置项赋值到 Java 实体类中。有关示例代码，如图 5-2 所示。

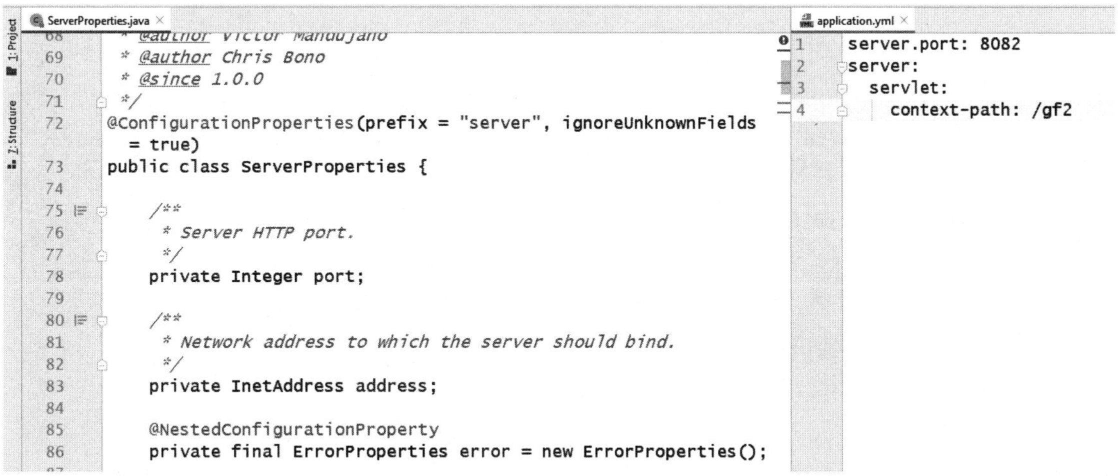

图 5-2 YAML 文件配置项

在 YAML 配置文件中，可以添加多个 profile 配置块，如图 5-3 所示。

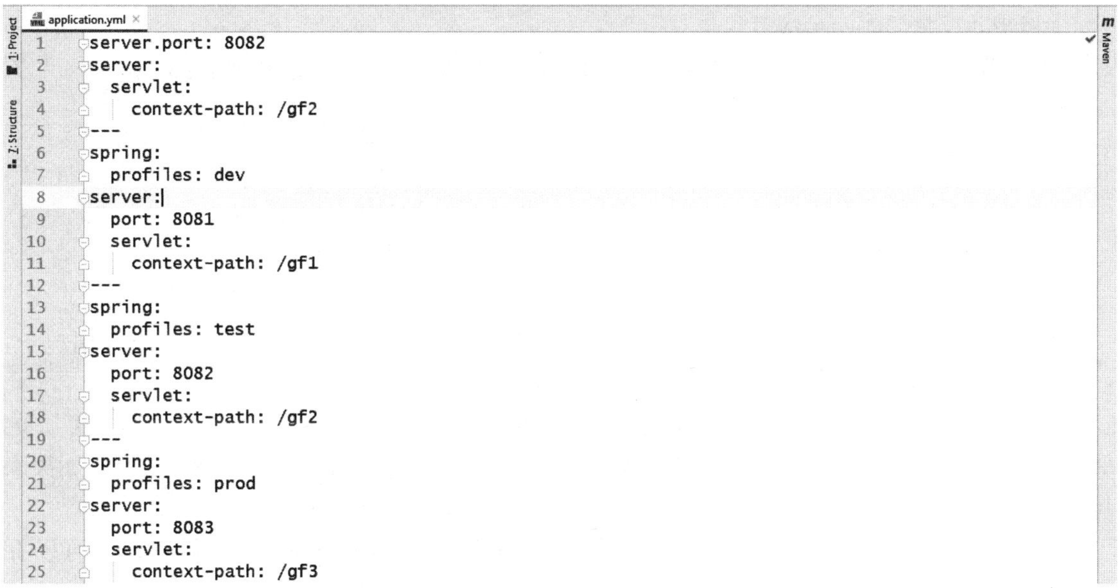

图 5-3 profile 配置块

5.4 Profiles

Spring Profiles 提供了一种把应用程序配置分离成为多个部分的方法，使某个部分仅在特定环境中可用。任何 @Component、@Configuration 或 @ConfigurationProperties 都可以标记为 @Profile，以限制何时加载相应的内容，如下所示：

```
package com.gf.sample.config;
import org.springframework.context.annotation.Bean;
import org.springframework.context.annotation.Configuration;
import org.springframework.context.annotation.Profile;
/**
 * 只有当 profile 的 dev 处于活动状态时，其中的 bean 才会生效。
 */
@Profile("dev")
@Configuration
public class DevConfig {
    @Bean
    public Object getMyObject() {
        //TODO: 编写具体的代码。
        return null;
    }
}
```

可以通过在 application.properties 配置文件中添加配置项来指定激活哪个 profile，如下所示：

spring.profiles.active=dev,gf

如果希望在任何情况下都可以加载某个 profile 配置块，则请在 application.properties 配置文件中添加配置项，如下所示：

spring.profiles.include=log

5.5 国际化

Spring Boot 支持本地化消息，以便应用程序可以适合不同语言偏好的用户。默认情况下，Spring Boot 会在类路径的根位置查找 messages 资源包的存在。

当配置的资源包的默认属性文件可用时，自动配置将应用（例如，默认的 messages.properties）。如果资源包仅包含特定语言的属性文件，则需要添加默认文件。如果未找到与任何配置的基名称匹配的属性文件，则不会自动配置 MessageSource。

想要查看更多可用的配置项，请查看 MessageSourceProperties 类的源代码。可以在 application.properties 配置文件中指定资源包的基名称，如下所示：

spring.messages.basename=config.i18n.msg

其中,"config.i18n"是多个以点号分隔的文件目录,"msg"是文件名的前缀,如图 5-4 所示。

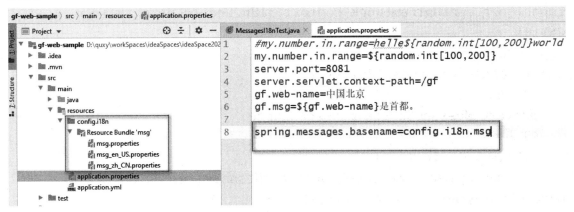

图 5-4 配置资源包文件目录

msg_zh_CN.properties 文件的内容,如图 5-5 所示。

图 5-5 msg_zh_CN.properties 文件内容

添加完各个文件后,就可以编写测试代码了,如下所示:

```
package com.gf.sample;
import lombok.extern.slf4j.Slf4j;
import org.junit.jupiter.api.Test;
import org.springframework.beans.factory.annotation.Autowired;
```

```java
import org.springframework.boot.test.context.SpringBootTest;
import org.springframework.context.MessageSource;
import org.springframework.context.i18n.LocaleContextHolder;
@Slf4j
@SpringBootTest
public class MessagesI18nTest {
    @Autowired
    private MessageSource messageSource;
    @Test
    public void test() {
        String msg = messageSource.getMessage("gf.china", null, LocaleContextHolder.getLocale());
//        msg = messageSource.getMessage("gf.china", null, Locale.CHINA);
        log.info(msg);
    }
}
```

在实际开发中,决定语言的参数大都是从前端应用传递到后端服务的,然后,后端服务接收到该参数之后就根据其值获取对应的属性值。

5.6 本章小结

本章介绍了 Spring Boot 应用程序的配置如何进行外部化,其中涵盖的内容有:生成随机值、Properties 文件、YAML 文件和 Profiles。最后,介绍了让 Spring Boot 应用程序支持多语言环境的国际化功能。

第六章 公共模块

总有一些工具类、开发时用到的便利配置和插件，它们是为了提高开发者的工作效率、编程质量而生。有了这些公共模块的支持，开发者将如虎添翼，将更加愿意在这优良的编程环境中编织自己的人生精彩。

6.1 公共类

公共类经常用于处理常见的代码逻辑固定的业务块，这些业务块被抽取出来形成一或多个方法进而形成公共类。公共类避免了开发者的重复编程，节省了开发者的编程时间。

6.1.1 GfResponseEntity 类

GfResponseEntity 类是基于 ResponseEntity 类构建的，它用于对响应数据的包装，从而使响应数据具有统一的 JSON 格式，示例代码，如下所示：

```java
package com.gf.common;
import org.springframework.http.ResponseEntity;
import java.io.Serializable;
import java.util.Collections;
import java.util.Optional;
/**
 * 本类用于对响应数据的包装，从而使响应数据具有统一的 JSON 格式。
 */
public class GfResponseEntity<T> implements Serializable {
    /**
     * HTTP 状态码
     */
    private int status;
    /**
     * 提示信息
     */
    private String msg = "";
    private long time = System.currentTimeMillis();
    /**
     * 业务数据
     */
    private T data;
```

```java
public GfResponseEntity() {}
public GfResponseEntity(T data) {
    this.data = data;
}
/**
 * 对象实例化
 */
public static <T> GfResponseEntity<T> self() {
    return new GfResponseEntity();
}
/**
 * 1、用于包装响应的对象数据。例如：根据 ID 查找详情信息。<br/>
 * 2、并且设置 HTTP 状态、内容长度、内容类型等信息。
 */
public ResponseEntity<GfResponseEntity<T>> build(T data, ResponseEntity.BodyBuilder bodyBuilder) {
    return build(data, (T) Collections.EMPTY_MAP, bodyBuilder);
}
/**
 * 1、用于包装响应的列表数据。例如：根据上级 ID 查找所有下级。<br/>
 * 2、并且设置 HTTP 状态、内容长度、内容类型等信息。
 */
public ResponseEntity<GfResponseEntity<T>> buildArr(T data, ResponseEntity.BodyBuilder bodyBuilder) {
    return build(data, (T) Collections.EMPTY_LIST, bodyBuilder);
}
/**
 * 1、用于包装响应数据。若没有响应数据，则可以返回默认值。<br/>
 * 2、并且设置 HTTP 状态、内容长度、内容类型等信息。
 */
public ResponseEntity<GfResponseEntity<T>> build(T data, T defaultValue, ResponseEntity.BodyBuilder bodyBuilder) {

    this.setData(Optional.ofNullable(data).orElse(defaultValue));
    ResponseEntity responseEntity = bodyBuilder.body(this);
    this.setStatus(responseEntity.getStatusCodeValue());
    return responseEntity;
}
/**
```

```
 * 用于请求执行成功后响应的数据。例如：创建数据成功后。
 */
public ResponseEntity ok() {
    return ok(null, (T) Collections.EMPTY_MAP);
}
/**
 * 用于包装响应的对象数据。例如：根据 ID 查找详情信息。
 */
public ResponseEntity<GfResponseEntity<T>> ok(T data) {
    return ok(data, (T) Collections.EMPTY_MAP);
}
/**
 * 用于包装响应的列表数据。例如：根据上级 ID 查找所有下级。
 */
public ResponseEntity<GfResponseEntity<T>> okArr(T data) {
    return ok(data, (T) Collections.EMPTY_LIST);
}
/**
 * 用于包装响应数据。若没有响应数据，则可以返回默认值。
 * @param data         业务数据
 * @param defaultValue 默认值
 */
public ResponseEntity<GfResponseEntity<T>> ok(T data, T defaultValue) {

this.setData(Optional.ofNullable(data).orElse(defaultValue));
    ResponseEntity responseEntity = ResponseEntity.ok(this);
    this.setStatus(responseEntity.getStatusCodeValue());
    return responseEntity;
}
/**
 * 设置提示信息
 */
public GfResponseEntity msg(String msg) {
    this.setMsg(msg);
    return this;
}
/**
 * 设置时间：毫秒数
```

```java
         */
        public GfResponseEntity time(long time) {
            this.setTime(time);
            return this;
        }
        public int getStatus() {
            return status;
        }
        private void setStatus(int status) {
            this.status = status;
        }
        public String getMsg() {
            return msg;
        }
        private void setMsg(String msg) {
            this.msg = msg;
        }
        public long getTime() {
            return time;
        }
        private void setTime(long time) {
            this.time = time;
        }
        public T getData() {
            return data;
        }
        private void setData(T data) {
            this.data = data;
        }
}
```

GfResponseEntity 类的定义代码只在此处列出，本书中其他代码块用到 GfResponseEntity 类的话，就不再重复列出定义代码。

6.1.2 StringUtil 类

StringUtil 类主要用于字符串的处理，是对已有的字符串处理类的补充。在各个框架中基本都有处理字符串的类，如下所示：

org.apache.commons.lang3.StringUtils

org.springframework.util.StringUtils

```java
import com.fasterxml.jackson.core.JsonProcessingException;
import com.fasterxml.jackson.databind.ObjectMapper;
import org.slf4j.Logger;
import org.slf4j.LoggerFactory;
public class StringUtil {
    private static final Logger log =
LoggerFactory.getLogger(StringUtil.class);
    /**
     * 将任意对象转换为字符串
     */
    public static String toString(ObjectMapper objectMapper, Object object) {
        try {
            return objectMapper.writeValueAsString(object);
        } catch (JsonProcessingException e) {
            log.error("StringUtil 报错：{}", e.getMessage());
        }
        return "";
    }
}
```

6.2 热加载

Spring Boot 包括一套增强的工具，可以使应用程序开发体验更加愉快。spring-boot-devtools 模块可以包含在任何项目中，以提供增强的开发时功能。要包括 devtools 支持，请将模块依赖项添加到 pom.xml 配置文件中。本书只讲解在 IntelliJ IDEA 中的热加载配置，有关 Eclipse 中的相关配置，还请自行查找资料。请添加依赖项，如下所示：

```xml
<dependency>
    <groupId>org.springframework.boot</groupId>
    <artifactId>spring-boot-devtools</artifactId>
    <optional>true</optional>
</dependency>
```

6.2.1 热加载静态资源

接下来,配置 IntelliJ IDEA,单击 File -> Settings... -> Compiler,如图 6-1 所示。

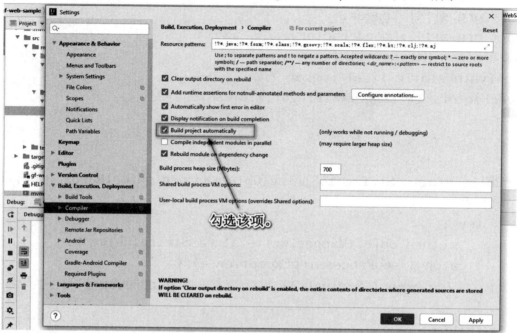

图 6-1 IntelliJ IDEA 配置

执行快捷键:Ctrl+Shift+Alt+/,并选择 "Registry...",如图 6-2 所示。

图 6-2 静态资源注册 1

然后，在弹出框中，勾选 compiler.automake.allow.when.app.running，如图 6-3 所示。

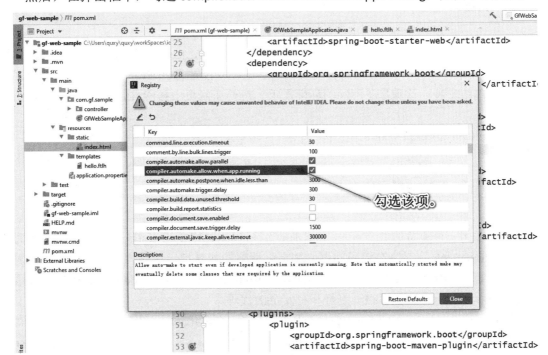

图 6-3 静态资源注册 2

经过以上配置后，即可实现不用重新启动应用而热加载静态资源和模板文件（有时候需要稍等片刻，再次刷新）。

6.2.2 热加载 Java 代码

经过上面的配置之后，修改 Java 代码会引发应用的重启启动。若希望禁用应用的重启启动，请在 application.yml 配置文件中添加配置项，如下所示：

spring.devtools.restart.enabled=false

配置后，应用只会在静态资源有修改时动态加载它们。更多详细的信息，请访问以下网址：

https://docs.spring.io/spring-boot/docs/2.3.1.RELEASE/reference/html/using-spring-boot.html#using-boot-devtools

另外一个值得一提的功能就是远程调试，可以在开发环境和测试环境中使用它。比较常见的应用场景是，远程调试测试环境中的问题，因为该问题不易在开发环境中重现。请记住，尽可能不要在生产环境中开启远程调试功能，以避免出现难以预料的问题。有关远程调试功能的详细信息，请访问以下网址：

https://docs.spring.io/spring-boot/docs/2.3.1.RELEASE/reference/html/using-spring-boot.html#using-boot-devtools-remote

6.3 IDEA 插件

每一款优秀的 IDE 工具都应该拥有丰富且优质的插件库，IDEA 就是这样的一款 IDE 工具。在 IDEA 插件库中，可以找到几乎所有你需要的一切辅助功能插件。IDEA 插件有两种安装方式：在线安装和离线安装，这满足了不同的环境需求。想要在网页中搜索 IDEA 插件的话，请访问以下网址：

https://plugins.jetbrains.com

6.3.1 SonarLint

SonarLint 是一个免费的 IDE 插件，可让你在编写代码时修复错误和漏洞。就像一个拼写检查器一样，SonarLint 实时高亮编码问题，并提供清晰的修复指导，这样你就可以在代码提交之前修复它们。在流行的 IDE（Eclipse、IntelliJ、Visual Studio 和 VS Code）和流行的编程语言中，SonarLint 帮助所有开发者编写更好和更安全的代码。

如果项目在 SonarQube 或 SonarCloud 上进行分析，SonarLint 可以连接到服务器以检索该项目的适当的质量配置文件和设置。运行 SonarLint 需要 Java 8，另外，JavaScript 和 TypeScript 的分析需要 Node.js >= 8。有两种方式可以安装 SonarLint：在线安装、离线安装。前者在 IDE 的插件管理中安装，后者需要提前从网站下载安装包。

启动 IDEA 后，依次单击：File -> Settings... -> Plugins，打开插件相关的弹出框，并搜索"SonarLint"，如图 6-4 所示。

图 6-4 搜索"SonarLint"插件

单击按钮[Install]，即可开始插件的安装，如图 6-5 所示。

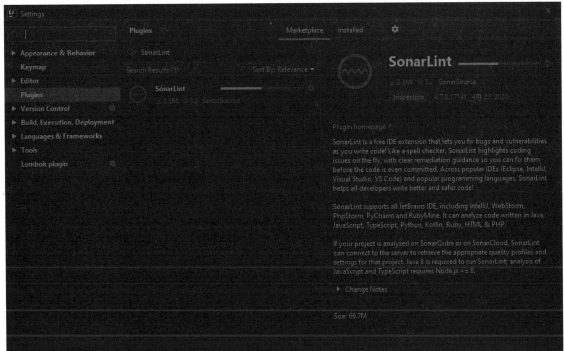

图 6-5 安装 "SonarLint" 插件

下载并安装后，需要重启 IDEA，如图 6-6 所示。

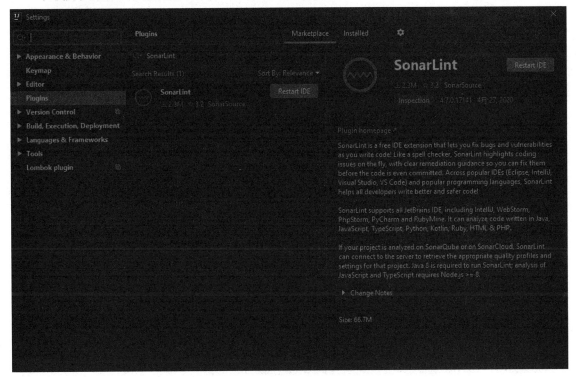

图 6-6 重启 IDEA

单击按钮[Restart IDE]，重启后插件即可生效，如图 6-7 所示。

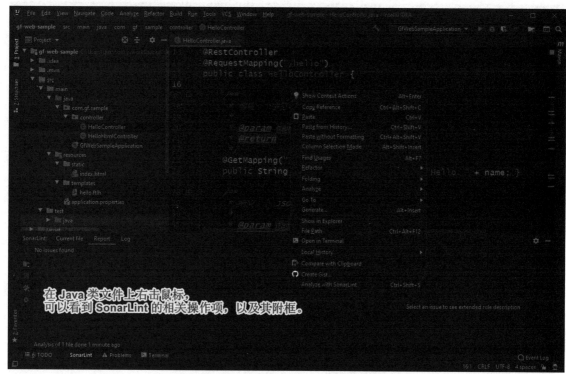

图 6-7 "SonarLint" 插件生效

也可以在项目名称上右击，对整个项目进行分析。执行分析操作后，我们看一下结果，如图 6-8 所示。

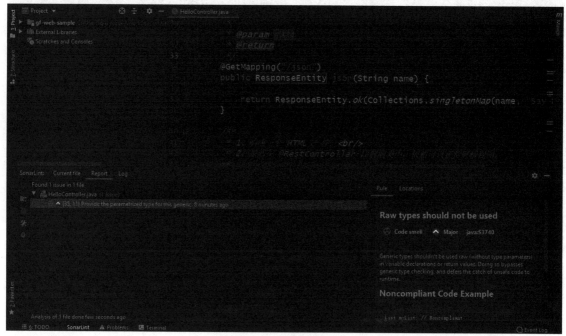

图 6-8 项目分析

在上图中列出需要调优的代码所在的行,并在右侧给出修改意见。各版本安装包见图6-9。

https://plugins.jetbrains.com/plugin/7973-sonarlint/versions

图 6-9 各版本安装包

下载离线安装包时,要注意插件的版本应与 IDEA 的版本相匹配。下载好后,我们可以把安装包直接拖拽到 IDEA 界面进行安装,还可以打开刚才的插件弹框,选择"从磁盘安装插件",如图 6-10 所示。

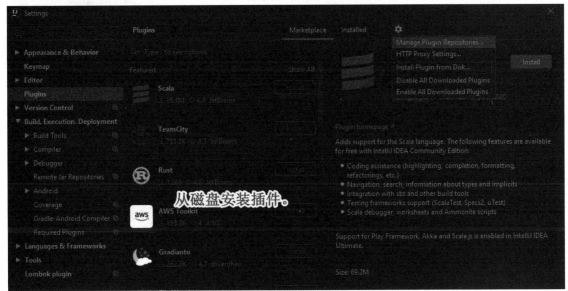

图 6-10 离线安装包下载

6.3.2 阿里 Java 规范插件

阿里 Java 规范插件是一款比较受欢迎的 Java 规范插件，许多公司的项目都在用。有了这款插件的支持，特别是对初级开发者来说，可以避免绝大部分低级规范问题（例如：类的首字母小写）。开始安装插件，打开 IDEA 的插件弹框后，搜索"alibaba"，即可找到本插件，如图 6-11 所示。

图 6-11 阿里 Java 规范插件

第一个就是阿里 Java 规范插件，请认准官方插件，其他类似的插件你有兴趣的话，也可以自行研究一下。插件离线安装包见图 6-12。

https://plugins.jetbrains.com/plugin/10046-alibaba-java-coding-guidelines/versions

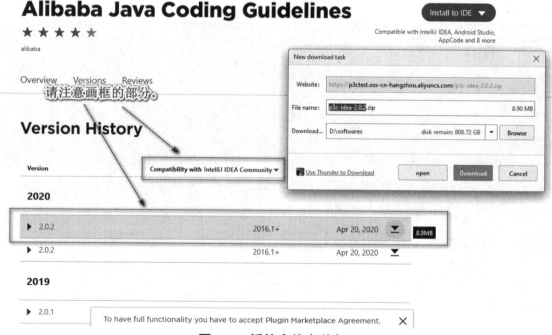

图 6-12 插件离线安装包

下载好离线安装包后，打开插件弹出框，并选择下载的安装包，如图 6-13 所示。

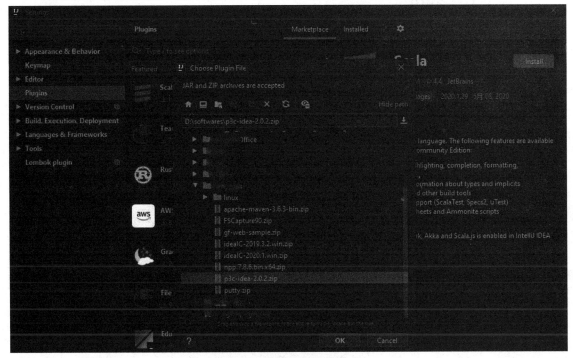

图 6-13 选择插件安装包

单击按钮[OK]，即可开始安装。安装完成后，记得重启 IDEA，才能使插件生效。右击 Java 类文件，可以看到相关的操作项，如图 6-14 所示。

图 6-14 安装插件

6.3.3 Lombok

有了这个插件，IntelliJ IDEA 就可以识别 lombok 项目中生成的所有 getter、setter 和其他一些东西，这样你就可以完成代码，并且可以在不出错的情况下使用这些方法。说得更直白一些就是，使用了 Lombok 插件及其依赖项，并在实体类上添加 @Data 注解，就可以免写 getter、setter 方法了，但在其他类中仍可以正常调用该实体类的 getter、setter 方法。Lombok 的神奇不止如此，希望你多加探索。离线安装包的下载网址，如图 6-15 所示。

https://plugins.jetbrains.com/plugin/6317-lombok/versions

图 6-15 Lombok 插件

6.4 本章小结

本章讲解了几个公共类、热加载功能和几个 IDEA 插件，它们都是为高效并优雅地编写代码服务的。本章所涵盖的内容只是冰山一角，还有大量的这样的类和工具等待大家去发现去使用。

第二部分：中级篇

本篇主要讲解 Spring MVC、Spring WebFlux、数据库操作、缓存和安全等技术，这涵盖了一个完整的 Spring Boot web 应用程序的绝大部分知识点。另外，针对有些知识点还给出相应的扩展链接，以此来带给大家更多的外延内容。

第七章　Spring MVC

　　Spring Web MVC 是基于 Servlet API 构建的原始 web 框架，从一开始就包含在 Spring Framework 中，但通常称其为"Spring MVC"。Spring MVC 是富"模型-视图-控制器"的 web 框架，它允许使用控制器 bean（带有 @Controller 和 @RestController 注解的类）处理 HTTP 请求。其中，MVC 是 Model-View-Controller 的缩写，即：模型-视图-控制器。通过使用 @RequestMapping、@PostMapping 和 @GetMapping 等注解来将控制器 bean 中的方法映射到相应的 HTTP 请求路径。执行方法中的逻辑代码并返回相应的结果数据来完成对 HTTP 请求的处理。

　　通过 Spring MVC 的相关注解将 HTTP 请求路径与控制器中方法关联起来，示例代码，如下所示：

```java
package com.gf.mvc.controller;
import com.fasterxml.jackson.core.JsonProcessingException;
import com.fasterxml.jackson.databind.ObjectMapper;
import com.gf.mvc.common.*;
import com.gf.mvc.dto.MenuDTO;
import com.gf.mvc.vo.MenuVO;
import org.slf4j.Logger;
import org.slf4j.LoggerFactory;
import org.springframework.beans.factory.annotation.Autowired;
import org.springframework.web.bind.annotation.*;
import java.util.Arrays;
import java.util.List;
@RestController
@RequestMapping({"/mvc/menu", "/mvc/menu/v2"})
public class MenuController {
    private Logger log = LoggerFactory.getLogger(this.getClass());
    @Autowired
    private ObjectMapper objectMapper;
    @PostMapping("/create")
    public JsonResponse create(MenuDTO params) throws JsonProcessingException {
        //TODO: 将新创建的菜单项，保存到数据库中。
        log.info("成功创建菜单项：", objectMapper.writeValueAsString(params));
        return GfResponse.success();
    }
    @GetMapping("/view/{id}")
```

```java
    public JsonResponse view(@PathVariable("id") String id) {
        MenuVO menuVO = null;
        if ("1".equals(id)) {
            menuVO = new MenuVO();
            menuVO.setId("1");
            menuVO.setName("用户管理");
            menuVO.setParentId("0");
            menuVO.setSorted(20);
        }
        return GfResponse.success(menuVO);
    }
    @RequestMapping("/list")
    public JsonResponse list() {
        List<String> list = Arrays.asList("用户管理", "菜单管理");
        return GfResponse.success(list);
    }
}
```

针对以上示例代码，给出如下解释：

1) @RestController 注解：表明本类是一个 RESTfull 风格的控制器类。
2) @RequestMapping 注解：它既可以用在类上，也可以用在方法上。它用在方法上时可以处理 POST 和 GET 等类型的请求。但是，我们建议只将其用在类上，在方法上只应用具体请求类型的注解。提示一下：它用在方法上时，代码质量检查工具 SonarQube 会将其判定为需要优化修改的项。它可以接受以 "{}" 括号包裹的路径数组，通过使用这种数组形式，我们可以在升级 API 的同时保证原有的 API 可用。
3) @PostMapping 和 @GetMapping 注解：这个两个注解只接受对应类型的请求，否则控制台会给出警告：不支持的请求类型，如下所示：

[org.springframework.web.HttpRequestMethodNotSupportedException: Request method 'GET' not supported]

本书的侧重点是前后端分离后的 Spring MVC 后端（非 UI 渲染）技术，在此不讲解模板引擎的技术。若需要查看相关技术，还请访问 Spring Boot 参考指南的相关网页，如下所示：

https://docs.spring.io/spring-boot/docs/2.3.1.RELEASE/reference/html/spring-boot-features.html#boot-features-spring-mvc-template-engines

7.1 DispatcherServlet

与许多其他 web 框架一样，Spring MVC 是围绕前控制器模式设计的，其中一个中心 Servlet DispatcherServlet 为请求处理提供了一个共享算法，而实际工作则由可配置的委托组件执行。该模型灵活，支持多种工作流。

DispatcherServlet 和任何 Servlet 一样，需要使用 Java 配置或 web.xml 并根据 Servlet 规范来声明和映射。反过来，DispatcherServlet 使用 Spring 配置来发现请求映射、视图解析、异常处理等所需的委托组件。

7.2 上下文层次结构

根 WebApplicationContext 通常包含基础架构 bean，例如，需要在多个 Servlet 实例之间共享的数据存储库和业务服务。这些 bean 是有效继承的，并且可以在特定 Servlet 的子 WebApplicationContext 中被覆盖（即重新声明），该子 WebApplicationContext 通常包含给定 Servlet 的本地 bean。这种关系的图形化表示，如图 7-1 所示。

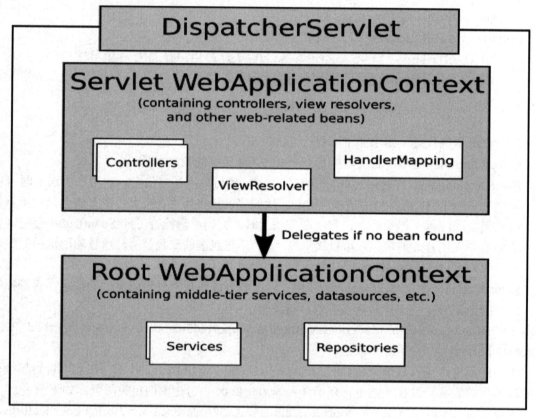

图 7-1 上下文层次结构图

7.3 拦截器

所有 HandlerMapping 实现都支持处理程序拦截器，当想要将特定功能应用于某些请求时，这些拦截器很有用。拦截器必须实现 HandlerInterceptor 接口，该接口提供了 3 个方法，它们提供足够的灵活性来进行各种预处理和后处理，如下所示：
1) preHandle(...)：本方法在执行实际处理程序之前执行；

2) postHandle(...)：本方法在执行实际处理程序之后执行；
3) afterCompletion(...)：本方法在完成请求后执行，它用得比较少。

有关定义拦截器和注册拦截器的示例代码，如下所示：

```java
package com.gf.mvc.common;
import org.slf4j.Logger;
import org.slf4j.LoggerFactory;
import org.springframework.web.servlet.HandlerInterceptor;
import org.springframework.web.servlet.ModelAndView;
import javax.servlet.http.HttpServletRequest;
import javax.servlet.http.HttpServletResponse;
/**
 * 定义一个拦截器
 */
public class GfHandlerInterceptor implements HandlerInterceptor {
    private static final Logger log =
LoggerFactory.getLogger(GfHandlerInterceptor.class);
    @Override
    public boolean preHandle(HttpServletRequest request,
HttpServletResponse response, Object handler) throws Exception {
        log.info("在业务代码执行之前执行。");
        return true;
    }
    @Override
    public void postHandle(HttpServletRequest request,
HttpServletResponse response, Object handler, ModelAndView
modelAndView) throws Exception {
        log.info("在业务代码执行之后执行。");
    }
    @Override
    public void afterCompletion(HttpServletRequest request,
HttpServletResponse response, Object handler, Exception ex) throws
Exception {
        log.info("在完成请求后执行。");
    }
}


package com.gf.mvc.config;
import com.gf.mvc.common.GfHandlerInterceptor;
```

```java
import org.springframework.context.annotation.Configuration;
import org.springframework.web.servlet.config.annotation.InterceptorRegistry;
import org.springframework.web.servlet.config.annotation.WebMvcConfigurer;
/**
 * 注册一个拦截器
 */
@Configuration
public class WebConfig implements WebMvcConfigurer {
    @Override
    public void addInterceptors(InterceptorRegistry registry) {
        registry.addInterceptor(new GfHandlerInterceptor());
    }
}
```

preHandle(...) 方法返回一个布尔值。你可以使用此方法来中断或继续执行链的处理。当此方法返回 true 时，处理程序执行链将继续。当它返回 false 时，DispatcherServlet 假设拦截器本身已经处理了请求（例如，呈现了一个适当的视图）并且不会继续处理执行链中的其他拦截器和实际处理程序。

注意，postHandle 在 @ResponseBody 和 ResponseEntity 的方法中用处不大，响应是在 HandleRapter 内和 postHandle 之前编写和提交的。这意味着对响应进行任何更改（例如：添加额外的头信息）都为时已晚。对于此类场景，可以实现 ResponseBodyAdvice 接口，并将其声明为 @ControllerAdvice bean，或者直接在 RequestMappingHandlerAdapter 上配置它。

在响应返回到客户端之前，添加头信息，并实现自定义 ContentType 的类型，如下所示：

```java
package com.gf.mvc.common;
import org.springframework.core.MethodParameter;
import org.springframework.http.HttpHeaders;
import org.springframework.http.MediaType;
import org.springframework.http.converter.HttpMessageConverter;
import org.springframework.http.server.ServerHttpRequest;
import org.springframework.http.server.ServerHttpResponse;
import org.springframework.util.MimeType;
import org.springframework.web.bind.annotation.RestControllerAdvice;
import org.springframework.web.servlet.mvc.method.annotation.ResponseBodyAdvice;
@RestControllerAdvice
```

```java
public class GfResponseBodyAdvice implements
ResponseBodyAdvice<String> {
    @Override
    public boolean supports(MethodParameter methodParameter, Class<? extends HttpMessageConverter<?>> aClass) {
        return true;
    }
    @Override
    public String beforeBodyWrite(String body, MethodParameter methodParameter, MediaType mediaType, Class<? extends HttpMessageConverter<?>> aClass, ServerHttpRequest serverHttpRequest, ServerHttpResponse serverHttpResponse) {
        HttpHeaders httpHeaders = serverHttpResponse.getHeaders();
        // 添加头信息
        httpHeaders.add("org", "gf.com");
        // 自定义 ContentType
        httpHeaders.setContentType(MediaType.asMediaType(MimeType.valueOf("application/json;charset=utf-8")));
        return body;
    }
}
```

7.4 文件上传

要启用 multipart 处理，需要在 DispatcherServlet 的 Spring 相应配置中声明一个名为 multipartResolver 的 MultipartResolver bean。关于这，Spring MVC 已经为我们提供了一个默认的 StandardServletMultipartResolver。DispatcherServlet 检测它并将其应用于传入请求。当收到内容类型为 multipart/form-data 的 POST 请求时，解析器解析内容并将当前的 HttpServletRequest 包装为 MultipartHttpServletRequest，以提供对解析部分的访问，此外还将它们作为请求参数公开。

在 application.yml 配置文件中可以配置的文件上传的键，如表 7-1 所示。

表格 7-1 可配置键

键	默认值	描述
spring.servlet.multipart.enabled	true	是否支持 multipart 上传。
spring.servlet.multipart.file-size-threshold	0 B	文件写入磁盘的阈值。
spring.servlet.multipart.location		上传文件的中间位置。
spring.servlet.multipart.max-file-size	1 MB	单个文件的最大大小。
spring.servlet.multipart.max-request-size	10 MB	一次请求中多个文件总的最大大小。

| spring.servlet.multipart.resolve-lazily | false | 是否在文件或参数访问时延迟解析 multipart 请求。 |

一个或多个文件上传的示例代码,如下所示:

```java
package com.gf.mvc.controller;
import org.slf4j.Logger;
import org.slf4j.LoggerFactory;
import org.springframework.web.bind.annotation.PostMapping;
import org.springframework.web.bind.annotation.RequestMapping;
import org.springframework.web.bind.annotation.RestController;
import org.springframework.web.multipart.MultipartFile;
import javax.servlet.http.HttpServletRequest;
@RestController
@RequestMapping("/file")
public class FileUploadController {
    private static final Logger log = LoggerFactory.getLogger(FileUploadController.class);
    /**
     * 同时上传多个文件
     */
    @PostMapping("/uploads")
    public String uploads(HttpServletRequest request,MultipartFile[] files) {
        if (files != null && files.length > 0) {
            for (MultipartFile f : files) {
                log.info("文件名称:" + f.getName() + ",文件大小:" + f.getSize());
            }
        }
        return "success";
    }
    /**
     * 上传一个文件
     */
    @PostMapping("/upload")
    public String upload(HttpServletRequest request,MultipartFile file) {
        log.info("文件名称:" + file.getName() + ",文件大小:" + file.getSize());
```

```
    return "success";
  }
}
```

7.5 静态资源

在前后端代码分离的应用模式中，html、css、js 和图片等静态资源是与后端代码分开放置的，放在不同的项目中。web 项目是从前后端代码和资源放在一个项目中逐步发展到前后端完全分离的。

单体应用的发展路径：Servlet、JSP、后端模板技术和前后端分离（前后端代码和资源放在一个项目，但前后端代码不放在一个文件中）。再之后，就是完全的前后端分离，后端代码放在一或多个项目中，前端代码放在另一或多个静态项目（借助 Nginx 服务器、Node.js 技术等构建成一个完整的应用服务，当然也可以使用 Spring Boot 存放各种静态资源）。前端应用通过后端应用提供的 API 来加载各种各样的数据，并与前端资源结合，最终向用户呈现视图界面。

Spring MVC 可以提供后端模板技术及两种前后端分离，这三种应用模式下的应用程序开发。从随着业务模块和各项功能的不断增加，单体应用变得越来越庞大臃肿，并且难以维护，bug 排查比较困难。小的 bug 修复就需要重启整个应用，这不利于应用服务的稳定，并且会降低最终用户的使用体验。庞大的单体应用，其开发工作也变得比较复杂，不同开发者的代码会相互影响，存在误改等行为，并且容易造成代码提交冲突。庞大的单体应用，其中的业务模块不能分别逐步上线，必须要等到其他业务模块开发完成才能一起上线。这样会导致单体应用迭代周期比较长。由于单体应用的以上种种问题，将其按照业务模块拆分成多个小微型应用，这成为了一种趋势。这种趋势最终演变成今天的微服务模式：网关服务、注册中心、配置中心、业务服务等，它们共同组成了一个完整有序的小型应用服务群，并且每个服务都会有多个节点，以此来增加各个服务的稳定性和容错性，并且提供相应的负载均衡和熔断等功能。

将前端项目包装在 Spring Boot 的静态资源目录中，这样就可以将其改造为一个只包含静态资源的 Spring Boot 项目了。这样做的好处是便于 Spring Cloud 中的注册中心、配置中心等管理该前端项目。

默认情况下，Spring Boot 从类路径中的 /static 目录提供静态资源，也可以是目录：/public、resources 或 /META-INF/resources。这些目录可以一起使用，默认情况下，这些目录中的资源都被映射到 /**。放在这些目录中的资源可以直接通过 HTTP 请求访问，如下所示：

http://localhost:8080/static.txt。

这些静态资源目录在项目中的具体位置，如图 7-2 所示。

图 7-2 静态资源目录

项目编译后，这些静态资源目录被放置在 classes 目录（类路径根目录）下，如图 7-3 所示。

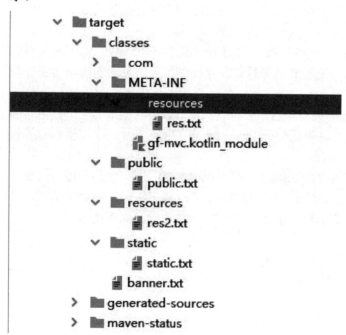

图 7-3 静态资源所在位置

可以通过 application.yml 配置文件中的 spring.mvc.static-path-pattern 属性调整静态资源映

射到 URL 的路径，它的默认值为：/**。

可以通过 application.yml 配置文件中的 spring.resources.static-locations 属性自定义静态资源的位置，它的默认值为：

classpath:/META-INF/resources/,classpath:/resources/,classpath:/static/,classpath:/public/

spring.mvc.static-path-pattern 属性用于匹配一个 URL 是否是静态资源。若是，则从 URL 中提取静态资源的相对路径（包含文件名）。接下来，spring.resources.static-locations 属性定义了在服务器的哪个位置查找该静态资源，与该相对路径一起，最终找到相应的资源。

若控制器类中方法的 URL 与静态资源对应的 URL 相同，则该 URL 优先应用于前者。

7.6 全局异常处理

默认情况下，Spring Boot 提供了一个 /error 映射，该映射以合理的方式处理所有错误，并且在 servlet 容器中注册为"全局"错误页面。对于机器客户端，它会生成一个 JSON 响应，其中包含错误、HTTP 状态和异常消息的详细信息。对于浏览器客户端，有一个"白标签"错误视图，它以 HTML 格式呈现相同的数据（要自定义它，请添加一个可解析为 error 的 View）。要完全替换默认行为，可以实现 ErrorController 接口并注册该类型的 bean 定义，或者添加 ErrorAttributes 类型的 bean 以使用现有机制，但替换内容。在 Spring MVC 中有三种处理全局异常的方式。

7.6.1 只替换响应内容

下面的类实现了 ErrorAttributes 接口，添加了响应数据字段：time、data 和 msg。下面的类的具体方法代码可以参考 org.springframework.boot.web.servlet.error.DefaultErrorAttributes 的方法代码。下面的类中未列出的方法，可以直接从 DefaultErrorAttributes 类中复制，并根据需要做适当修改。

```
import org.springframework.boot.web.servlet.error.DefaultErrorAttributes;
import org.springframework.boot.web.servlet.error.ErrorAttributes;
import org.springframework.http.HttpStatus;
import org.springframework.util.StringUtils;
import org.springframework.validation.BindingResult;
import org.springframework.web.bind.MethodArgumentNotValidException;
import org.springframework.web.context.request.RequestAttributes;
import org.springframework.web.context.request.WebRequest;
import javax.servlet.ServletException;
import java.io.PrintWriter;
import java.io.StringWriter;
```

```java
import java.util.*;

public class GfErrorAttributes implements ErrorAttributes {
    private static final String ERROR_ATTRIBUTE =
DefaultErrorAttributes.class.getName() + ".ERROR";
    private final boolean includeException;
    public GfErrorAttributes() {this(false);}
    public GfErrorAttributes(boolean includeException) {
        this.includeException = includeException;
    }
    @Override
    public Map<String, Object> getErrorAttributes(WebRequest
webRequest, boolean includeStackTrace) {
        Map<String, Object> errorAttributes = new LinkedHashMap();
        errorAttributes.put("time", System.currentTimeMillis());
        errorAttributes.put("data", Collections.EMPTY_MAP);
        this.addStatus(errorAttributes, webRequest);
        this.addErrorDetails(errorAttributes, webRequest,
includeStackTrace);
        this.addPath(errorAttributes, webRequest);
        return errorAttributes;
    }

    private void addErrorDetails(Map<String, Object> errorAttributes,
WebRequest webRequest, boolean includeStackTrace) {
        Throwable error = this.getError(webRequest);
        if (error != null) {
            while (true) {
                if (!(error instanceof ServletException) ||
error.getCause() == null) {
                    if (this.includeException) {
                        errorAttributes.put("exception",
error.getClass().getName());
                    }

                    this.addErrorMessage(errorAttributes, error);
                    if (includeStackTrace) {
                        this.addStackTrace(errorAttributes, error);
                    }
```

```
                break;
            }
            error = error.getCause();
        }
    }

    Object message = this.getAttribute(webRequest,
"javax.servlet.error.message");
    if ((!StringUtils.isEmpty(message) ||
errorAttributes.get("msg") == null) && !(error instanceof
BindingResult)) {
        errorAttributes.put("msg", StringUtils.isEmpty(message) ?
"No message available" : message);
    }
  }
  ……
}
```

```
import org.springframework.boot.web.servlet.FilterRegistrationBean;
import org.springframework.boot.web.servlet.error.ErrorAttributes;
import org.springframework.context.annotation.Bean;
import org.springframework.context.annotation.Configuration;
@Configuration
public class WebConfiguration {
    /**
     * 添加 ErrorAttributes bean，以替换默认的 DefaultErrorAttributes
     */
    @Bean
    public ErrorAttributes errorAttributes() {
        return new GfErrorAttributes(true);
    }
}
```

7.6.2 根据响应内容类型添加处理程序

BasicErrorController 可以用作实现 ErrorController 接口的自定义类的基类。如果想为新的内容类型（ContentType）添加处理程序（默认是专门处理 text/html 并为其他所有内容提供回退），这一点特别有用。为此，需要创建新类型的 bean，并继承 BasicErrorController，添加具有 produces 属性的 @RequestMapping 的公共方法，如下所示：

```
import org.springframework.boot.autoconfigure.web.ErrorProperties;
import org.springframework.boot.autoconfigure.web.servlet.error.BasicErrorController;
import org.springframework.boot.autoconfigure.web.servlet.error.ErrorViewResolver;
import org.springframework.boot.web.servlet.error.ErrorAttributes;
import org.springframework.http.HttpStatus;
import org.springframework.http.MediaType;
import org.springframework.http.ResponseEntity;
import org.springframework.stereotype.Controller;
import org.springframework.web.bind.annotation.ExceptionHandler;
import org.springframework.web.bind.annotation.RequestMapping;
import javax.servlet.http.HttpServletRequest;
import java.util.List;
import java.util.Map;
@Controller
public class GfErrorController extends BasicErrorController {

    public GfErrorController(ErrorAttributes errorAttributes, List<ErrorViewResolver> errorViewResolvers) {
        super(errorAttributes, new ErrorProperties(), errorViewResolvers);
    }
    /**
     * 如果需要更改现有的处理方法，则可以将其覆盖即可。
     */
    @RequestMapping(produces = MediaType.APPLICATION_JSON_VALUE)
    @Override
    public ResponseEntity<Map<String, Object>> error(HttpServletRequest request) {
        HttpStatus status = this.getStatus(request);
        if (status == HttpStatus.NO_CONTENT) {
            return new ResponseEntity(status);
        } else {
```

```
            Map<String, Object> body = this.getErrorAttributes(request,
this.isIncludeStackTrace(request, MediaType.APPLICATION_JSON));
            return new ResponseEntity(body, status);
        }
    }
    /**
     * 为新的内容类型（ContentType）添加处理程序
     */
    @RequestMapping(produces = MediaType.APPLICATION_PDF_VALUE)
    public ResponseEntity<Map<String, Object>>
errorPdf(HttpServletRequest request) {
        HttpStatus status = this.getStatus(request);
        if (status == HttpStatus.NO_CONTENT) {
            return new ResponseEntity(status);
        } else {
            Map<String, Object> body = this.getErrorAttributes(request,
this.isIncludeStackTrace(request, MediaType.APPLICATION_PDF));
            return new ResponseEntity(body, status);
        }
    }
    @ExceptionHandler(RuntimeException.class)
    public ResponseEntity<String>
mediaTypeNotAcceptable(HttpServletRequest request) {
        HttpStatus status = getStatus(request);
        return ResponseEntity.status(status).build();
    }
}
```

7.6.3 根据异常类型添加处理程序

还可以定义一个用 @ControllerAdvice 注解的类，为不同的控制器和（或）异常类型响应的不同的 JSON 数据，如下所示：

```
import com.gf.mvc.controller.MenuController;
import org.springframework.http.HttpStatus;
import org.springframework.http.ResponseEntity;
import org.springframework.web.bind.annotation.ControllerAdvice;
import org.springframework.web.bind.annotation.ExceptionHandler;
import org.springframework.web.bind.annotation.ResponseBody;
```

```java
import org.springframework.web.context.request.RequestAttributes;
import org.springframework.web.context.request.ServletWebRequest;
import org.springframework.web.context.request.WebRequest;
import org.springframework.web.servlet.mvc.method.annotation.ResponseEntityExceptionHandler;
import javax.servlet.http.HttpServletRequest;
@ControllerAdvice(basePackageClasses = MenuController.class)
public class GfControllerAdvice extends ResponseEntityExceptionHandler {
    @ExceptionHandler(NullPointerException.class)
    @ResponseBody
    ResponseEntity<?> handleControllerException(HttpServletRequest request, Throwable ex) {
        HttpStatus status = getStatus(request);
        String path = request.getRequestURI();
        return new ResponseEntity<>(new CustomErrorType(status.value(), ex.getMessage(), path), status);
    }
    private HttpStatus getStatus(HttpServletRequest request) {
        WebRequest webRequest = new ServletWebRequest(request);
        Integer statusCode = (Integer) this.getAttribute(webRequest, "javax.servlet.error.status_code");
        if (statusCode == null) {
            return HttpStatus.INTERNAL_SERVER_ERROR;
        }
        return HttpStatus.valueOf(statusCode);
    }

    private <T> T getAttribute(RequestAttributes requestAttributes, String name) {
        return (T) requestAttributes.getAttribute(name, RequestAttributes.SCOPE_REQUEST);
    }
}
```

在上面的示例中,如果与 MenuController 控制器类在同一包中定义的控制器抛出了异常的类型 NullPointerException,则向客户端响应 CustomErrorType 实体类的 JSON 数据而不是 ErrorAttributes 的数据。ErrorController 接口的现实类处理所有未被处理的任何异常。

7.7 跨域资源共享

出于安全考虑,浏览器禁止 AJAX 调用当前来源之外的资源。例如,A 站点中 JS 脚本向 B 站点的 API 发起请求,这是被禁止的。

跨域资源共享(CORS)是被大多数浏览器实现的 W3C 规范,可让你指定授权的跨域请求类型,而不是使用基于 IFRAME 或 JSONP 的不太安全和不太强大的解决方法。从 4.2 版本开始,Spring MVC 支持跨域资源共享。

7.7.1 为指定 API 添加 CORS

可以通过为控制器中的方法添加 @CrossOrigin 注解来实现单个 API 的 CORS 功能。这种方式只需要添加该注解即可,无需其他任何特定配置,如下所示:

```
import org.springframework.http.ResponseEntity;
import org.springframework.web.bind.annotation.CrossOrigin;
import org.springframework.web.bind.annotation.GetMapping;
import org.springframework.web.bind.annotation.RequestMapping;
import org.springframework.web.bind.annotation.RestController;
import java.util.Collections;
@RestController
@RequestMapping("/mvc/cross")
public class CrossOriginController {
    @GetMapping("/view")
    @CrossOrigin
    public ResponseEntity view() {
        return ResponseEntity.ok(Collections.singletonMap("gf", "中国"));
    }
}
```

还可以在类级别添加 @CrossOrigin 注解,这样该控制器中的所有方法都会继承 CORS 特性,如下所示:

```
import org.springframework.http.ResponseEntity;
import org.springframework.web.bind.annotation.CrossOrigin;
import org.springframework.web.bind.annotation.GetMapping;
import org.springframework.web.bind.annotation.RequestMapping;
import org.springframework.web.bind.annotation.RestController;
import java.util.Collections;
@CrossOrigin(origins = "http://localhost:8082",maxAge =1800 )
```

```
@RestController
@RequestMapping("/mvc/cross2")
public class CrossOrigin2Controller {
    @GetMapping("/view")
    public ResponseEntity view() {
        return ResponseEntity.ok(Collections.singletonMap("gf", "中国-类级别 CORS"));
    }
}
```

1) origins 属性可以指定一或多个允许的 URL。
2) maxAge 属性是预检响应的缓存持续时间的最大期限（以秒为单位，默认值：1800 秒，即 30 分钟）。它控制预检请求的 Access-Control-Max-Age 响应头的值。将它设置为合理的值可以减少浏览器所需的预检请求/响应交互次数。负值表示未定义。

在 JS 代码中，发出未设置 CORS 的 URL 请求，在浏览器的控制台中会报错误，从而导致该请求无法完成，如图 7-4 所示。

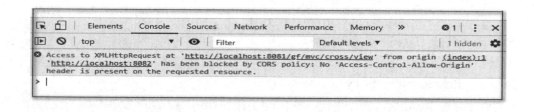

图 7-4 浏览器控制台

7.7.2 全局配置

自定义一个 WebMvcConfigurer 接口的实现类，并重写 addCorsMappings(CorsRegistry) 方法，然后将该类注册为 bean，即可为匹配的 URL 配置 CORS。示例代码，如下所示：
```
import org.springframework.context.annotation.Bean;
import org.springframework.context.annotation.Configuration;
```

```java
import org.springframework.web.servlet.config.annotation.CorsRegistry;
import org.springframework.web.servlet.config.annotation.WebMvcConfigurer;
@Configuration
public class WebConfiguration {
    /**
     * 为匹配的 URL 配置 CORS
     */
    @Bean
    public WebMvcConfigurer corsConfigurer() {
        return new WebMvcConfigurer() {
            @Override
            public void addCorsMappings(CorsRegistry registry) {
                registry.addMapping("/api/**");
            }
        };
    }
}
```

添加 CORS 全局配置后，所有与 /api/** 匹配的 URL 都支持 CORS 请求。请注意上下文路径（server.servlet.context-path 属性指定的值）不参与模式匹配，本例中 /gf 不参与匹配，如图 7-5 所示。

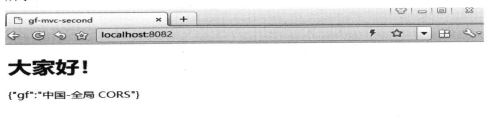

图 7-5 不参与模式匹配的上下文路径

在过滤器章节也有关于 CORS 的内容，以过滤器的方式实现了 CORS。有关详细信息，请查看该章节。

7.8 过滤器

Spring MVC 使用过滤器来处理：表单数据、转发头、弱 ETag 和 CORS。在本书中，只讲自定义过滤器和 CORS 过滤器。

过滤器和拦截器在功能方面比较相近，但具体实现技术是不同的。两者的区别，如下所示：
1) 过滤器是依赖于 Servlet 容器，是 Servlet 规范的一部分，而拦截器则是独立存在的；
2) 过滤器的执行由 Servlet 容器回调完成，而拦截器通常通过动态代理的方式来执行；
3) 过滤器的生命周期由 Servlet 容器管理，而拦截器则可以通过 IoC 容器来管理。

7.8.1 自定义过滤器

自定义过滤器最常用的方式是实现 Filter 接口。定义过滤器和注册过滤器的示例代码，如下所示：

```java
package com.gf.mvc.config;
import org.slf4j.Logger;
import org.slf4j.LoggerFactory;
import javax.servlet.*;
import java.io.IOException;
/**
 * 定义 web 过滤器
 */
public class GfFilter implements Filter {
    private Logger log = LoggerFactory.getLogger(this.getClass());
    @Override
    public void doFilter(ServletRequest servletRequest,
ServletResponse servletResponse, FilterChain filterChain) throws
IOException, ServletException {
        log.info("执行前：逻辑处理代码");
        filterChain.doFilter(servletRequest, servletResponse);
        log.info("执行后：逻辑处理代码");
    }
}
```

```java
package com.gf.mvc.config;
import org.springframework.boot.web.servlet.FilterRegistrationBean;
```

```
import org.springframework.context.annotation.Bean;
import org.springframework.context.annotation.Configuration;
@Configuration
public class WebConfiguration {
    /**
     * 注册 web 过滤器
     */
    @Bean
    public FilterRegistrationBean filterRegistrationBean() {
        FilterRegistrationBean registrationBean = new FilterRegistrationBean(new GfFilter());
        registrationBean.addUrlPatterns("/*");
        registrationBean.setName("gfFilter");
        return registrationBean;
    }
}
```

7.8.2 @WebFilter 注解

在 Spring Boot 项目中，也可以使用 @WebFilter 注解配置过滤器，这样就可以不用手动添加过滤器注册的代码。示例代码，如下所示：

```
package com.gf.mvc.config;
import lombok.extern.slf4j.Slf4j;
import org.springframework.stereotype.Component;
import javax.servlet.*;
import javax.servlet.annotation.WebFilter;
import java.io.IOException;
@Slf4j
@WebFilter
@Component
public class MyWebFilter implements Filter {
    @Override
    public void doFilter(ServletRequest servletRequest, ServletResponse servletResponse, FilterChain filterChain) throws IOException, ServletException {
        log.info("执行前：逻辑处理代码（WebFilter）");
        filterChain.doFilter(servletRequest, servletResponse);
        log.info("执行后：逻辑处理代码（WebFilter）");
```

            }
        }

在不加任何属性的情况下，@WebFilter 注解会匹配所有的 URL，要想只匹配指定的 URL，需要为它设置 urlPatterns 属性。@WebFilter 注解的常用属性，如表 7-2 所示。

表 7-2 @WebFilter 注解的常用属性

属性	类型	说明
asyncSupported	boolean	指定过滤器是否支持异步模式
dispatcherTypes	DispatcherType[]	指定过滤器对哪种方式的请求进行过滤。支持的属性：ASYNC、ERROR、FORWARD、INCLUDE 和 REQUEST；默认过滤所有方式的请求
filterName	String	过滤器的名称
initParams	WebInitParam[]	配置参数
servletNames	String[]	指定对哪些 Servlet 进行过滤
urlPatterns	String[]	指定过滤的路径，可以使用"*"通配符
value	String[]	指定过滤的路径，与 urlPatterns 属性相同

7.8.3 CORS 过滤器

Spring MVC 通过对控制器的注解（@CrossOrigin）为 CORS 配置提供细粒度的支持。然而，当与 Spring Security 一起使用时，我们建议依赖内置的 CorsFilter，该过滤器必须在 Spring Security 的过滤器链之前进行排序。示例代码，如下所示：

```
import org.springframework.boot.web.servlet.FilterRegistrationBean;
import org.springframework.context.annotation.Bean;
import org.springframework.context.annotation.Configuration;
import org.springframework.web.cors.CorsConfiguration;
import org.springframework.web.cors.UrlBasedCorsConfigurationSource;
import org.springframework.web.filter.CorsFilter;
import org.springframework.web.servlet.config.annotation.CorsRegistry;
@Configuration
public class WebConfiguration {
    /**
     * 注册 CorsFilter 过滤器，以支持 CORS
     */
    @Bean
    public FilterRegistrationBean corsFilterRegistrationBean() {
        CorsConfiguration config = new CorsConfiguration();
```

```
        config.setAllowCredentials(true);
        config.addAllowedOrigin("http://localhost:8082");
        config.addAllowedHeader("*");
        config.addAllowedMethod("*");
        UrlBasedCorsConfigurationSource source = new
UrlBasedCorsConfigurationSource();
        source.registerCorsConfiguration("/**", config);
        FilterRegistrationBean registrationBean = new
FilterRegistrationBean(new CorsFilter(source));
        registrationBean.addUrlPatterns("/*");
        registrationBean.setName("corsFilter");
        return registrationBean;
    }
}
```

7.9 函数式端点

Spring Web MVC 包括 WebMvc.fn，这是一种轻量级的函数式编程模型，其中函数用于路由和处理请求，契约是为不变性而设计的。它是基于注解的编程模型的替代方案，但在其他方面运行在同一个 DispatcherServlet 上。

在 WebMvc.fn 中，HTTP 请求使用 HandlerFunction 来处理：它是一个接受 ServerRequest 并返回 ServerResponse 的函数。作为响应对象的请求都具有不可变的契约，这些契约提供了对 HTTP 请求和响应的 JDK 8 的友好访问。HandlerFunction 相当于基于注解的编程模型中 @RequestMapping 方法的主体。

传入的请求被路由到具有 RouterFunction 的处理函数：接受 ServerRequest 并返回可选 HandlerFunction（即 Optional<HandlerFunction>）。当路由器函数匹配时，返回处理器函数；否则为空 Optional。RouterFunction 相当于 @RequestMapping 注解，但主要区别在于路由器函数不仅提供数据，还提供行为。

定义处理器函数

下面类中的每个方法就是一个处理器函数，相当于控制器类中的方法。相比于控制器类，处理器类中没有与路由相关的注解，它实现了业务处理代码与路由的完全分离。

```
import com.fasterxml.jackson.databind.ObjectMapper;
import com.gf.mvc.dto.MenuDTO;
import com.gf.mvc.vo.MenuVO;
import org.slf4j.Logger;
import org.slf4j.LoggerFactory;
import org.springframework.beans.factory.annotation.Autowired;
import org.springframework.http.MediaType;
```

```java
import org.springframework.stereotype.Component;
import org.springframework.web.servlet.function.ServerRequest;
import org.springframework.web.servlet.function.ServerResponse;
import javax.servlet.ServletException;
import java.io.IOException;
import java.util.Arrays;
import java.util.Collections;
import java.util.List;
@Component
public class MenuHandler {
    private Logger log = LoggerFactory.getLogger(this.getClass());
    @Autowired
    private ObjectMapper objectMapper;
    public ServerResponse create(ServerRequest request) throws IOException, ServletException {
        MenuDTO params = request.body(MenuDTO.class);
        //TODO: 将新创建的菜单项，保存到数据库中。
        log.info("成功创建菜单项：", objectMapper.writeValueAsString(params));
        return ServerResponse.ok().body(Collections.EMPTY_MAP);
    }
    public ServerResponse view(ServerRequest request) {
        String id = request.pathVariable("id");
        MenuVO menuVO = null;
        if ("1".equals(id)) {
            menuVO = new MenuVO();
            menuVO.setId("1");
            menuVO.setName("用户管理");
            menuVO.setParentId("0");
            menuVO.setSorted(20);
        }
        return ServerResponse.ok()
                .contentType(MediaType.APPLICATION_JSON)
                .body(menuVO);
    }
    public ServerResponse list(ServerRequest request) {
        List<String> list = Arrays.asList("用户管理", "菜单管理", "系统日志");
        return ServerResponse.ok()
```

```
                .contentType(MediaType.APPLICATION_JSON)
                .body(list);
    }
}
```

定义路由器函数

我们使用 @Configuration 类来注册路由器函数的 bean。每个 bean 都可以指定一或多个路由信息，及其对应的处理器函数。

RouterFunctions.route() 提供了一个路由器构建器，它有助于创建路由器，如下所示：

```
import com.gf.mvc.handler.MenuHandler;
import org.springframework.context.annotation.Bean;
import org.springframework.context.annotation.Configuration;
import org.springframework.web.servlet.function.RouterFunction;
import static org.springframework.http.MediaType.*;
import static org.springframework.web.servlet.function.RequestPredicates.*;
import static org.springframework.web.servlet.function.RouterFunctions.*;
import org.springframework.web.servlet.function.ServerResponse;
@Configuration
public class MenuRouter {
    private final MenuHandler handler;
    public MenuRouter(MenuHandler handler) {
        this.handler = handler;
    }
    @Bean
    public RouterFunction<ServerResponse> routerOne() {
        return route().path("/mvc/menu/r", builder -> builder
                .POST("/create", accept(APPLICATION_JSON), handler::create)
                .GET("view/{id}", accept(APPLICATION_JSON), handler::view)
                .POST("/list", accept(APPLICATION_JSON), handler::list)).build();
    }
}
```

7.10 异步请求

Spring MVC 与 Servlet 3.0 异步请求处理有广泛的集成：
1) 在控制器方法中 DeferredResult 和 Callable 的返回值，并为单个异步返回值提供基本支持；
2) 控制器可以流式传输多个值，包括 SSE 和原始数据；
3) 控制器可以使用反应式客户端并返回反应式类型进行响应处理。

有关以上内容的更多信息，请访问以下网址：
https://docs.spring.io/spring/docs/5.2.7.RELEASE/spring-framework-reference/web.html#mvc-ann-async

Spring 框架从 3.0 开始，提供了 @Async 注解，使用这个注解可以优雅地进行异步调用。这个注解可以用在无返回值的方法上，也可以用在有返回值的方法上，但是前者用得较多。示例代码，如下所示：

```
package com.gf.mvc.service.impl;
import com.baomidou.mybatisplus.extension.service.impl.ServiceImpl;
import com.gf.mvc.entity.Menu;
import com.gf.mvc.mapper.MenuMapper3;
import com.gf.mvc.service.MenuService3;
import org.springframework.scheduling.annotation.Async;
import org.springframework.stereotype.Service;
import java.util.List;
@Service
public class MenuServiceImpl3 extends ServiceImpl<MenuMapper3, Menu>
implements MenuService3 {
    /**
     * 修改完成菜单后，异步向指定用户发送消息。
     */
    @Async
    public void sendMsg(String userId) {
        // TODO: 编写具体的代码……
    }
}
```

7.11 HTTP/2

需要 Servlet 4 容器来支持 HTTP/2，Spring Framework 5 与 Servlet API 4 兼容。从编程模型的角度来看，应用程序无需做任何特定的事情，但是，有与服务器配置相关的注意事项。有关更多详细信息，请参见 HTTP/2 wiki 页面，如下所示：
https://github.com/spring-projects/spring-framework/wiki/HTTP-2-support

Servlet API 确实公开了一个与 HTTP/2 相关的构造。可以使用 javax.servlet.http.PushBuilder 主动将资源推送到客户端，并且它被支持作为 @RequestMapping 方法的参数。

7.12 验证

只要类路径上有一个 JSR-303 实现（如 Hibernate 验证器），Bean Validation 1.1 支持的方法验证功能就会自动启用。这使得 bean 方法可以在其参数和（或）返回值上使用 javax.validation 约束进行注解。具有此注解方法的目标类需要在类型级别使用 @Validated 注解进行注解，以便搜索其方法以获得内联约束注解。添加依赖项，以便使用各种验证注解，如下所示：

```xml
<dependency>
    <groupId>org.springframework.boot</groupId>
    <artifactId>spring-boot-starter-validation</artifactId>
</dependency>
```

我们为 MenuController 控制器中的 create 方法的参数添加验证，代码示例，如下所示：

```java
import com.fasterxml.jackson.core.JsonProcessingException;
import com.fasterxml.jackson.databind.ObjectMapper;
import com.gf.mvc.dto.MenuDTO;
import org.slf4j.Logger;
import org.slf4j.LoggerFactory;
import org.springframework.beans.factory.annotation.Autowired;
import org.springframework.http.ResponseEntity;
import org.springframework.web.bind.annotation.*;
import javax.validation.Valid;
import java.util.Collections;
@RestController
@RequestMapping({"/mvc/menu", "/mvc/menu/v2"})
public class MenuController2 {
    private Logger log = LoggerFactory.getLogger(this.getClass());
    @Autowired
    private ObjectMapper objectMapper;
    /**
     * 为 params 添加验证
     */
    @PostMapping("/create")
    public ResponseEntity create(@RequestBody @Valid MenuDTO params)
            throws JsonProcessingException {
        //TODO: 将新创建的菜单项，保存到数据库中。
```

```java
        log.info("成功创建菜单项：", objectMapper.writeValueAsString(params));
        return ResponseEntity.ok(Collections.EMPTY_MAP);
    }
}
```

定义用于接收参数的 DTO 类，示例代码，如下所示：

```java
import javax.validation.constraints.Min;
import javax.validation.constraints.NotEmpty;
import javax.validation.constraints.NotNull;
public class MenuDTO {
    @NotEmpty // 不能为空，空字符串也不可以。
    private String name;
    @NotNull // 不能为 null，空字符串是允许的。
    private String parentId;
    @Min(0) // 不能小于 0
    private int sorted;
    // 省略 setter、getter ...
}
```

上面的示例代码，只是使用验证功能，我们更希望在调用 API 时，若出现验证不通过的字段，在 API 的响应中应给出提示。这时，还需要添加一个 MethodArgumentNotValidException 类型的全局异常处理，如下所示：

```java
import org.slf4j.Logger;
import org.slf4j.LoggerFactory;
import org.springframework.http.*;
import org.springframework.validation.BindingResult;
import org.springframework.validation.FieldError;
import org.springframework.validation.ObjectError;
import org.springframework.web.bind.MethodArgumentNotValidException;
import org.springframework.web.bind.annotation.ControllerAdvice;
import org.springframework.web.bind.annotation.ExceptionHandler;
import org.springframework.web.bind.annotation.ResponseBody;
import org.springframework.web.context.request.RequestAttributes;
import org.springframework.web.context.request.ServletWebRequest;
import org.springframework.web.context.request.WebRequest;
import javax.servlet.http.HttpServletRequest;
import java.util.List;
@ControllerAdvice
public class ExceptionControllerAdvice2 {
```

```java
private final Logger log = LoggerFactory.getLogger(getClass());
/**
 * 请求参数验证
 */
@ResponseBody
@ExceptionHandler(MethodArgumentNotValidException.class)
public ResponseEntity<?> validationBodyException(HttpServletRequest request, MethodArgumentNotValidException ex) {
    StringBuilder strBuilder = new StringBuilder();
    BindingResult result = ex.getBindingResult();
    if (result.hasErrors()) {
        List<ObjectError> errors = result.getAllErrors();
        errors.forEach(p -> {
            FieldError fieldError = (FieldError) p;
            log.warn("请求参数验证: object{" +
                    fieldError.getObjectName() + "},field{" +
                    fieldError.getField() + "},errorMessage{" +
                    fieldError.getDefaultMessage() + "}");
            strBuilder.append(", ");
            strBuilder.append(fieldError.getField());
            strBuilder.append(": ");
            strBuilder.append(fieldError.getDefaultMessage());
        });
    }
    if (strBuilder.length() > 0) {
        strBuilder.deleteCharAt(0);}
    HttpStatus status = getStatus(request);
    String path = request.getRequestURI();
    return new ResponseEntity<>(new CustomErrorType(status.value(), strBuilder.toString(), path), status);
}
private HttpStatus getStatus(HttpServletRequest request) {
    WebRequest webRequest = new ServletWebRequest(request);
    Integer statusCode = (Integer) this.getAttribute(webRequest, "javax.servlet.error.status_code");
    if (statusCode == null) {
        return HttpStatus.INTERNAL_SERVER_ERROR;}
    return HttpStatus.valueOf(statusCode);
```

```
    }
    private <T> T getAttribute(RequestAttributes requestAttributes,
String name) {
        return (T) requestAttributes.getAttribute(name,
RequestAttributes.SCOPE_REQUEST);
    }
}
```

表中列出一些验证注解及其说明，如表 7-3 所示。:

表 7-3 验证注解及其说明

注解	说明
@NotNull	值不能为 NULL
@Null	值必须为 NULL
@Pattern(regex=)	字符串必须匹配正则表达式
@Size(min, max)	集合元素的值必须在 min 和 max 之间
@CreditCardNumber(ignoreNonDigitCharacters=)	字符串必须是信用卡号，默认按照美国的标准验证
@Email	字符串必须是 Email 地址
@Length(min, max)	字符串的长度必须在 min 和 max 之间
@NotBlank	不能为空白字符串
@NotEmpty	字符串不能为 NULL 和空字符串；对于集合类型，其中必须有元素
@Range(min, max)	数值必须在 min 和 max 之间
@SafeHtml	字符串必须是安全的 HTML
@URL	字符串必须是合法的 URL
@AssertFalse	值必须是 false
@AssertTrue	值必须是 true
@DecimalMax(value=, inclusive=)	值必须小于等于（inclusive=true）/小于（inclusive=false）属性指定的值，也可以注解在字符串类型的属性上。
@DecimalMin(value=, inclusive=)	值必须大于等于（inclusive=true）/大于（inclusive=false）属性指定的值，也可以注解在字符串类型的属性上。
@Digist(integer=,fraction=)	数字格式检查。integer 指定整数部分的最大长度，fraction 指定小数部分的最大长度
@Future	时间必须是未来的
@Past	时间必须是过去的
@Max(value=)	值必须小于等于 value 指定的值。不能注解在字符串类型属性上。
@Min(value=)	值必须小于等于 value 指定的值。不能注解在字符串类型属性上。

7.13 本章小结

本章讲解了 Spring MVC 的各个知识点：拦截器、过滤器、文件上传、静态资源、全局异常处理和跨域资源共享等等。当然，还有一些知识点并未涉及到，希望大家在自己的开发实践中掌握更多更全面的技能。

第八章 Spring WebFlux

Spring 框架中包含的原始 web 框架 Spring Web MVC 是专门为 Servlet API 和 Servlet 容器构建的。反应式堆栈 web 框架 Spring WebFlux 是在 5.0 版本中添加的。它是完全非阻塞的,支持反应式流背压,并可以在 Netty、Undertow 和 Servlet 3.1+ 容器等服务器上运行。

这两个 web 模块在 Spring 框架中是共存的,还有就是每个模块都是可选的。应用程序可以选择使用其中一个,或者在某些情况下同时使用,例如,带有反应式 WebClient 的 Spring MVC 控制器。

8.1 为什么创建 Spring WebFlux

答案之一是需要一个非阻塞的 web 堆栈来处理具有少量线程的并发,并用较少的硬件资源进行扩展。Servlet 3.1 确实提供了一个用于非阻塞 I/O 的 API,但是,使用它会使 Servlet API 的其余部分有所不同,后者的契约是同步的(Filter 和 Servlet)或阻塞的(getParameter 和 getPart)。这是新的公共 API 作为任何非阻塞运行时的基础的动机。还有就是,Netty 之类的服务器在异步、非阻塞空间中已经很好地建立起来了。

答案之二是函数式编程。正如 Java 5 中添加注解创造了机会(例如,带注解的 REST 控制器或单元测试),Java 8 中添加的 lambda 表达式也为 Java 中的函数式 API 创造了机会。这对于允许异步逻辑的声明式组合的非阻塞应用程序和延续风格的 API(CompletableFuture 和 ReactiveX 所普及的)来说是一个福音。在编程模型级别,Java 8 使 Spring WebFlux 能够提供函数式的 web 端点和带注解的控制器。

8.2 选择 Spring MVC 还是 WebFlux

这是一个很自然的问题,但却建立了一个不合理的二分法。实际上,两者共同努力扩大了可用选项的范围。这两种设计都是为了彼此的连续性和一致性,它们是并排提供的,并且来自每一方的反馈对双方都有好处。有关两者之间的关系,如图 8-1 所示。

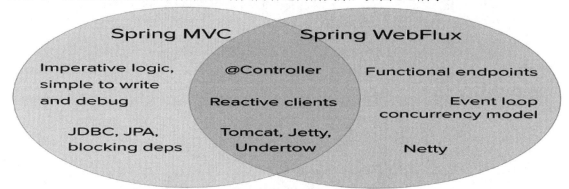

图 8-1 Spring MVC 与 WebFlux 的关系

如果已经有一个运行良好的 Spring MVC 应用程序，那么就没有必要进行更改。命令式编程是编写、理解和调试代码的最简单方法。对于 Spring MVC 应用程序，可以选择最多的库，因为从历史上看，大多数库都是阻塞的。

WebFlux 提供了服务器（Netty、Tomcat、Jetty、Undertow 和 Servlet 3.1+ 容器）的选择和编程模型（带注解的控制器和函数式 web 端点）的选择，以及反应式的选择库（Project Reactor 和 RxJava 等）。

如果对使用 Java 8 Lambda 的轻量级和功能性 web 框架感兴趣，可以使用 WebFlux 函数式 web 端点。对于较小的应用程序或要求较不复杂的微服务来说，这也是一个不错的选择，这些应用程序或微服务可以从更高的透明度和控制中受益。

在微服务架构中，可以混合使用 Spring MVC 或 WebFlux 控制器或 WebFlux 功能端点的应用程序。在两个框架中都支持相同的基于注解的编程模型，可以更轻松地重用知识，同时还可以为正确的工作选择正确的工具。

评估应用程序的一个简单方法是检查其依赖关系。如果要使用阻塞持久性 API（JPA 和 JDBC）或网络 API，Spring MVC 至少是通用架构的最佳选择。从技术上讲，Project Reactor 和 RxJava 都可以在单独的线程上执行阻塞调用，但这样不会充分利用非阻塞 web 堆栈。

如果有一个 Spring MVC 应用程序调用远程服务，请尝试使用反应式 WebClient。可以直接从 Spring MVC 控制器方法返回反应式类型（Project Reactor 和 RxJava 等）。每次呼叫的延迟越大或呼叫之间的相互依赖性越大，好处就越显著。

如果你有一个庞大的团队，请记住在向非阻塞、函数式和声明式编程转变的过程中的陡峭学习曲线。在没有完全切换的情况下启动的一个实用方法是使用反应式 WebClient。除此之外，从小事做起，衡量收益。我们预计，对于广泛的应用，这种转变是不必要的。如果你不确定要寻找什么好处，请从了解非阻塞 I/O 的工作原理开始（例如，单线程 Node.js 上的并发）以及它的影响。

8.3 简单示例

想要在项目中使用 Spring WebFlux 的编程风格及其相应的反应式容器，请先添加相应的依赖项，如下所示：

```
<dependency>
    <groupId>org.springframework.boot</groupId>
    <artifactId>spring-boot-starter-webflux</artifactId>
</dependency>
```

然后，编写主类，并在类级别添加 @EnableWebFlux 注解，如下所示：

```
package com.gf.webflux;
import org.springframework.boot.SpringApplication;
import org.springframework.boot.autoconfigure.SpringBootApplication;
import org.springframework.web.reactive.config.EnableWebFlux;
```

```
@EnableWebFlux
@SpringBootApplication
public class WebFluxApplication {
    public static void main(String[] args) {
        SpringApplication.run(WebFluxApplication.class, args);
    }
}
```

请注意，在未引入 spring-boot-starter-web 依赖项的情况下，可以省略 @EnableWebFlux 注解。完成了以上工作后，我们使用 Mono 和 Flux 就可以开始编写具体的业务代码了，如下所示：

```
package com.gf.webflux.controller;
import com.gf.webflux.entity.User;
import org.springframework.http.ResponseEntity;
import org.springframework.web.bind.annotation.*;
import reactor.core.publisher.Flux;
import reactor.core.publisher.Mono;
import java.util.ArrayList;
import java.util.Collections;
import java.util.List;
import static com.gf.common.GfResponseEntity.self;
import static org.springframework.http.ResponseEntity.status;
@RestController
@RequestMapping("/gf/webflux")
public class HelloReactiveController {
    @GetMapping("/hello")
    public Mono<ResponseEntity> hello() {
        return Mono.just(self().ok(Collections
                .singletonMap("webflux", "Hello World!")));
    }
    @GetMapping("/num/list")
    public Flux<Integer> list() {
        return Flux.range(1, 10)
                // 元素的值加一
                .map(num -> num++)
                // 过滤掉非偶数的值
                .filter(num -> num % 2 == 0);
    }
    @PostMapping("/list")
```

```
    public Mono<ResponseEntity> list(@RequestBody Mono<User>
monoParams) {
        return monoParams
                .map(user -> {
                    List<String> menus = new ArrayList<>();
                    menus.add("菜单管理");
                    menus.add("系统管理");
                    menus.add("日志管理");
                    return Collections.singletonMap(user.getId(),
menus);
                })
                .map(data -> self().msg("中国")
                        .build(data, status(200).header("org", "gf")));
    }
}
```

Flux 和 Mono 是 Project Reactor 中的两个基本概念。Flux 表示的是包含 0~N 个元素的异步序列，Mono 表示的是包含 0~1 个元素的异步序列，两者之间是可以相互转换的。有关 Project Reactor 的更多详细信息，请访问以下网址：

https://projectreactor.io

反应式编程 Reactor 3 参考指南中文文档，如下所示：

https://github.com/jijicai/ProjectReactor/tree/master/book/Reactor3

8.4 WebClient

Spring WebFlux 包括一个针对 HTTP 请求的反应式、非阻塞的 WebClient。它具有函数式的、流畅的 API，其中包含用于声明式组合的反应式类型。WebFlux 客户端和服务器依靠相同的非阻塞编解码器对请求和响应内容进行编码和解码。

在内部，WebClient 把具体的请求逻辑委托给一个具体的 HTTP 客户端库。默认情况下，它使用 Reactor Netty，对 Jetty 反应式 HttpClient 有内置支持，其他可以通过 ClientHttpConnector 插入。有关 WebClient 实例化的示例代码，如下所示：

```
package com.gf.webflux;
import io.netty.channel.ChannelOption;
import org.junit.jupiter.api.Test;
import org.springframework.http.client.reactive.ReactorClientHttpConnector
;
```

```java
import org.springframework.web.reactive.function.client.ExchangeStrategies;
import org.springframework.web.reactive.function.client.WebClient;
import reactor.netty.http.client.HttpClient;
/**
 * 实例化 WebClient
 */
public class WebClientTest {
    @Test
    public void test1() {
        WebClient webClient = WebClient.create();
    }
    @Test
    public void test2() {
        String baseUrl = "http://localhost:8080";
        WebClient webClient = WebClient.create(baseUrl);
    }
    @Test
    public void test3() {
        // 配置在编解码器中缓冲的内存数据的限制，以避免应用程序出现内存问题。
        // 默认值为 256 KB
        ExchangeStrategies exchangeStrategies = ExchangeStrategies
                .builder().codecs(codecs -> {
                    codecs.defaultCodecs().maxInMemorySize(2 * 1024 * 1024);
                }).build();
        WebClient client = WebClient.builder()
                .exchangeStrategies(exchangeStrategies).build();
    }
    @Test
    public void test4() {
        // 配置连接超时
        HttpClient httpClient = HttpClient.create()
                .tcpConfiguration(client -> client.option(ChannelOption.CONNECT_TIMEOUT_MILLIS, 10000));
        WebClient webClient = WebClient.builder()
                .clientConnector(new ReactorClientHttpConnector(httpClient)).build();
```

        }
    }
    StepVerifier 接口位于 reactor-test 模块中，用于验证反应式流中的元素，相应的依赖项，如下所示：

```xml
<dependency>
    <groupId>io.projectreactor</groupId>
    <artifactId>reactor-test</artifactId>
</dependency>
```

使用 WebClient 的实例对象调用 API 的示例代码，如下所示：

```java
package com.gf.webflux;
import com.fasterxml.jackson.core.JsonProcessingException;
import com.fasterxml.jackson.databind.ObjectMapper;
import com.gf.common.GfResponseEntity;
import org.junit.jupiter.api.Test;
import org.springframework.http.MediaType;
import org.springframework.util.LinkedMultiValueMap;
import org.springframework.util.MultiValueMap;
import org.springframework.web.reactive.function.client.WebClient;
import reactor.core.publisher.Flux;
import reactor.core.publisher.Mono;
import reactor.test.StepVerifier;
/**
 * 使用 webClient 对象
 */
public class WebClientUseTest {
    private ObjectMapper objectMapper = new ObjectMapper();
    private String baseUrl = "http://localhost:8080";
    private WebClient webClient = WebClient.create(baseUrl);
    @Test
    public void test1() {
        // 声明具体的 API 调用和响应的处理逻辑
        Flux<Integer> resultFlux = webClient.get()
                .uri("/gf/webflux/num/list")
                .accept(MediaType.APPLICATION_JSON).exchange()
                .flatMapMany(response ->
response.bodyToFlux(Integer.class));
        // 打印出 API 返回的数据，并进行相应的转化，以便用于下一步。
        Mono<Boolean> resultMono = resultFlux.collectList()
                .doOnNext(System.out::println).hasElement();
```

```
            // 触发 API 的调用
        StepVerifier.create(resultMono)
                // 验证转化后的结果：若有响应数据，则通过验证。
                .expectNext(true).verifyComplete();
    }
    @Test
    public void test2() throws JsonProcessingException {
        // 和 API 一起的表单数据
        MultiValueMap<String, String> formData = new LinkedMultiValueMap<>();
        formData.add("name", "123");
        // 声明具体的 API 调用和响应的处理逻辑
        Mono<GfResponseEntity> result = webClient.post()
                .uri("/gf/webflux/data").retrieve()
                .bodyToMono(GfResponseEntity.class);
        // 转化为同步调用 API
        GfResponseEntity responseEntity = result.block();
System.out.println(objectMapper.writeValueAsString(responseEntity));
    }
}
```

8.5 RSocket

RSocket 是一种应用协议，用于通过 TCP、WebSocket 和其他字节流传输进行多路复用和双工通信，使用以下交互模型之一：
1) Request-Response：发送一条消息，然后接收一条消息；
2) Request-Stream：发送一条消息并接收一条消息流；
3) Channel：向两个方向发送消息流；
4) Fire-and-Forget：发送单向消息。

一旦建立了初始连接，"客户端"与"服务器"的区别就会消失，因为双方都变得对称，并且每一方都可以发起上述交互之一。这就是为什么在协议中将参与方称为"请求者"和"响应者"，而上述交互被称为"请求流"或简称为"请求"。以下是 RSocket 协议的主要功能和优点：

1) 跨网络边界的反应式流语义对于流式请求（如请求流和通道），背压信号在请求者和响应者之间传输，允许请求者在源端减慢响应器的速度，从而减少对网络层拥塞控制的依赖，以及对网络级缓冲的需要或者在任何级别。
2) 请求限制：此功能在可以从每一端发送的 LEASE 帧之后命名为"Leasing"，以限制另一端在给定时间内允许的请求总数。
3) 会话恢复：这是为连接失去而设计的，需要保持一些状态。状态管理对于应用程序是透明的，并且与背压相结合可以很好地工作，可以在可能的情况下停止生产者并减少所需的状态量。
4) 大消息的碎片化和重新组装。
5) 保持活力（心跳）。

Java 实现

RSocket 的 Java 实现构建在 Project Reactor 上。TCP 和 WebSocket 的传输建立在 Reactor Netty 上。Reactor 作为一个反应式流库，简化了协议的实现工作。对于应用程序来说，使用 Flux 和 Mono 与声明式操作符和透明的背压支持是很自然的。

RSocket Java 中的 API 有意是最小和基本的。它专注于协议的功能特性，并将应用程序编程模型（例如：RPC 代码生成和其他）作为一个更高级别的独立关注点。

Spring 支持

Spring Boot 2.2、Spring Security 5.2、Spring Integration 5.2 和 Spring Cloud Gateway，它们都提供了对 RSocket 的支持。

Spring Boot 2.2 支持通过 TCP 或 WebSocket 建立 RSocket 服务器，包括在 WebFlux 服务器中通过 WebSocket 公开 RSocket 的选项。Spring Boot 2.2 还提供了客户端支持和自动配置 RSocketRequest.Builder 以及 RSocketStrategies。

有关 RSocket 的更多详细信息，请访问以下网址：
https://docs.spring.io/spring/docs/5.2.7.RELEASE/spring-framework-reference/web-reactive.html#rsocket

8.6 反应式库

Spring WebFlux 是基于 reactor-core 构建的，并在内部使用它来组成异步逻辑并提供反应式流支持。一般来说，WebFlux API 返回 Flux 或 Mono，并且宽容地接受任何反应式流 Publisher 实现作为输入。Flux 和 Mono 是很重要的，它们用于表示基数（例如，预期的是一个还是多个异步值），这对于决策（例如，编码或解码 HTTP 消息）至关重要。

对于带注解的控制器，WebFlux 透明地适应应用程序选择的反应库。这是在 ReactiveAdapterRegistry 的帮助下完成的，它为反应式库和其他异步类型提供了可插拔的支持。注册中心内置了对 RxJava 和 CompletableFuture 的支持，但是也可以注册其他的。

对于函数式 API（例如，函数式端点和 WebClient 等），WebFlux API 的一般规则：将 Flux 和 Mono 作为返回值，将反应式流 Publisher 作为输入。当提供了一个 Publisher（无论是自定义的还是来自另一个反应式库）时，它只能被视为具有未知语义的流（0~N）。但是，如果语义是已知的，可以用 Flux 或 Mono.from(Publisher) 而不是传递原始 Publisher。

8.7 本章小结

本章讲解了 Spring WebFlux，它用于搭建异步非阻塞的反应式 web 应用程序。还进一步讲解了 WebClient、RSocket 和反应式库等相关知识点。

第九章 单元测试和集成测试

测试是企业软件开发不可分割的一部分。Spring 框架和 Spring Boot 分别为各种测试（例如：web 测试、JDBC 测试）提供了多种多样的工具类、注解和模拟环境等。本书所讲的测试的类型大体上分为两类：单元测试和集成测试。由于这两块包含的内容比较多，我们只能选取其中的一部分进行讲解，有关测试的更多详细信息，还请查看官方文档。

9.1 单元测试

真正的单元测试通常运行得非常快，因为没有要设置的运行时基础设施。应用程序中的 POJO 应该在 JUnit 或 TestNG 中是可测试的，它的对象通过使用 new 运算符实例化，而不需要 Spring 或任何其他容器。

对于某些单元测试场景，Spring 框架提供了模拟对象和测试支持类。可以使用模拟对象在隔离中测试代码。遵循 Spring 的架构建议：代码库的干净分层和组件化，这将有助于简化单元测试。例如，可以通过模拟 DAO 或存储库接口来测试服务层对象，而无需在运行单元测试时访问持久数据。另外，依赖注入可以降低代码对容器的依赖程度。

要对业务代码进行单元测试，可以使用 JUnit 5 这个单元测试框架，添加依赖项，如下所示：

```xml
<dependency>
    <groupId>org.junit.jupiter</groupId>
    <artifactId>junit-jupiter</artifactId>
    <scope>test</scope>
</dependency>
```

添加完成依赖项后，我们编写一个 UnitTest 测试类，如下所示：

```java
package com.gf.test;
import org.junit.jupiter.api.*;
@DisplayName("gf 单元测试")
public class UnitTest {
    @BeforeAll
    public static void init() {
        System.out.println("【测试开始】执行初始化代码……");
    }
    @BeforeEach
    public void beforeEach() {
        System.out.println("在每个测试方法之前执行的代码……");
    }
    @AfterEach
    public void afterEach() {
```

```java
        System.out.println("在每个测试方法之后执行的代码......");
    }
    @DisplayName("测试 1")
    @Tag("test")
    @Test
    public void test1() {
        System.out.println("执行测试 1 的逻辑");
    }
    @DisplayName("测试 2")
    @Tag("test")
    @Test
    public void test2() {
        System.out.println("执行测试 2 的逻辑");
    }
    @AfterAll
    public static void destroy() {
        System.out.println("【测试结束】执行清理资源的代码......");
    }
}
```

模拟对象

Spring 包含许多专门用于模拟的包：Environment、JNDI、Servlet API 和 Spring Web Reactive。

MockEnvironment 和 MockPropertySource 对于某些代码的容器外测试非常有用，这些代码依赖于特定环境的属性。

org.springframework.mock.web 包中有一组全面的 Servlet API 模拟对象，这些对象对于测试 web 上下文、控制器和过滤器非常有用。

MockServerHttpRequest 和 MockServerHttpResponse 都从与服务器特定的实现相同的抽象基类继承，并与它们共享行为。例如，模拟请求一旦创建就不可更改，但是可以使用 ServerHttpRequest 中的 mutate() 方法创建修改后的实例。

WebTestClient 构建在模拟请求和响应的基础上，为测试没有 HTTP 服务器的 WebFlux 应用程序提供支持。它还可以用于运行服务器的端到端测试。

单元测试支持类

AopTestUtils 是与 AOP 相关的工具方法的集合。可以使用这些方法来获取对底层目标对象的引用，这些对象隐藏在一个或多个 Spring 代理后面的。例如，使用 EasyMock 或 Mockito 等库将 bean 配置为动态模拟，并且该模拟被包装在 Spring 代理中，在这种情况下，可能需要直接访问底层模拟以配置对其的期望并执行验证。

Spring Boot 提供了许多工具和注解，可在测试应用程序时提供帮助。测试支持由两个模块提供：spring-boot-test 包含核心功能，而 spring-boot-test-autoconfigure 支持测试的自动配置。

在 Spring Boot 应用程序中一般使用 spring-boot-starter-test 依赖项，它既导入了 Spring Boot 测试模块，也导入了 JUnit Jupiter、AssertJ、Hamcrest 和许多其他有用的库。该依赖项还带来了老式引擎，这样就可以同时运行 JUnit 4 和 JUnit 5 测试，如下所示：

```
<dependency>
    <groupId>org.springframework.boot</groupId>
    <artifactId>spring-boot-starter-test</artifactId>
    <scope>test</scope>
</dependency>
```

spring-boot-starter-test 依赖项（在测试范围内）包含以下提供的库：

1) JUnit 5：它是 Java 应用程序单元测试的事实标准，其包括与 JUnit4 向后兼容的老式引擎；
2) Spring Test 和 Spring Boot Test：它们提供了适用于 Spring Boot 应用程序的工具和集成测试支持；
3) AssertJ：一个流畅的断言库；
4) Hamcrest：匹配器的对象库（也称为约束或谓词）；
5) Mockito：一个 Java 模拟框架；
6) JSONassert：JSON 的断言库；
7) JsonPath：用于 JSON 的 XPath。

我们通常发现这些常见的库在编写测试时很有用。如果这些库都不适合需要，还可以添加其他的测试依赖项。

9.2 测试 Spring 应用程序

依赖注入的一个主要优点是它可以使代码更容易进行单元测试。在使用依赖注入方式编写代码的情况下，可以使用 new 操作符实例化对象，甚至不需要涉及 Spring，还可以使用模拟对象而不是实际的依赖项。

通常情况下，需要超越单元测试并开始集成测试（使用 Spring ApplicationContext）。能够在不需要部署应用程序或连接到其他基础设施的情况下执行集成测试非常有用。Spring 框架包括有用于这种集成测试的专用测试模块。可以直接声明一个 spring-test 依赖项，或者使用 spring-boot-starter-test 依赖项以传递的方式引入它。spring-test 依赖项，如下所示：

```
<dependency>
    <groupId>org.springframework</groupId>
    <artifactId>spring-test</artifactId>
</dependency>
```

有关 Spring 框架的测试模块的更多信息，请访问以下网址：
https://docs.spring.io/spring/docs/5.2.7.RELEASE/spring-framework-reference/testing.html

9.3 测试 Spring Boot 应用程序

Spring Boot 应用程序使用的是 Spring ApplicationContext，因此除了使用普通的 Spring 上下文通常会做的事情之外，不需要做任何特别的事情来测试它。

Spring Boot 提供 @SpringBootTest 注解，当需要 Spring Boot 功能时，可以将其用作 spring-test 模块的 @ContextConfiguration 注解的替代者。默认情况下，@SpringBootTest 注解不会启动服务器。但是，可以使用 @SpringBootTest 注解的 webEnvironment 属性指定测试的运行方式：

1) MOCK（默认）：加载 web ApplicationContext 并提供模拟 web 环境。使用此注解时，未启动嵌入式服务器。如果类路径上没有可用的 web 环境，则此模式透明地退回到创建常规的非 web ApplicationContext。在这种模式下，也可以与 @AutoConfigureMockMvc 注解或 @AutoConfigureWebTestClient 注解结合使用，用于基于模拟的 web 应用程序测试。

2) RANDOM_PORT：加载 WebServerApplicationContext 并提供真实的 web 环境。嵌入式服务器启动并在随机端口上监听。

3) DEFINED_PORT：加载 WebServerApplicationContext 并提供真实的 web 环境。将启动嵌入式服务器，并在定义的端口（来自 application.properties）或默认端口 8080 上监听。

4) NONE：使用 SpringApplication 加载 ApplicationContext，但不提供任何 web 环境（mock 或其他）。

如果测试中包含事务，则默认情况下，会在每个测试方法结束时回滚事务。但是，在 RANDOM_PORT 或 DEFINED_PORT 模式下，在服务器上启动的任何事务都不会回滚。因为，这两种模式都提供了真实的 servlet 环境，HTTP 客户端和服务器运行在单独的线程和单独的事务中。

使用 @SpringBootTest 注解进行接口测试，示例代码，如下所示：

```
package com.gf.test;
import org.junit.jupiter.api.Test;
import org.springframework.beans.factory.annotation.Autowired;
import org.springframework.boot.test.context.SpringBootTest;
import org.springframework.boot.test.context.SpringBootTest.WebEnvironment;
import org.springframework.boot.test.web.client.TestRestTemplate;
import static org.assertj.core.api.Assertions.assertThat;
@SpringBootTest(webEnvironment = WebEnvironment.RANDOM_PORT)
class SpringBootTest1 {
    @Test
    void test1(@Autowired TestRestTemplate restTemplate) {
        String body = restTemplate.getForObject("/gf/city", String.class);
```

```
        assertThat(body).isEqualTo("北京");
    }
}
```

MongoDB 测试

可以使用 @DataMongoTest 测试 MongoDB 应用程序。默认情况下，它配置内存中的嵌入式 MongoDB（如果可用），配置 MongoTemplate，扫描 @Document 类，并配置 Spring Data MongoDB 存储库。常规 @Component bean 不会被加载到 ApplicationContext 中。要使用嵌入式 MongoDB，请添加依赖项，如下所示：

```xml
<dependency>
    <groupId>de.flapdoodle.embed</groupId>
    <artifactId>de.flapdoodle.embed.mongo</artifactId>
    <scope>test</scope>
</dependency>
```

使用嵌入式 MongoDB 进行测试的示例代码，如下所示：

```java
package com.gf.test;
import com.fasterxml.jackson.core.JsonProcessingException;
import com.fasterxml.jackson.databind.ObjectMapper;
import com.gf.test.entity.Menu;
import org.junit.jupiter.api.Assertions;
import org.junit.jupiter.api.Test;
import org.springframework.beans.factory.annotation.Autowired;
import org.springframework.boot.test.autoconfigure.data.mongo.DataMongoTest;
import org.springframework.data.mongodb.core.MongoTemplate;
@DataMongoTest
class EmbeddedMongoTest {
    @Autowired
    private MongoTemplate mongoTemplate;
    private ObjectMapper objectMapper = new ObjectMapper();
    @Test
    public void test1() throws JsonProcessingException {
        Menu menu = new Menu();
        menu.setName("系统菜单");
        mongoTemplate.insert(menu);
        System.out.println(objectMapper.writeValueAsString(menu));
        // menu 的 id 字段有值，代表数据插入成功，表示测试通过。
        Assertions.assertTrue(menu.getId() != null);
```

        }
}

内存嵌入式 MongoDB 通常适用于测试,因为它的速度快,不需要任何开发者安装。但是,如果希望对真实的 MongoDB 服务器运行测试,则应排除嵌入式 MongoDB 自动配置,如下所示:

```
import org.springframework.boot.autoconfigure.mongo.embedded.EmbeddedMongoAutoConfiguration;
import org.springframework.boot.test.autoconfigure.data.mongo.DataMongoTest;
@DataMongoTest(excludeAutoConfiguration = EmbeddedMongoAutoConfiguration.class)
public class MongoTest {}
```

有关针对不同技术的测试内容的文档,请访问以下网址:

https://docs.spring.io/spring-boot/docs/2.3.1.RELEASE/reference/html/spring-boot-features.html#boot-features-testing-spring-boot-applications

有关 Spring Boot 的测试内容的文档,请访问以下网址:

https://docs.spring.io/spring-boot/docs/2.3.1.RELEASE/reference/html/spring-boot-features.html#boot-features-testing

9.4 测试工具

一些测试工具类被打包为 spring-boot 的一部分,它们在测试应用程序时是有用的。这些测试工具类,举例如下:

ConfigFileApplicationContextInitializer

ConfigFileApplicationContextInitializer 是 ApplicationContextInitializer 接口的一个实现类,可以将其应用于测试,以便加载 Spring Boot 应用程序中的 application.properties 配置文件。可以在不需要 @SpringBootTest 提供的全套功能时使用它,如下所示:

@ContextConfiguration(classes = Config.class,
initializers = ConfigFileApplicationContextInitializer.class)

单独使用 ConfigFileApplicationContextInitializer 并不支持 @Value("${...}") 注入,它唯一的任务就是确保 application.properties 文件被加载到 Spring 的 Environment 中。对于 @Value 支持,需要另外配置一个 PropertySourcesPlaceholderConfigurer 或使用 @SpringBootTest。

TestPropertyValues

TestPropertyValues 可快速向 ConfigurableEnvironment 或 ConfigurableApplicationContext 添

加属性。可以使用 key = value 字符串调用它，如下所示：
TestPropertyValues.of("org=Spring", "name=Boot").applyTo(env);

OutputCapture

OutputCapture 是一个 JUnit 测试框架的扩展，可以用来捕获 System.out 和 System.err 输出。要使用它，请添加 @ExtendWith(OutputCaptureExtension.class)，并将 CapturedOutput 作为参数注入到测试类构造器或测试方法，如下所示：

```java
package com.gf.test;
import org.junit.jupiter.api.Test;
import org.junit.jupiter.api.extension.ExtendWith;
import org.springframework.boot.test.system.CapturedOutput;
import org.springframework.boot.test.system.OutputCaptureExtension;
import static org.assertj.core.api.Assertions.assertThat;
@ExtendWith(OutputCaptureExtension.class)
public class OutputCaptureTest {
    /**
     * 若控制台的输出中包含字符串"World"，则表示测试通过。
     * @param output
     */
    @Test
    public void test1(CapturedOutput output) {
        System.out.println("Hello World!");
        assertThat(output).contains("World");
    }
}
```

TestRestTemplate

TestRestTemplate 是 Spring RestTemplate 的一个方便的替代者，它在集成测试中很有用。通过它，可以获得一个普通模板，或者一个发送基本 HTTP 身份验证（带有用户名和密码）的模板。在这两种情况下，模板都以一种测试友好的方式运行，不会在服务器端错误上引发异常。可以在集成测试中直接实例化 TestRestTemplate，如下所示：

```java
package com.gf.test;
import org.junit.jupiter.api.Test;
import org.springframework.boot.test.web.client.TestRestTemplate;
import org.springframework.http.HttpHeaders;
import static org.assertj.core.api.Assertions.assertThat;
public class TestRestTemplateTest {
    private TestRestTemplate template = new TestRestTemplate();
    @Test
```

```java
    public void test1() throws Exception {
        String url = "https://www.baidu.com";
        HttpHeaders headers = this.template.
                getForEntity(url, String.class).getHeaders();
        assertThat(headers.getLocation()).hasHost("www.baidu.com");
    }
}
```

或者，使用 @SpringBootTest 注解，可以注入完全配置的 TestRestTemplate。如有必要，可以通过 RestTemplateBuilder bean 应用其他自定义。任何未指定主机和端口的 URL 都会自动连接到嵌入式服务器，如下所示：

```java
package com.gf.test;
import org.junit.jupiter.api.Test;
import org.springframework.beans.factory.annotation.Autowired;
import org.springframework.boot.test.context.SpringBootTest;
import org.springframework.boot.test.context.TestConfiguration;
import org.springframework.boot.test.web.client.TestRestTemplate;
import org.springframework.boot.web.client.RestTemplateBuilder;
import org.springframework.context.annotation.Bean;
import org.springframework.http.ResponseEntity;
import java.time.Duration;
import static org.assertj.core.api.Assertions.assertThat;
@SpringBootTest(webEnvironment =
SpringBootTest.WebEnvironment.RANDOM_PORT)
class TestRestTemplateTest2 {
    @Autowired
    private TestRestTemplate template;
    @Test
    void test1() {
        ResponseEntity<String> result = this.template.
                getForEntity("/name", String.class);
        System.out.println(result.getBody());
        assertThat(result.getBody()).isNotEmpty();
    }
    @TestConfiguration(proxyBeanMethods = false)
    static class Config {
        @Bean
        RestTemplateBuilder restTemplateBuilder() {
            return new
RestTemplateBuilder().setConnectTimeout(Duration.ofSeconds(1))
```

```
                .setReadTimeout(Duration.ofSeconds(1));
        }
    }
}
```

9.5 本章小结

　　本章讲解了代码测试的相关知识点，测试分为单元测试和集成测试，并且每个分类又有不同的技术和框架来支撑。编写规范的代码，才能更便于进行测试，才能使代码更加健壮。希望大家能多学测试技术，编写高质量的代码。

第十章 日 志

Spring Boot 使用 Commons Logging 进行所有内部日志记录，但底层日志实现保持打开状态。为 Java Util Logging、Log4J2 和 Logback 提供了默认配置。在每种情况下，日志器都预先配置为使用控制台输出，并提供可选的文件输出。

默认情况下，如果使用 starter，则 Logback 被用于日志记录。还包括适当的 Logback 路由，以确保使用 Java Util Logging、Commons Logging、Log4J 或 SLF4J 的依赖库都能正常工作。在使用 ELK 或 Elastic Stack 之类的日志分析系统时，用的较多的是 Logback。

10.1 日志格式

在一般情况下，我们使用 Spring Boot 的默认日志输出即可，如图 10-1 所示。

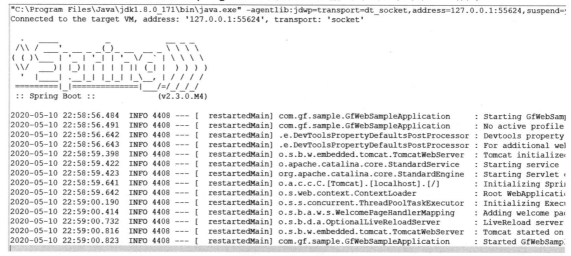

图 10-1 输出日志

对于上图中的日志输出，每行的日志信息包含的内容，如下所示：
1) 日期和时间：毫秒精度，易于分类；
2) 日志级别：ERROR、WARN、INFO、DEBUG 或 TRACE；
3) 进程 ID；
4) "---"分隔符，用于区分实际日志消息的开头；
5) 线程名称：用方括号括起来（控制台输出时可能会被截断）；
6) 日志器名称：这通常是源类名（通常缩写）；
7) 日志信息。

10.2 文件输出

默认的日志配置在消息写入时将消息回显到控制台。默认情况下，会记录 ERROR、WARN

和 INFO 级别的消息。如果要在控制台输出之外写入日志文件，则需要设置 logging.file 或 logging.path 属性（例如，在 application.properties 配置文件中），如表 10-1 所示。

表 10-1 设置 logging.file 或 logging.path 属性

logging.file	logging.path	示例	描述
（无）	（无）		只有控制台日志
特定的文件	（none）	my.log	写到指定的日志文件。名称可以是确切的位置或相对于当前目录。
（无）	特定的目录	/var/log	将 spring.log 文件写入指定目录。名称可以是确切的位置或相对于当前目录。

所有受支持的日志系统都可以使用 logging.level.*=在 Spring Environment（例如，在 application.properties 配置文件中）中设置日志级别，其中 level 可以是其中之一：TRACE、DEBUG、INFO、WARN、ERROR、FATAL 或 OFF。展示了 application.properties 配置文件中可能的日志设置，如下所示：

logging.level.root=warn
logging.level.org.springframework.web=debug
logging.level.org.hibernate=error

10.3 日志组

能够将相关的日志器组合在一起，以便可以同时对它们进行配置，这通常是很有用的。例如，通常可以更改所有与 Tomcat 相关的日志器的日志级别，但是你不容易记住顶级包。为了实现这一点，Spring Boot 允许在 Spring Environment 中定义日志组。通过将"tomcat"组添加到 application.properties 配置文件中来定义日志组，如下所示：

logging.group.tomcat=org.apache.catalina, org.apache.coyote, org.apache.tomcat
定义后，可以用一行配置项更改组中所有日志器的级别：
logging.level.tomcat=TRACE
Spring Boot 包括预定义的日志组，可以开箱即用，如表 10-2 所示。

表 10-2 日志组

名称	日志器列表
web	org.springframework.core.codec、org.springframework.http、org.springframework.web
sql	org.springframework.jdbc.core、org.hibernate.SQL

10.4 Logback

Logback 旨在作为流行的 log4j 项目的继承者，从 log4j 退出的地方开始。Logback 的体系结构足够通用，以便在不同的情况下应用。目前，Logback 分为三个模块：logback-core、 logback-classic 和 logback-access。

　　logback-core 模块为另外两个模块奠定了基础。logback-classic 模块可以被视为 log4j 的显著改进版本。此外，logback-classic 自身实现了 SLF4J API，因此可以轻松地在 logback 和其他日志框架（如 log4j 或 java.util.logging(JUL)）之间来回切换。

　　logback-access 模块与 Servlet 容器（如 Tomcat 和 Jetty）集成，以提供 HTTP 访问日志功能。请注意，可以在 logback-core 之上轻松构建自己的模块。Logback 的官网，如下所示：

　　http://logback.qos.ch

　　Spring Boot 包含许多 Logback 扩展，可以帮助进行高级配置。可以在 logback-spring.xml 配置文件中使用这些扩展。由于标准 logback.xml 配置文件加载太早，因此不能在其中使用扩展。需要使用 logback-spring.xml 或定义 logging.config 属性。

　　只要引入 spring-boot-starter-web 模块，就间接引入了 Logback 的依赖项。因此，想要使用 Logback 无需添加依赖项，添加 logback-spring.xml 配置文件后即可使用它。logback-spring.xml 配置文件，如下所示：

```xml
<?xml version="1.0" encoding="UTF-8"?>
<configuration scan="true" scanPeriod="10 seconds">
    <contextName>logback</contextName>
    <property name="log.path" value="D:/log/gf"/>
    <property name="PATTERN" value="%d{yyyy-MM-dd HH:mm:ss.SSS} [%thread] %-5level %logger{50} - %msg%n"/>
    <property name="CHARSET" value="UTF-8"/>
    <property name="LEVEL" value="trace"/>
    <appender name="CONSOLE" class="ch.qos.logback.core.ConsoleAppender">
        <!--这个级别决定了最低打印（或输出）的日志级别。控制台输出的日志级别是大于或等于此级别的-->
        <filter class="ch.qos.logback.classic.filter.ThresholdFilter">
            <level>${LEVEL}</level>
        </filter>
        <encoder>
            <Pattern>${PATTERN}</Pattern>
            <charset>${CHARSET}</charset>
        </encoder>
    </appender>
    <appender name="FILE" class="ch.qos.logback.core.rolling.RollingFileAppender">
        <file>${log.path}/gf-web.log</file>
        <encoder>
            <pattern>${PATTERN}</pattern>
```

```xml
            <charset>${CHARSET}</charset>
        </encoder>
        <!-- 日志记录器的滚动策略，按日期，按大小记录 -->
        <rollingPolicy class="ch.qos.logback.core.rolling.TimeBasedRollingPolicy">
            <!-- 每天日志归档路径以及格式 -->
            <fileNamePattern>${log.path}/gf-web-%d{yyyy-MM-dd}.%i.log</fileNamePattern>
            <timeBasedFileNamingAndTriggeringPolicy class="ch.qos.logback.core.rolling.SizeAndTimeBasedFNATP">
                <maxFileSize>100MB</maxFileSize>
            </timeBasedFileNamingAndTriggeringPolicy>
            <maxHistory>7</maxHistory>
        </rollingPolicy>
        <filter class="ch.qos.logback.classic.filter.ThresholdFilter">
            <level>${LEVEL}</level>
        </filter>
        <!-- 只记录 info 级别的 -->
        <!--<filter class="ch.qos.logback.classic.filter.LevelFilter">
            <level>info</level>
            <onMatch>ACCEPT</onMatch>
            <onMismatch>DENY</onMismatch>
        </filter>-->
    </appender>

    <springProfile name="dev">
        <root level="info"><!--这个级别决定了最低产生的日志级别-->
            <appender-ref ref="CONSOLE"/>
        </root>
    </springProfile>

    <springProfile name="test">
        <root level="info">
            <appender-ref ref="CONSOLE"/>
            <appender-ref ref="FILE"/>
        </root>
    </springProfile>
```

```
</configuration>
```
　　使用 property 元素可以定义其值，然后在其他元素中使用，这样可以减少重复的配置项。使用 springProfile 元素可以为开发、测试和生产环境配置不同的日志输出。

10.5　本章小结

　　本章讲解了日志的各个知识点：日志格式、文件输出、日志组和 Logback。有关其他日志系统的内容，各个官网和相应的博客文章都提供了丰富的示例。

第十一章　分布式 ID

ID 是数据的唯一标识，在单库单表的情况下，可以使用 UUID 或数据库的自增 ID。随着数据量的增加，需要对数据进行分库分表操作，之前的 ID 方案已不能满足要求。因此，提出了分布式 ID 的概念，其特点有：全局唯一、趋势递增、单调递增和信息安全。其中，中间两者是互斥的，无法应用于同一场景。

ID 的值类型有两种：数值类型和字符串类型，其中前者通常是 long 型数值。

11.1　UUID

UUID（Universally Unique Identifier）的标准形式包含 32 个 16 进制数字，并以 "-" 分为五段，格式为 8-4-4-4-12。所以，每个 UUID 有 36 个字符。

优点

性能非常高：本地生成，没有网络消耗。

缺点

1) 不易于存储：它有 36 个字符，太长，并且无序；
2) 信息不安全：它是基于 MAC 地址生成的，容易造成该地址的泄露；
3) 作为主键会降低性能。

从 Java 1.5 开始提供了 UUID 类，示例代码，如下所示：

```java
import lombok.extern.slf4j.Slf4j;
import org.junit.jupiter.api.Test;
import java.util.UUID;
@Slf4j
public class UUIDTest {
    @Test
    public void generateUUID() {
        //生成 UUID 并转换为字符串
        String uuid = UUID.randomUUID().toString();
        log.info(uuid);
    }
}
```

11.2　数据库自增 ID

单节点

基于单个的 MariaDB 实例的自增 ID，可以构建分布式 ID 服务，但存在以下致命缺点：

1) 性能低：每次需要 ID 时都需要访问数据库，性能受限于单个数据库实例的读写性能；

2) 不可靠：若该数据库实例掉线了，会影响到所有用到此 ID 的业务系统。

这种方式满足了全局唯一、单调递增的特性。相应的创建数据库和表的 SQL 语句，如下所示：

```sql
CREATE DATABASE `id_generators` /*!40100 DEFAULT CHARACTER SET utf8mb4 */;
USE `id_generators`;

CREATE TABLE `order_id` (
  `id` bigint(20) unsigned NOT NULL AUTO_INCREMENT,
  `stub` char(10) NOT NULL DEFAULT '',
  PRIMARY KEY (`id`),
  UNIQUE KEY `stub` (`stub`)
) ENGINE=InnoDB DEFAULT CHARSET=utf8mb4;

#获取新的 ID
begin;
replace into order_id(stub) values('gf');
select last_insert_id();
commit;
```

多主模式

运行多个数据库实例,给每个实例中的自增 ID 设置不同的初始值,其步长和实例数相等。这种方式构建的分布式 ID 服务，有以下缺点：
1) 水平扩展比较困难；
2) 效率低：虽然相较于单个数据库实例来说，性能有所提升，但是每次获取 ID 还是要访问数据库的。

这种方式满足了全局唯一、趋势递增的特性。在 MariaDB 中，查看步长，请执行以下语句：

```
show variables like 'auto_increment%';
```

在 MariaDB 中，步长是系统变量，是全局的，被应用于数据库实例的每个数据库。设置步长，请执行以下语句：

```
set @@auto_increment_increment=1;
```

11.3 Leaf 号段方案

本方案是美团公司开源的分布式 ID 生成方案，它解决了上一个方案的缺点，本方案采取了添加代理 server 以批量获取 ID 号段的方式。本方案利用代理 server 每次批量从数据库中获取 ID 号段（如每次 1000 个），用完之后再从数据库获取新的。这样就可以避免每次都访问数据的麻烦，大大减轻了数据库的压力。本方案就不再利用数据库的自增 ID 了。

优点

1) 代理 server 可以很方便地扩展，性能完全能够支撑大多数业务场景；
2) 容灾性高：代理 server 中有号段缓存，即使数据库实例宕机，短时间内代理 server 仍能正常对外提供服务。

缺点

1) ID 不够随机，会泄露发号数量的信息；
2) 号段消耗完时取下一号段，会造成临界点 ID 获取时间较长，进而造成线程阻塞；
3) 数据库宕机会造成整个系统不可用。

双 buffer 优化

对于上面的第二个缺点，优化方式就是采用双 buffer。我们希望从数据库取号段的过程不会阻塞请求线程，提前取号段可以做到这一点。也就是说，在剩余 ID 达到某个临界值时，就异步地把下一个号段加载到代理 server。

高可用容灾

对于上面的第三个缺点，数据库可以采用主从模式，并通过中间件进行主从切换。Leaf 号段方案生成的 ID 是呈现趋势递增特点的，同时是可计算的。这种方案不适用于订单 ID 场景，因为可以通过差值估算一定时间段的订单量，这造成了信息泄露。

11.4 Leaf 雪花方案

本方案是美团公司开源的分布式 ID 生成方案，它沿用了推特公司的雪花方案的比特位设计，以"1+41+10+12"的方式组装 ID。本方案使用 ZooKeeper 配置 workerID。

本方案解决了雪花算法中的时钟回拨问题：机器的时钟回拨有可能会导致生成重复的 ID。有关本方案的源代码，请访问以下网址：

https://github.com/Meituan-Dianping/Leaf

有关本方案的中文文档及其服务搭建，请访问以下网址：

https://github.com/Meituan-Dianping/Leaf/blob/master/README_CN.md

11.5 雪花算法

Snowflake 是推特公司开源的用于生成分布式 ID 的算法。其生成 ID 的结构，如下所示：
0 - 0000000000 0000000000 0000000000 0000000000 0 - 00000 - 00000 - 000000000000

1) 从左至右，第 1 位是符号位，0 表示正数；
2) 41 位为毫秒级时间，一般会是时间戳的差值（当前时间-固定的起始时间），这样可以使 ID 从更小值开始，最大可以表示 69 年；
3) 10 位工作机器 ID：可以使用前 5 位作为数据中心机房标识，后 5 位作为机器标识，最多支持部署 1024 个节点；
4) 最后 12 位是毫秒内的计数，支持每个节点每毫秒产生 4096 个 ID。

有关雪花算法的源代码，请访问以下网址：
https://github.com/twitter-archive/snowflake

11.6 其他方案

UidGenerator

UidGenerator 是百度公司开源的用 Java 语言（Java 8+）实现的，基于 Snowflake 算法的分布式 ID 生成器。UidGenerator 以组件形式工作在应用项目中，支持自定义 workerId 位数和初始化策略。在实现上，UidGenerator 通过借用未来时间来解决 sequence 天然存在的并发限制。UidGenerator 的分布式 ID 产出速率在单机情况下可达 600 万/秒。

想要查看 UidGenerator 的源代码和文档等更多有用信息，请访问以下网址：
https://github.com/baidu/uid-generator
UidGenerator 中文文档，如下所示：
https://github.com/baidu/uid-generator/blob/master/README.zh_cn.md
在 UidGenerator 中文文档中，有快速入门的步骤，请结合自己项目的实际需求做出相应的调整。另外提一下，UidGenerator 是基于 MariaDB 构建的，所以在使用之前记得安装好数据库。

Tinyid

Tinyid 是用 Java 语言（JDK 1.7+）开发的一款分布式 ID 生成系统，是基于数据库号段算法实现的。它支持 ID 生成本地化，以便获得更好的性能与可用性。它是滴滴公司的非官方项目。

Tinyid 适用于：ID 是趋势递增的，可以容忍 ID 不连续、有浪费的场景；它不适用的场景：类似订单 ID 的业务，会引起安全问题（例如：被扫库、估测订单量）。

想要查看 Tinyid 的源代码和文档等更多有用信息，请访问以下网址：
https://github.com/didi/tinyid
Tinyid 中文文档，请访问以下网址：
https://github.com/didi/tinyid/wiki

11.7 本章小结

本章一共介绍了 7 种分布式 ID 生成方案，它们中有的是基于雪花算法构建的。不同的业务场景需要不同的 ID 生成方案，有些场景需要 ID 是单调递增的，有些场景需要 ID 不能泄露系统信息（如：订单数量）。UUID 不太适合作为分布式 ID，数据库自增 ID 适合作为小规模非核心业务数据的 ID；Leaf 号段方案适合于需要单调递增 ID 的场景，基于雪花算法的方案适合于要求信息安全的场景。

第十二章 数据库 MariaDB

MariaDB 是由 MySQL 原始开发者创建的开源免费的数据库。它在用户级别上兼容 MySQL 的主流版本，包括 API 和命令行，所以，可以很轻松地从 MySQL 切换到 MariaDB。目前，MariaDB 主要由开源社区在维护。本章所有的数据库操作都是基于 MariaDB 数据库的。

12.1 JDBC

Spring Data JDBC 的目标是在概念上更加简单，包括以下设计决策：
1) 如果加载实体，则执行 SQL 语句。完成后，将拥有一个完全加载的实体。不会执行延迟加载或缓存。
2) 如果保存实体，它将被保存。如果没有保存，它就没有。没有脏跟踪和会话。
3) 对于如何将实体映射到表，有一个简单的模型。它可能只适用于非常简单的情况。如果不喜欢，则可以制定自己的策略。Spring Data JDBC 对使用注解定制策略的支持非常有限。

在接下来，我们会一步一步地进行 SQL 数据库相关的配置，并编写简单的代码，完成一个增删改查的示例。首先，添加 SQL 数据库相关的依赖项，如下所示：

```xml
<dependency>
    <groupId>org.springframework.boot</groupId>
    <artifactId>spring-boot-starter-jdbc</artifactId>
</dependency>
<dependency>
    <groupId>mysql</groupId>
    <artifactId>mysql-connector-java</artifactId>
</dependency>
```

然后，新建 application-dev.yml 配置文件，并添加配置项，如下所示：

```yaml
spring:
  datasource:
    url: jdbc:mysql://localhost:3306/gf-platform2?serverTimezone=UTC&useUnicode=true&characterEncoding=utf8
    username: root
    password: 123456
    #driver-class: com.mysql.jdbc.Driver # 可以省略
    type: # 配置数据连接池
```

在 Spring Boot 项目中，可以省略 spring.datasource.driver-class-name，根据 url 即可推断驱动器类。当然，对于一些不常见的数据库，最好还是加上这个属性。

spring.datasource.type 属性用于指定数据库连接池。若容器是 Tomcat（非嵌入式），则会提供 tomcat-jdbc 作为默认值。该属性在未设置的情况下，Spring Boot 自动从类路径中查找，一般是由 HikariPool 提供连接池。

要想在应用启动时能够加载 application-dev.yml 配置文件，需要在 application.yml 配置文件中添加配置项，如下所示：

spring.profiles.active: dev

spring.profiles.active 属性指定了当前活动的配置文件为 application-dev.yml，其中，"application-"为前缀。该属性的值可以为逗号分隔的列表。另外，spring.profiles.include 属性与前一个属性类似，但区别在于，它是无条件激活指定的以逗号分隔的配置文件列表。上述内容配置完毕后，我们就可以启动应用了，主类代码，如下所示：

```
import org.springframework.boot.SpringApplication;
import org.springframework.boot.autoconfigure.SpringBootApplication;
@SpringBootApplication
public class MvcApplication {
    public static void main(String[] args) {
        SpringApplication.run(MvcApplication.class,args);
    }
}
```

启动时，若数据库配置正确，控制台会产生以下输出（也能会产生数据库相关的报错，还请仔细查看报错提示，以便找到问题原因并解决它）：

2020-04-27 20:40:48.925　INFO 2336 --- [main] com.zaxxer.hikari.HikariDataSource　　　　　: HikariPool-1 - Starting...
2020-04-27 20:40:49.365　INFO 2336 --- [main] com.zaxxer.hikari.HikariDataSource　　　　　: HikariPool-1 - Start completed.

接下来，我们继续编写代码：controller 层、service 层和 dao 层。

controller 层

本层负责将 URL 映射到具体方法，并接受请求参数，以及参数校验。相应业务处理完毕后，它负责响应请求。

service 层

本层接受上层的参数输入，然后调用 dao 层的方法，从数据库获取相应的数据（可能会分多次调用）。之后，本层对数据进行各种处理及过滤，以形成最终的数据结构，并返回给上层。

dao 层

本层负责对数据库的增删改查。大部分的操作是针对单表的，涉及到多表查询的也有一些。JdbcTemplate、Hibernate 和 Mybatis Plus 等持久层框架一般在本层使用。借助这些框架技术，可以更快捷地进行数据库的开发。当然，也可以借助所用的 JDBC 实现直接操作数据库，这样做可能比较费时费事。

JdbcTemplate 类通过模板设计模式帮助我们消除了冗长的代码，只做需要做的事情（即可变部分），并且帮我们做那些固定部分，如连接的创建及关闭。JdbcTemplate 主要提供以下几类方法：

1) execute 方法：可以用于执行任何 SQL 语句，一般用于执行 DDL 语句；
2) update 方法及 batchUpdate 方法：update 方法用于执行新增、修改和删除等语句；batchUpdate 方法用于执行批处理相关语句；
3) query 方法及 queryForXXX 方法：用于执行查询相关语句；
4) call 方法：用于执行存储过程、函数相关的语句。

有关 controller 层的代码，这里只列出一部分，示例如下：

```java
import com.fasterxml.jackson.core.JsonProcessingException;
import com.fasterxml.jackson.databind.ObjectMapper;
import com.gf.mvc.dto.MenuDTO;
import com.gf.mvc.service.MenuService;
import org.slf4j.Logger;
import org.slf4j.LoggerFactory;
import org.springframework.beans.factory.annotation.Autowired;
import org.springframework.http.ResponseEntity;
import org.springframework.web.bind.annotation.*;
import javax.validation.Valid;
@RestController
@RequestMapping({"/mvc/menu", "/mvc/menu/v2"})
public class MenuController2 {
    private Logger log = LoggerFactory.getLogger(this.getClass());
    @Autowired
    private ObjectMapper objectMapper;
    @Autowired
    private MenuService menuService;
    /**
     * 为 params 添加验证
     */
    @PostMapping("/create")
    public ResponseEntity create(@RequestBody @Valid MenuDTO params)
throws JsonProcessingException {
        boolean isOK = menuService.create(params);
        if (isOK) {
            log.info("成功创建菜单项：", objectMapper.writeValueAsString(params));
        } else {
```

```
            log.error("创建菜单项失败："，
objectMapper.writeValueAsString(params));
        }
        return ResponseEntity.ok(isOK);
    }
}
```

有关service层的代码，这里只列出一部分，示例如下：

```
package com.gf.mvc.service.impl;
import com.gf.mvc.dao.MenuDao;
import com.gf.mvc.dto.MenuDTO;
import com.gf.mvc.entity.Menu;
import com.gf.mvc.service.MenuService;
import org.springframework.stereotype.Service;
import java.util.List;
@Service
public class MenuServiceImpl implements MenuService {
    private final MenuDao menuDao;
    public MenuServiceImpl(MenuDao menuDao) {
        this.menuDao = menuDao;
    }
    @Override
    public boolean create(MenuDTO params) {
        Menu entity = new Menu();
        entity.setName(params.getName());
        entity.setParentId(params.getParentId());
        entity.setSorted(params.getSorted());
        entity = menuDao.create2(entity);
        return entity.getId() > -1;
    }
    ……
}
```

对于创建数据后需要返回数据的情况，在如下代码中也有示例给出。有关dao层的代码，示例如下：

```
import com.fasterxml.jackson.databind.ObjectMapper;
import com.gf.mvc.dao.MenuDao;
import com.gf.mvc.entity.Menu;
import com.gf.mvc.util.StringUtil;
import org.slf4j.Logger;
import org.slf4j.LoggerFactory;
```

```java
import org.springframework.jdbc.core.JdbcTemplate;
import org.springframework.jdbc.core.PreparedStatementCreator;
import org.springframework.jdbc.support.GeneratedKeyHolder;
import org.springframework.jdbc.support.KeyHolder;
import org.springframework.stereotype.Repository;
import java.io.Serializable;
import java.sql.Connection;
import java.sql.PreparedStatement;
import java.sql.SQLException;
@Repository
public class MenuDaoImpl implements MenuDao {
    private final Logger log = LoggerFactory.getLogger(getClass());
    private final ObjectMapper objectMapper = new ObjectMapper();
    private final JdbcTemplate jdbcTemplate;
    private final String sqlCreate = "insert into s_menu(name,parent_id,status) values(?,?,?)";
    private final String sqlView = "select id,name,parent_id,status,sorted,remark from s_menu where id=?";
    private final String sqlUpdate = "update from s_menu set name=?,parent_id=? where id=?";
    private final String sqlDelete = "delete from s_menu where id=?";
    /**
     * Spring Boot 提供 JdbcTemplate bean 初始化。
     */
    public MenuDaoImpl(JdbcTemplate jdbcTemplate) {
        this.jdbcTemplate = jdbcTemplate;
    }
    /**
     * 新建菜单
     */
    @Override
    public int create(Menu entity) {
        return jdbcTemplate.update(sqlCreate, entity.getName(), entity.getParentId(), entity.getSorted());
    }
    /**
     * 新建菜单：插入数据成功后，返回带 ID 的对象。
     */
    public Menu create2(Menu entity) {
```

```java
        KeyHolder keyHolder = new GeneratedKeyHolder();
        int num = jdbcTemplate.update(new PreparedStatementCreator() {
            @Override
            public PreparedStatement createPreparedStatement(Connection connection) throws SQLException {
                PreparedStatement psst = connection.prepareStatement(sqlCreate, new String[]{"id"});
                psst.setString(1, entity.getName());
                psst.setLong(2, entity.getParentId());
                psst.setInt(3, entity.getSorted());
                return psst;
            }
        }, keyHolder);
        // 设置生成的主键 ID
        entity.setId(keyHolder.getKey().longValue());
        log.info("新建菜单成功，获取 ID：{}", StringUtil.toString(objectMapper, entity));
        return entity;
    }
    /**
     * 根据 ID 获取相应的菜单详细数据
     */
    @Override
    public Menu view(Serializable id) {
        return jdbcTemplate.queryForObject(sqlView, new Serializable[]{id}, (rs, paramInt) -> {
            Menu menu = new Menu();
            menu.setId(rs.getLong("id"));
            menu.setName(rs.getString("name"));
            menu.setParentId(rs.getLong("parent_id"));
            menu.setStatus(rs.getInt("status"));
            menu.setSorted(rs.getInt("sorted"));
            menu.setRemark(rs.getString("remark"));
            return menu;
        });
    }
    /**
     * 更新指定 ID 的数据
     */
```

```java
    @Override
    public int update(Menu entity) {
        return jdbcTemplate.update(sqlUpdate, entity.getName(),
entity.getParentId(), entity.getId());
    }
    /**
     * 删除指定 ID 的数据
     */
    @Override
    public int delete(Serializable id) {
        return jdbcTemplate.update(sqlDelete, id);
    }
}
```

有关菜单的数据表（用于 MariaDB 数据库），如下所示：

```
DROP TABLE IF EXISTS `s_menu`;
CREATE TABLE `s_menu` (
  `Id` bigint(11) NOT NULL AUTO_INCREMENT,
  `name` varchar(200) DEFAULT NULL COMMENT '菜单名称',
  `parent_id` bigint(20) DEFAULT NULL COMMENT '父级',
  `status` int(11) DEFAULT 0 COMMENT '状态：0=已启用，1=已禁用',
  `sorted` int(11) DEFAULT NULL COMMENT '排序',
  `remark` varchar(500) DEFAULT '' COMMENT '备注',
  PRIMARY KEY (`Id`)
) ENGINE=InnoDB AUTO_INCREMENT=7 DEFAULT CHARSET=utf8mb4 COMMENT='菜单表';
```

12.2 连接池

HikariCP

HikariCP 是一个几乎零开销的 JDBC 连接池，它是快速、简单、可靠的，它非常轻，大约 130kb。Spring Boot 默认使用这个连接池库。有关 HikariCP、Dbcp2、C3P0 和 Vibur 的性能对比测试详情，请访问以下网址：

https://github.com/brettwooldridge/HikariCP/wiki/Bad-Behavior:-Handling-Database-Down

有关 HihariCP 的源代码等更多详细信息，请访问以下网址：

https://github.com/brettwooldridge/HikariCP

Druid

Druid 连接池是阿里巴巴公司开源的数据库连接池项目。Druid 连接池为监控而生，内置强大的监控功能，监控特性不影响性能。它功能强大，能防 SQL 注入，内置 Loging 能诊断 Hack 应用行为。关于各类连接池的对比表格，如表 12-1 所示。

表 12-1 各类连接池对比表格

功能类别	功能	Druid	HikariCP	DBCP	Tomcat-jdbc	C3P0
性能	PSCache	是	否	是	是	是
	LRU	是	否	是	是	是
	SLB 负载均衡支持	是	否	否	否	否
稳定性	ExceptionSorter	是	否	否	否	否
扩展	扩展	Filter			JdbcIntercepter	
监控	监控方式	jmx、log http	jmx metrics	jmx	jmx	jmx
	支持 SQL 级监控	是	否	否	否	否
	Spring/Web 关联监控	是	否	否	否	否
	诊断支持	LogFilter	否	否	否	否
	连接泄露诊断	logAbandoned	否	否	否	否
安全	SQL 防注入	是	无	无	无	无
	支持配置加密	是	否	否	否	否

以上表格数据来自 Druid Wiki，关于 Druid 的更多详细信息，请访问以下网址：
https://github.com/alibaba/druid/wiki/Druid%E8%BF%9E%E6%8E%A5%E6%B1%A0%E4%BB%8B%E7%BB%8D

要想在 Spring Boot 项目中引入 Druid 连接池，请添加依赖项，如下所示：

```
<dependency>
    <groupId>com.alibaba</groupId>
    <artifactId>druid-spring-boot-starter</artifactId>
    <version>1.1.22</version>
</dependency>
```

添加完依赖项后，便引入了 Druid 连接池，启动应用时，控制台的输出，如下所示：

```
2020-04-28 20:36:24.878  INFO 704 --- [main]
c.a.d.s.b.a.DruidDataSourceAutoConfigure : Init DruidDataSource
2020-04-28 20:36:25.329  INFO 704 --- [main]
com.alibaba.druid.pool.DruidDataSource   : {dataSource-1} inited
```

我们可以直接使用 Druid 连接池的默认配置，也可以根据实际的需要进行相应的配置调整。相关的具体参数，本书就不展开介绍了，有关详情，请访问以下网址：

https://github.com/alibaba/druid/tree/master/druid-spring-boot-starter

12.3 JPA

Spring JPA 在 org.springframework.orm.jpa 包下被提供，它以类似于与 Hibernate 集成的方式提供对 Java Persistence API 的全面支持，同时可以了解底层实现，以便提供额外的特性。Spring JPA 支持提供了三种设置 JPA EntityManagerFactory 的方法，应用程序使用这些方法来获取实体管理器，如下所示：
1) 使用 LocalEntityManagerFactoryBean
2) 从 JNDI 获得 EntityManagerFactory
3) 使用 LocalContainerEntityManagerFactoryBean

Java Persistence API（JPA）是一种标准技术，可以将对象映射到关系数据库中的表。spring-boot-starter-data-jpa POM 提供了一种快速入门的方法，它提供以下关键依赖项：
1) Hibernate：最流行的 JPA 实现之一；
2) Spring Data JPA：使实现基于 JPA 的存储库变得容易；
3) Spring ORM：Spring 框架的核心 ORM 支持。

在 Spring Boot 中，要想使用 JPA 来操作数据库，请添加依赖项，如下所示：

```
<dependency>
    <groupId>org.springframework.boot</groupId>
    <artifactId>spring-boot-starter-data-jpa</artifactId>
</dependency>
```

由于 spring-boot-starter-data-jpa 中已经包含 spring-boot-starter-jdbc 依赖项，添加前者后，可以把后者给移除掉。添加上面的依赖项之后，就可以开始代码编写了。首先定义实体类，如下所示：

```
import javax.persistence.*;
import java.io.Serializable;
@Entity
@Table(name = "s_menu")
public class MenuEntity implements Serializable {
    @Id
    @GeneratedValue(strategy = GenerationType.IDENTITY)
    private Long id;
    @Column(nullable = false)
    private String name;
    private Long parentId;
    private int status;
    private int sorted;
    private String remark;
    // 省略 setter、getter ...
}
```

对上面实体类中的注解作说明，如下所示：
1) @Table 注解用于指定与实体类对应的表名。若实体类名与对应的表名相同的话，则可以省略 name 属性。
2) @Id 注解用于将该字段映射为数据表的主键。
3) @GeneratedValue 用于标注主键的生成策略，通过 strategy 属性指定。默认情况下，JPA 自动选择一个最适合底层数据库的主键生成策略。

在 javax.persistence.GenerationType 中定义了几种可供选择的策略，如表 12-2 所示。

表 12-2 可供选择策略

类型	说明
IDENTITY	采用数据库 ID 自增长的方式来自增主键字段。MySQL（MariaDB）和 SQL Server 等支持这种方式，Oracle 不支持这种方式
AUTO	JPA 自动选择合适的策略，这是默认选项
SEQUENCE	通过序列产生主键，通过 @SequenceGenerator 注解指定序列名。Oracle 支持这种方式，MySQL 不支持这种方式
TABLE	通过数据表产生主键，框架借由数据表模拟序列产生主键，使用该策略可以使应用更易于数据库移植。默认情况下，表结构如下所示： DROP TABLE IF EXISTS \`hibernate_sequence\`; CREATE TABLE \`hibernate_sequence\` (\`next_val\` bigint(20) DEFAULT 0) ENGINE=InnoDB DEFAULT CHARSET=utf8mb4 COMMENT='ID 表';

Spring Data 存储库通常从 Repository 或 CrudRepository 接口扩展。如果使用自动配置，将从包含主配置类（用 @EnableAutoConfiguration 或 @SpringBootApplication 注解的类）的包中搜索存储库。CrudRepository 接口可以满足基本的单表操作，对于更复杂的查询，可以使用 Spring Data 的 @Query 注解对方法进行注解。实现了 CrudRepository 接口的存储库，如下所示：

```
import com.gf.mvc.entity.MenuEntity;
import org.springframework.data.jpa.repository.Query;
import org.springframework.data.repository.CrudRepository;
import org.springframework.data.repository.query.Param;
import java.util.List;
/**
 * 菜单实体的存储库定义
 */
public interface MenuEntityCrudRepository extends
CrudRepository<MenuEntity, Long> {
    /**
     * 根据 name 精确查找对应的菜单数据。
     */
    @Query("select m from MenuEntity m where m.name =:name")
```

```java
    MenuEntity viewByName(@Param("name") String name);
    /**
     * 根据 name 模糊查找对应的多个菜单数据。
     */
    @Query("select m from MenuEntity m where m.name like %:name%")
    List<MenuEntity> listByName(@Param("name") String name);
    /**
     * 根据 name 模糊查找对应的多个菜单 ID 数据。
     */
    @Query("select m.id from MenuEntity m where m.name like %:name%")
    List<Long> idsByName(@Param("name") String name);
}
```

与之对应的 service 层代码，如下所示：

```java
import com.fasterxml.jackson.databind.ObjectMapper;
import com.gf.mvc.dao.MenuDao;
import com.gf.mvc.dto.MenuDTO;
import com.gf.mvc.entity.Menu;
import com.gf.mvc.entity.MenuEntity;
import com.gf.mvc.repository.MenuEntityCrudRepository;
import com.gf.mvc.repository.MenuEntityRepository;
import com.gf.mvc.service.MenuService;
import com.gf.mvc.util.StringUtil;
import org.slf4j.Logger;
import org.slf4j.LoggerFactory;
import org.springframework.stereotype.Service;
import java.util.ArrayList;
import java.util.List;
@Service
public class MenuServiceImpl implements MenuService {
    private final Logger log = LoggerFactory.getLogger(getClass());
    private final ObjectMapper objectMapper;
    private final MenuDao menuDao;
    private final MenuEntityRepository menuEntityRepository;
    private final MenuEntityCrudRepository menuEntityCrudRepository;
    public MenuServiceImpl(ObjectMapper objectMapper, MenuDao menuDao,
MenuEntityRepository menuEntityRepository,
MenuEntityCrudRepository menuEntityCrudRepository) {
        this.objectMapper = objectMapper;
        this.menuDao = menuDao;
```

```java
        this.menuEntityRepository = menuEntityRepository;
        this.menuEntityCrudRepository = menuEntityCrudRepository;
    }
    /**
     * 使用存储库的方式，新建菜单。
     */
    public boolean create4Crud(MenuDTO params) {
        MenuEntity entity = new MenuEntity();
        entity.setName(params.getName());
        entity.setParentId(params.getParentId());
        entity.setSorted(params.getSorted());
        entity = menuEntityCrudRepository.save(entity);
        log.info("成功创建菜单后，返回 ID：", StringUtil.toString(objectMapper, entity));
        return entity.getId() > -1;
    }
    @Override
    public List<MenuEntity> listAll() {
        List<MenuEntity> list = new ArrayList<>();
        menuEntityCrudRepository.findAll().forEach(list::add);
        return list;
    }
    /**
     * 根据菜单 ID 查找对应的菜单数据。
     */
    @Override
    public MenuEntity viewCrud(Long id) {
        return menuEntityCrudRepository.findById(id).orElse(new MenuEntity());
    }
    /**
     * 根据 name 精确查找对应的菜单数据。
     */
    @Override
    public MenuEntity viewByName(String name) {
        return menuEntityCrudRepository.viewByName(name);
    }
}
```

在 application-dev.yml 配置文件中，添加配置项可以实现在控制台中打印 SQL 语句，如下所示：
```
spring.jpa.show-sql: true
spring.properties.hibernate.format_sql: true
```

12.4 事务

Spring TX 模块负责在 Spring 框架中实现事务管理功能。以 AOP 切面的方式将事务注入到业务代码中，并实现不同类型的事务管理器。

默认情况下，存储库实例上的 CRUD 方法是事务性的。对于读取操作，事务配置 readOnly 标志设置为 true。所有其他配置都只使用 @Transactional，以便应用默认事务配置。如果需要调整存储库中声明的方法之一的事务配置，请在存储库接口中重新声明该方法。

Spring 全面的事务管理支持是使用 Spring 框架的重要理由之一。Spring 框架为事务管理提供了一致的抽象，这样做的好处，如下所示：
1) 跨不同事务 API 的一致编程模型，例如：Java 事务 API（JTA）、JDBC、Hibernate 和 Java 持久性 API（JPA）；
2) 支持声明式事务管理；
3) 用于编程式事务管理的 API 比复杂事务 API（例如 JTA）更简单；
4) 与 Spring 的数据访问抽象完美集成。

在 service 层中，用到声明式事务管理的示例代码，如下所示：

```java
package com.gf.mvc.service.impl;
import com.baomidou.mybatisplus.extension.service.impl.ServiceImpl;
import com.gf.mvc.dto.MenuUpdateDTO;
import com.gf.mvc.entity.Menu;
import com.gf.mvc.mapper.MenuMapper3;
import com.gf.mvc.service.MenuService3;
import org.springframework.stereotype.Service;
import org.springframework.transaction.annotation.Transactional;
@Service
public class MenuServiceImpl3 extends ServiceImpl<MenuMapper3, Menu>
implements MenuService3 {
    @Transactional
    @Override
    public void updateMenu(MenuUpdateDTO params) {
        // 修改菜单
        // 添加菜单修改日志
    }
}
```

通常情况下，JAVA EE 开发者有两种事务管理选择：全局事务管理或本地事务管理，这两种方法都有很大的局限性。下面，我们回顾一下全局和本地事务管理，然后讨论 Spring 框架的事务管理支持如何解决全局和本地事务模型的局限性。

全局事务管理的缺点

全局事务管理使应用程序可以使用多个事务资源，通常是关系数据库和消息队列。应用程序服务器通过 JTA 管理全局事务，这是一个繁琐的 API（部分原因是其异常模型）。此外，JTA 用户事务通常需要依赖 JNDI。JTA 通常仅在应用程序服务器环境中使用，这会导致应用程序包含相关的代码，这降低了代码的通用性。

以前，使用全局事务的首选方法是通过 EJB CMT（容器托管事务）。CMT 是声明式事务管理的一种形式（区别于编程式事务管理）。它的缺点是 CMT 与 JTA 和应用服务器环境相关联，独立性比较差，所以它不是一个好的事务管理选择。

本地事务管理的缺点

本地事务管理是针对单个资源（数据库）的，应用程序服务器不参与事务管理。值得注意的是，大多数应用程序使用单个事务资源。本地事务管理易于使用，但它不能跨多个事务资源工作，另一个缺点是本地事务管理对编程模型具有侵入性。

Spring 方案

Spring 框架提供了声明式和编程式事务管理，这解决了全局和本地事务管理的缺点。它允许应用程序开发者在任何环境中使用一致的编程模型。大多数开发者更喜欢声明式事务管理，这也是被官方推荐的。

12.5 分布式事务

通过使用 Atomikos 嵌入式事务管理器，Spring Boot 支持跨多个 XA 资源的分布式 JTA 事务。当检测到 JTA 环境时，Spring 的 JtaTransactionManager 用于管理事务。自动配置的 JMS、DataSource 和 JPA bean 已升级以支持 XA 事务。可以使用标准的 Spring 注解等（如 @Transaction）参与分布式事务。如果在 JTA 环境中，但仍希望使用本地事务，可以将 spring.jta.enabled 属性设置为 false，以禁用 JTA 自动配置。

什么是 XA？XA 协议由 Tuxedo 首先提出的，并交给 X/Open 组织，作为资源管理器（如数据库）与事务管理器的接口标准。目前，Oracle、Informix、DB2 和 Sybase 等各大数据库厂家都提供对 XA 的支持。XA 协议采用两阶段提交方式来管理分布式事务。XA 接口提供资源管理器与事务管理器之间进行通信的标准接口。

XA 性能有局限性，它的效率较低，准备阶段的时间成本比较长，全局事务状态的成本持久，性能与本地事务相差 10 倍左右；提交前，出现故障难以恢复和隔离问题。

MySQL 从 5.0.3 开始支持 XA 分布式事务，且只有 InnoDB 存储引擎支持。MySQL Connector/J 从 5.0.0 版本开始直接提供对 XA 的支持。

在 MariaDB 中，XA 事务只能与支持的存储引擎一起使用，例如：InnoDB、TokuDB、SPIDER 和 MyRocks。对于 InnoDB，可以通过将 innodb_support_xa 服务器系统变量设置为 0 来禁用 XA 事务。有关 MariaDB 的 XA 事务的更多信息，请访问以下网址：

https://mariadb.com/kb/en/xa-transactions/

需要注意的是，业务开发者在编写代码时，不应该直接操作这些 XA 事务操作的接口。因为在 DTP 模型中，资源管理器上的事务分支的开启、结束、准备、提交、回滚等操作，都应该是由事务管理器来统一管理。

Atomikos

Atomikos 是一款流行的开源事务管理器，可以嵌入到 Spring Boot 应用程序中。可以使用 spring-boot-starter-jta-atomikos 依赖项引入相应的 Atomikos 库。Spring Boot 自动配置 Atomikos，并确保将适当的依赖设置应用于 Spring bean，以实现正确的启动和关闭顺序。在属性类 AtomikosProperties 中包含了可设置的各个属性。

为确保多个事务管理器可以安全地协调同一资源管理器，必须为每个 Atomikos 实例配置唯一的 ID。默认情况下，该 ID 是运行 Atomikos 的机器的 IP 地址。为了确保该 ID 在生产环境中的唯一性，应该为应用程序的每个实例配置具有不同值的 spring.jta.transaction-manager-id 属性。有关 Atomikos 事务管理器的更多信息，请访问以下网址：

https://www.atomikos.com

如果想在 Spring Boot 应用程序中添加自定义的嵌入式事务管理器，请实现两个接口，如下所示：

XAConnectionFactoryWrapper 和 XADataSourceWrapper 接口

有关对其他嵌入式事务管理器支持的更多信息，请访问以下网址：

https://docs.spring.io/spring-boot/docs/2.3.2.RELEASE/reference/html/spring-boot-features.html#boot-features-jta-supporting-alternative-embedded

12.6 本章小结

本章讲解了基于数据库 MariaDB 的各个数据处理知识点，包括：JDBC、连接池、自定义转换器、JPA、事务和分布式事务。其中，涉及到事务的内容只做了简略的概括总结，这方面的知识点是比较多的，希望大家利用好相关的 URL 链接来进一步学习。

第十三章 Redis

Redis 是一个开源（BSD 许可）的内存数据结构存储，用作数据库、缓存和消息代理。它支持数据结构，如字符串、哈希、列表、集合、带范围查询的排序集合、位图、超日志、带半径查询的地理空间索引和流。Redis 内置了复制、Lua 脚本、LRU 回收、事务和不同级别的磁盘持久化，并通过 Redis Sentinel 和带有 Redis 集群的自动分区提供高可用性。

13.1 搭建

我们在 CentOS 8 中下载并安装 redis-6.0.1.tar.gz。首先，下载 Redis，如下所示：
$ wget http://download.redis.io/releases/redis-6.0.1.tar.gz

解压并编译 Redis，如下所示：
$ tar xzf redis-6.0.1.tar.gz
$ cd redis-6.0.1
$ make

单节点运行 Redis 服务器：$ ./src/redis-server ./redis.conf。如图 13-1 所示。

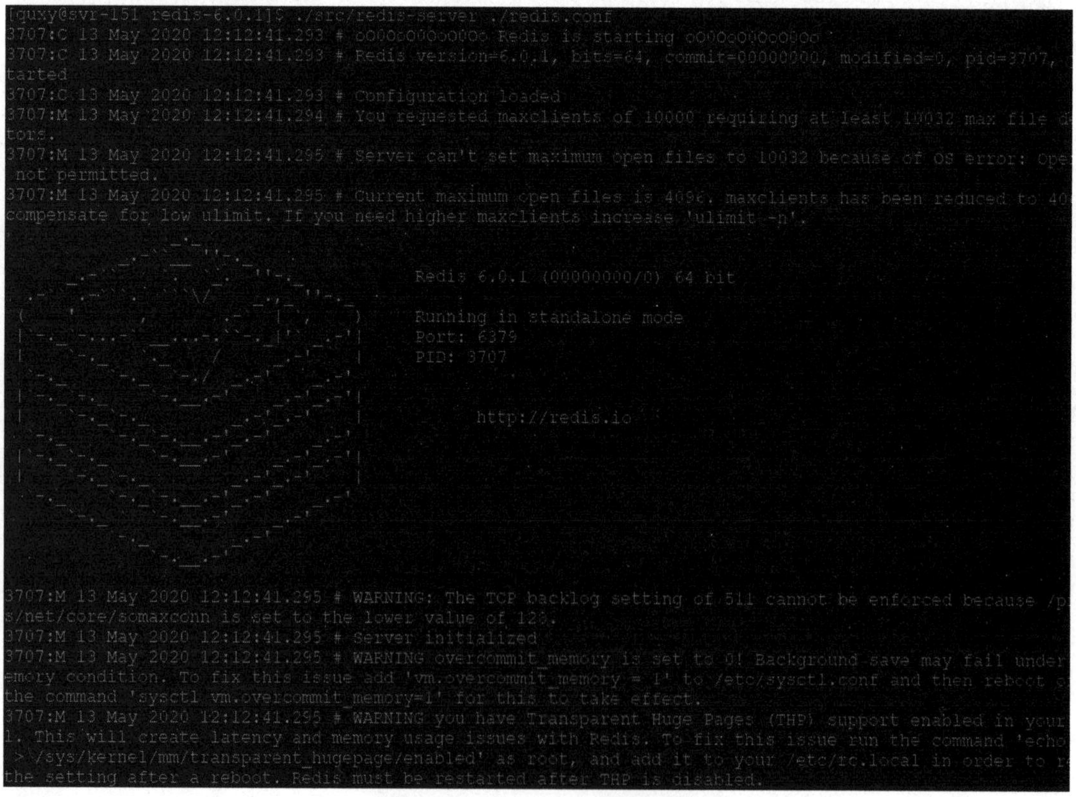

图 13-1 redis.conf 配置

使用内置的客户端与 Redis 服务器交互：$ src/redis-cli。如图 13-2 所示。

图 13-2 内置客户端与 Redis 服务器交互

上图中演示了字符串的存储和获取，对于中文字符会进行编码。

13.2 进一步配置

接下来，我们设置本地测试用局域网访问及密码。使用 vim 打开 redis.config 配置文件。将 bind 修改为：bind 0.0.0.0，如图 13-3 所示。

图 13-3 配置 redis.config（1）

将 protected-mode 修改为 protected-mode no，如图 13-4 所示。

图 13-4 配置 redis.config（2）

如上图所示，可以使用 port 更改端口，这里我们不做调整了。做了上述调整后，再次启动 Redis 服务器后，就可以在本地测试用局域网中的另一台机器上访问它了。执行命令：
$ ./src/redis-cli -h 192.168.31.151 -p 6379。如图 13-5 所示。

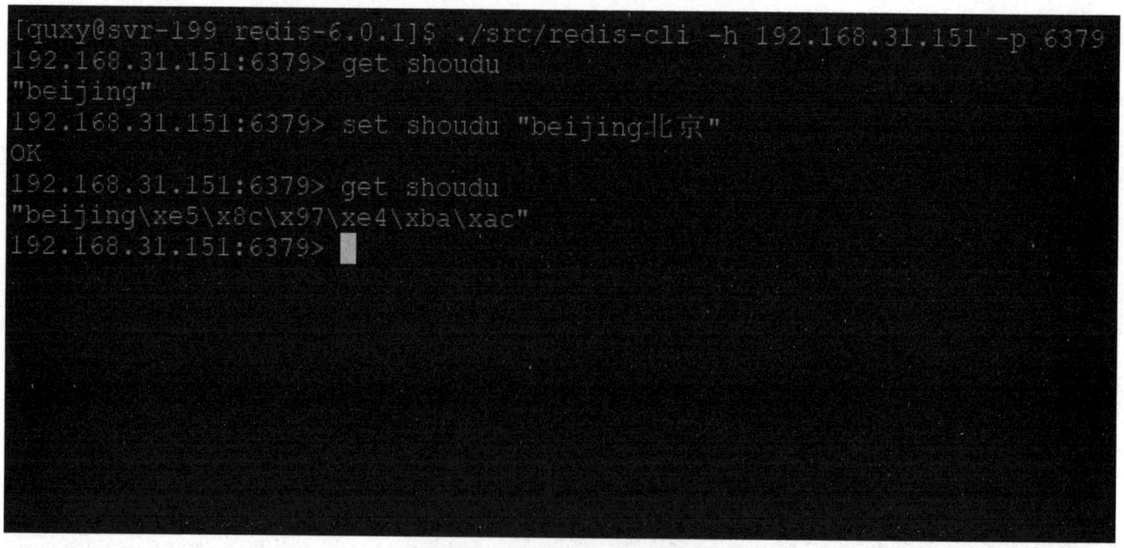

图 13-5 执行 Redis 服务

将 Redis 服务器作为后台进程来运行，相应的配置项为：daemonize yes。如图 13-6 所示。

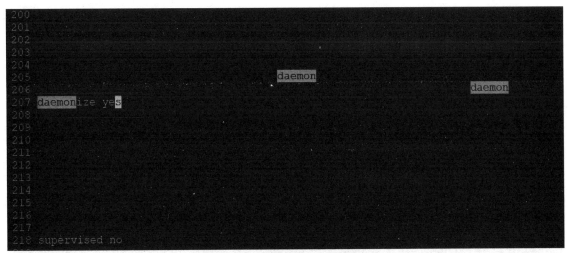

图 13-6 配置 Redis 服务器

配置后，仍然使用原来的启动命令，控制台的输出，如图 13-7 所示。

```
[quxy@svr-151 redis-6.0.1]$ ./src/redis-server ./redis.conf
8046:C 14 May 2020 07:41:46.101 # oO0OoO0OoO0Oo Redis is starting oO0OoO0OoO0Oo
8046:C 14 May 2020 07:41:46.102 # Redis version=6.0.1, bits=64, commit=00000000, modified=0, pid=8046, just started
8046:C 14 May 2020 07:41:46.102 # Configuration loaded
[quxy@svr-151 redis-6.0.1]$
```

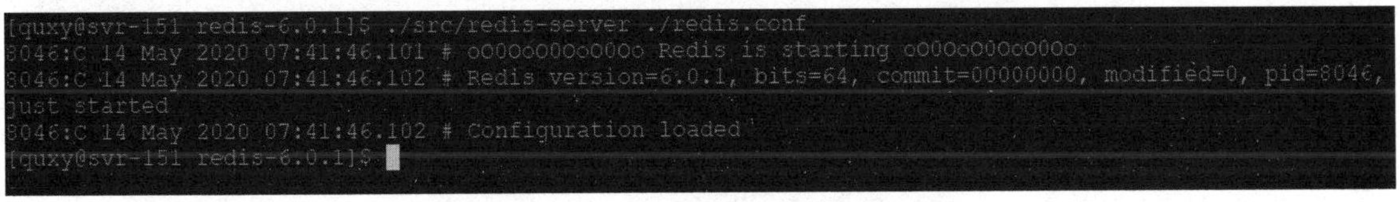

图 13-7 Redis 服务器控制台输出

对于生成环境来说，给 Redis 服务器添加密码是必不可少的。可以使用相应的命令设置密码，如下所示：

> CONFIG set requirepass "123456"

设置的密码对已经建立的会话连接不生效，只对之后建立的会话连接生效。具体情况，如图 13-8 所示。

```
192.168.31.151:6379> get shoudu
"beijing\xe5\x8c\x97\xe4\xba\xac"
192.168.31.151:6379> CONFIG set requirepass "123456"
OK
192.168.31.151:6379> get shoudu
"beijing\xe5\x8c\x97\xe4\xba\xac"
192.168.31.151:6379> set shoudu beijing
OK
192.168.31.151:6379>
[quxy@svr-199 redis-6.0.1]$ ./src/redis-cli -h 192.168.31.151 -p 6379
192.168.31.151:6379> get foo
(error) NOAUTH Authentication required.
192.168.31.151:6379> AUTH 123456
OK
192.168.31.151:6379> get foo
"bar"
192.168.31.151:6379>
```

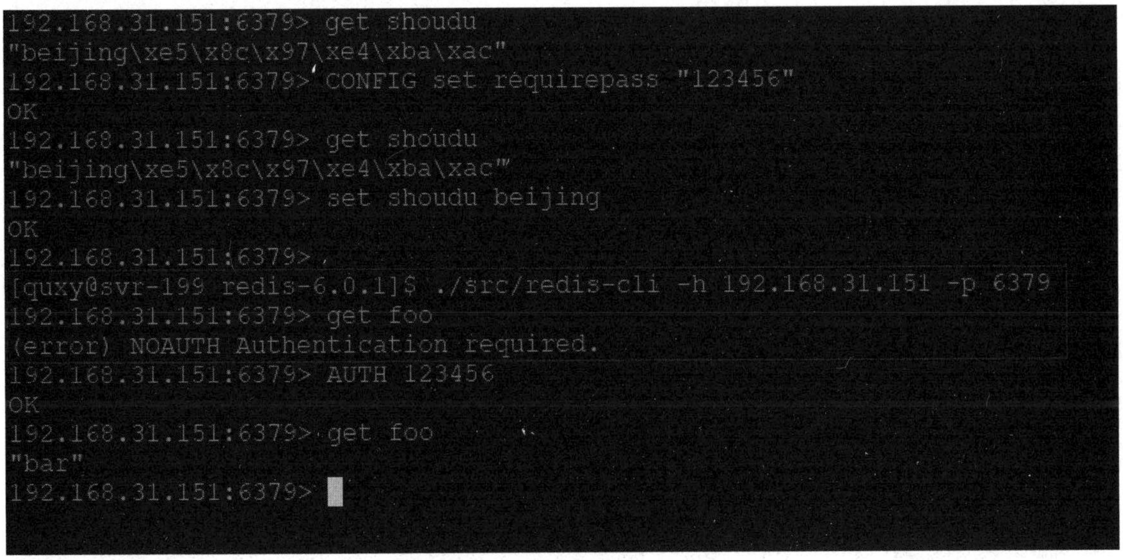

图 13-8 密码设置（1）

另外，要注意的是：通过上面的命令设置的密码，在 Redis 服务器重启后会失效的。要想永久生效，就需要在 redis.config 配置文件中添加密码，相应的配置项为：requirepass 123456。如图 13-9 所示。

```
1787 port 6379
1788 cluster-enabled yes
1789 cluster-config-file nodes.conf
1790 cluster-node-timeout 5000
1791 appendonly yes
1792
1793
1794 requirepass 123456
```

图 13-9 密码设置（2）

13.3 搭建集群

在这里，我们创建一个拥有 3 个主节点 0 个从节点的集群，具体规划，如表 13-1 所示。

表 13-1 集群节点列表

IP	端口	节点类型
192.168.31.11	6379	master
192.168.31.151	6379	master
192.168.31.199	6379	master

首先，修改每个节点的 redis.conf 配置文件，在文件的底部添加配置项，如图 13-10 所示。

```
1789 bind 0.0.0.0
1790 protected-mode no
1791 port 6379
1792 cluster-enabled yes
1793 cluster-config-file nodes.conf
1794 cluster-node-timeout 5000
1795 appendonly yes
```

图 13-10 各节点 redis.conf 文件配置内容

cluster-config-file 指定的文件是自动生成的，无需手动处理它。配置完成后，依次启动这个三个节点，执行命令，如下所示：

$ ./src/redis-server ./redis.conf

控制台的输出，如图 13-11 所示。

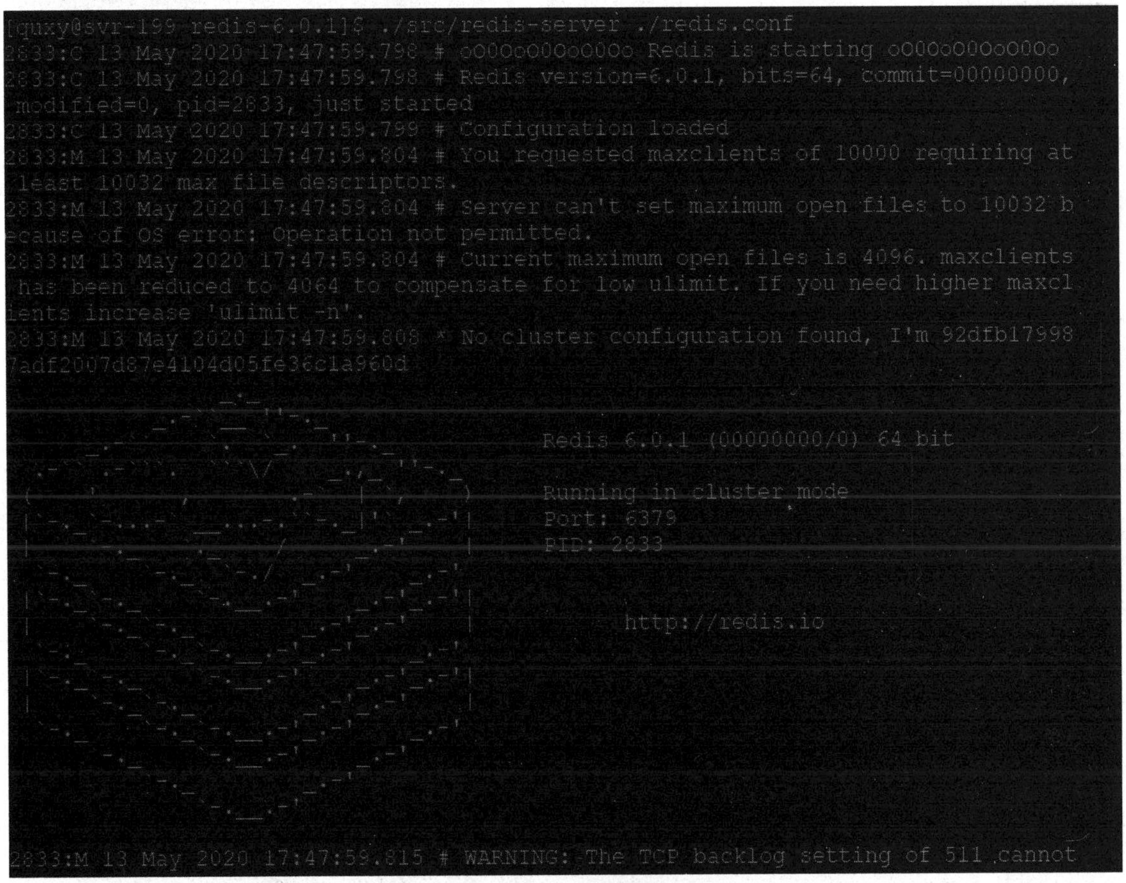

图 13-11 启动节点

进入任意一个主节点的主目录，执行命令，如下所示：

$ ./src/redis-cli --cluster create \
192.168.31.11:6379 192.168.31.151:6379 192.168.31.199:6379 \
  --cluster-replicas 0

上面的命令产生的输出，如图 13-12 所示。

```
[quxy@svr-151 redis-6.0.1]$ ./src/redis-cli --cluster create 192.168.31.11:6379
192.168.31.151:6379 192.168.31.199:6379 --cluster-replicas 0
>>> Performing hash slots allocation on 3 nodes...
Master[0] -> Slots 0 - 5460
Master[1] -> Slots 5461 - 10922
Master[2] -> Slots 10923 - 16383
M: 9420175697d6a073477c7906f648fb355e9fffd9 192.168.31.11:6379
   slots:[0-5460] (5461 slots) master
M: 54a0ddde2680f55279db2536fe4320fdb488e9b4 192.168.31.151:6379
   slots:[5461-10922] (5462 slots) master
M: 92dfb179987adf2007d87e4104d05fe36c1a960d 192.168.31.199:6379
   slots:[10923-16383] (5461 slots) master
Can I set the above configuration? (type 'yes' to accept): yes
>>> Nodes configuration updated
>>> Assign a different config epoch to each node
>>> Sending CLUSTER MEET messages to join the cluster
Waiting for the cluster to join

>>> Performing Cluster Check (using node 192.168.31.11:6379)
M: 9420175697d6a073477c7906f648fb355e9fffd9 192.168.31.11:6379
   slots:[0-5460] (5461 slots) master
M: 54a0ddde2680f55279db2536fe4320fdb488e9b4 192.168.31.151:6379
   slots:[5461-10922] (5462 slots) master
M: 92dfb179987adf2007d87e4104d05fe36c1a960d 192.168.31.199:6379
   slots:[10923-16383] (5461 slots) master
[OK] All nodes agree about slots configuration.
>>> Check for open slots...
>>> Check slots coverage...
[OK] All 16384 slots covered.
```

图 13-12 备份分片

在 Redis 的主目录中，我们找到 nodes.conf 配置文件，执行命令：cat nodes.conf 打开它，如图 13-13 所示。

```
[quxy@svr-151 redis-6.0.1]$ cat nodes.conf
9420175697d6a073477c7906f648fb355e9fffd9 192.168.31.11:6379@16379 master - 0 158
9363871596 1 connected 0-5460
92dfb179987adf2007d87e4104d05fe36c1a960d 192.168.31.199:6379@16379 master - 0 15
89363871697 3 connected 10923-16383
54a0ddde2680f55279db2536fe4320fdb488e9b4 192.168.31.151:6379@16379 myself,master
 - 0 0 2 connected 5461-10922
vars currentEpoch 3 lastVoteEpoch 0
```

图 13-13 修改 nodes.conf 配置

从上图中可以看到，其中，包含了：节点 ID、IP、端口等信息。要想查看所有节点的状态信息，请执行命令：$ ./src/redis-cli -p 6379 cluster nodes。如图 13-14 所示。

```
[quxy@svr-151 redis-6.0.1]$ ./src/redis-cli -p 6379 cluster nodes
92dfb179987adf2007d87e4104d05fe36c1a960d 192.168.31.199:6379@16379 master - 0 1589417062204 3 connected
 10923-16383
9420175697d6a073477c7906f648fb355e9fffd9 192.168.31.11:6379@16379 master - 0 1589417063229 1 connected
 0-5460
54a0ddde2680f55279db2536fe4320fdb488e9b4 192.168.31.151:6379@16379 myself,master - 0 1589417062000 2 co
nnected 5461-10922
```

图 13-14 集群节点

经过以上的配置、运行、测试之后，Redis 集群已经部署好了。接下来，我们用命令行客户端试用一下。执行命令：$ ./src/redis-cli -c。如图 13-15 所示。

```
[guxy@svr-151 redis-6.0.1]$ ./src/redis-cli -c
127.0.0.1:6379> set shoudu beijing
-> Redirected to slot [1941] located at 192.168.31.11:6379
OK
192.168.31.11:6379> get shoudu
"beijing"
192.168.31.11:6379> set shoudu beijing2
OK
192.168.31.11:6379> set henan zhengzhou
OK
192.168.31.11:6379> get henan
"zhengzhou"
192.168.31.11:6379>
```

图 13-15 集群测试

从上图中，我们可以看出：shoudu 的这个键及其值并没有存储到本地机器上而是存储在了另外一台服务器上。在存储后，命令行也随之跳转到另外一台机器上。

13.4 Redis 示例

Spring Data Redis（SDR）框架通过 Spring 出色的基础设施支持，消除了与存储交互所需的冗余任务和样板代码，使编写使用 Redis 键值存储的 Spring 应用程序变得容易。

Spring Data Redis 提供了从 Spring 应用程序轻松配置和访问 Redis 的功能。它提供了与存储交互的低级和高级抽象，将用户从基础设施问题中解放出来。要在项目中使用 Redis 服务器，请添加 Redis 依赖项，如下所示：

```xml
<dependency>
    <groupId>org.springframework.boot</groupId>
    <artifactId>spring-boot-starter-data-redis</artifactId>
</dependency>
```

在本依赖项中，默认使用 Lettuce 连接 Redis 服务器，如图 13-16 所示。

```
\- org.springframework.boot:spring-boot-starter-data-redis:jar:2.3.0.M4:compile
   +- org.springframework.data:spring-data-redis:jar:2.3.0.RC1:compile
   |  +- org.springframework.data:spring-data-keyvalue:jar:2.3.0.RC1:compile
   |  +- org.springframework:spring-oxm:jar:5.2.5.RELEASE:compile
   |  \- org.springframework:spring-context-support:jar:5.2.5.RELEASE:compile
   \- io.lettuce:lettuce-core:jar:5.2.2.RELEASE:compile
      +- io.netty:netty-common:jar:4.1.48.Final:compile
      +- io.netty:netty-handler:jar:4.1.48.Final:compile
      |  +- io.netty:netty-resolver:jar:4.1.48.Final:compile
      |  +- io.netty:netty-buffer:jar:4.1.48.Final:compile
      |  \- io.netty:netty-codec:jar:4.1.48.Final:compile
      +- io.netty:netty-transport:jar:4.1.48.Final:compile
      \- io.projectreactor:reactor-core:jar:3.3.4.RELEASE:compile
         \- org.reactivestreams:reactive-streams:jar:1.0.3:compile
```

图 13-16 Redis 依赖项

接着，在 application.yml 配置文件中添加配置项，如图 13-17 所示。

```yaml
spring:
  redis:
    password: 123456
    cluster:
      nodes:
        - 192.168.31.11:6379
        - 192.168.31.151:6379
        - 192.168.31.199:6379
    lettuce:
      shutdown-timeout: 100ms
      pool:
        max-active: 8
        max-idle: 8
        max-wait: -1ms
        min-idle: 0
```

图 13-17 修改 application.yml 配置

接下来，我们使用 RedisTemplate 接口操作 Redis 服务器，Spring Boot 提供的默认实现是 StringRedisTemplate 类。编写 controller 层的代码，如下所示：

```java
import com.fasterxml.jackson.databind.ObjectMapper;
import com.gf.common.GfResponseEntity;
import org.springframework.data.redis.core.RedisTemplate;
import org.springframework.http.ResponseEntity;
import org.springframework.web.bind.annotation.GetMapping;
import org.springframework.web.bind.annotation.PostMapping;
import org.springframework.web.bind.annotation.RequestMapping;
import org.springframework.web.bind.annotation.RestController;
import java.util.Collections;
import java.util.List;
import java.util.Map;
@RestController
@RequestMapping("/mvc/redis")
public class RedisController {
    private final RedisTemplate<String, String> redisTemplate;
    private final ObjectMapper objectMapper;
    public RedisController(RedisTemplate<String, String> redisTemplate, ObjectMapper objectMapper) {
        this.redisTemplate = redisTemplate;
        this.objectMapper = objectMapper;
    }
    /**
     * 存储一个关联 key 的字符串
     */
    @PostMapping("/create")
    public ResponseEntity create(String key, String val) {
        redisTemplate.opsForValue().set(key, val);
```

```java
        return GfResponseEntity.self().ok();
    }
    /**
     * 根据 key 获取对应的字符串
     */
    @GetMapping("/view")
    public ResponseEntity view(String key) {
        String val = redisTemplate.opsForValue().get(key);
        return GfResponseEntity.self().ok(Collections.singletonMap(key, val));
    }
    /**
     * 根据 key 获取对应的 Map 数据
     */
    @GetMapping("/viewHash")
    public ResponseEntity viewHash(String key) {
        Map<Object, Object> person = redisTemplate.boundHashOps(key).entries();
        return GfResponseEntity.self().ok(person);
    }
    /**
     * 存储一个关联 key 的列表
     */
    @PostMapping("/createList")
    public ResponseEntity createList(String key, String[] vals) {
        redisTemplate.opsForList().rightPushAll(key, vals);
        return GfResponseEntity.self().ok();
    }
    /**
     * 根据 key 获取对应的列表数据
     */
    @GetMapping("/viewList")
    public ResponseEntity viewList(String key) {
        List<String> list = redisTemplate.opsForList().range(key, 0, 10);
        return GfResponseEntity.self().ok(Collections.singletonMap(key, list));
    }
}
```

13.5 UI 客户端

在日常的开发中，除了使用命令行客户端外，使用 UI 客户端可以更方便快捷地进行查看、修改、删除数据等操作，Redis 有以下几种 UI 客户端可供选择：

RedisClient

RedisClient 是一个功能较为丰富的 Redis 客户端，能满足日常的开发要求。目前，它可以支持较低版本的 Redis。因为，RedisClient 已经很长时间未更新（在编写本书时，windows 分支的最后更新时间是 3 年前），所以不能支持 Redis 的新版本。要想让 RedisClient 支持新版本的 Redis，请下载源代码，提升 pom.xml 配置文件中 jedis 的版本号，本书中使用的版本号是：2.10.2。RedisClient 源代码，如下所示：

https://github.com/caoxinyu/RedisClient

启动 RedisClient 并连接 Redis，查看其中数据，如图 13-18 所示。

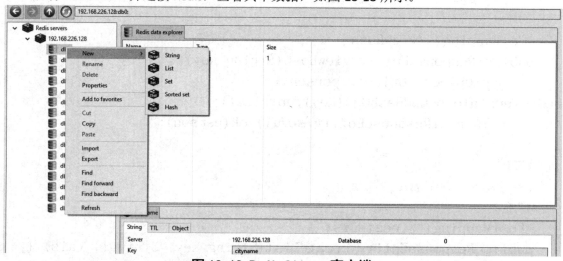

图 13-18 RedisClient 客户端

Another Redis DeskTop Manager

这是一个免费开源的 Redis 桌面管理器，它是更快、更好、更稳定的，在 Linux、Windows 和 Mac 操作系统中都可以运行。此外，当加载大量密钥时，它不会崩溃。它是使用 JavaScript 语言开发的，使用 npm 命令进行安装。要想查看源代码、下载安装和使用，请访问以下网址：

https://github.com/qishibo/AnotherRedisDesktopManager

RedisStudio

RedisStudio 是一个用 C++语言编写的简单的 Redis 客户端，它是运行在 Windows 环境中的，主要用于查看 Redis 中的数据。RedisStudio 的 exe 程序是免安装的，直接双击运行即可，启动后，连接 Redis，界面如图 13-19 所示。

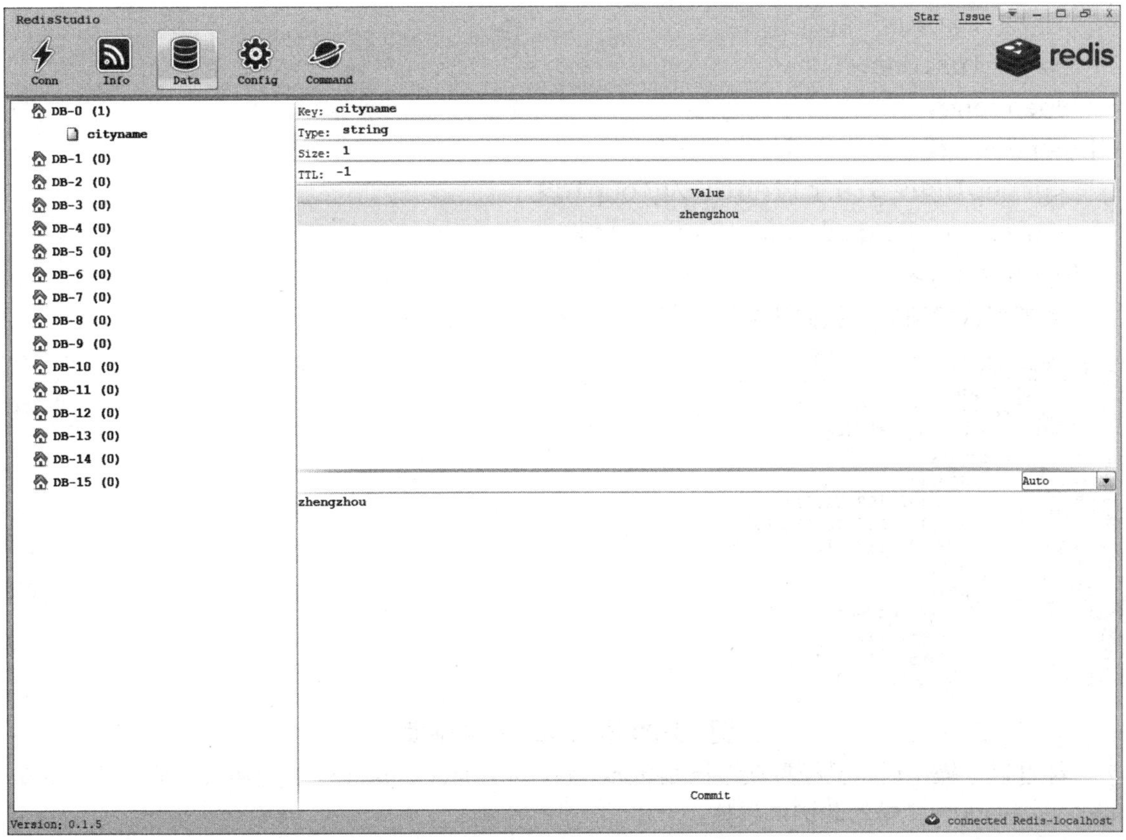

图 13-19 RedisStudio 客户端

查看源代码、下载安装和使用，请访问以下网址：
https://github.com/cinience/RedisStudio

13.6 Jedis

Jedis 是一个非常小而健全的 Redis Java 客户端，它非常易于使用。有关 Jedis 的源代码等更多信息，请访问以下网址：

https://github.com/xetorthio/jedis

在 Spring Boot 应用程序中，想要把默认的 Lettuce 替换为 Jedis，不需要调整 Java 代码，只需要排除 Lettuce 依赖项，添加 Jedis 依赖项，如下所示：

```
<dependency>
    <groupId>org.springframework.boot</groupId>
    <artifactId>spring-boot-starter-data-redis</artifactId>
    <exclusions>
        <exclusion>
            <groupId>io.lettuce</groupId>
            <artifactId>lettuce-core</artifactId>
```

```xml
        </exclusion>
    </exclusions>
</dependency>
<dependency>
    <groupId>redis.clients</groupId>
    <artifactId>jedis</artifactId>
</dependency>
```

调整完依赖项后，我们还需要调整一下配置项，如图 13-20 所示。

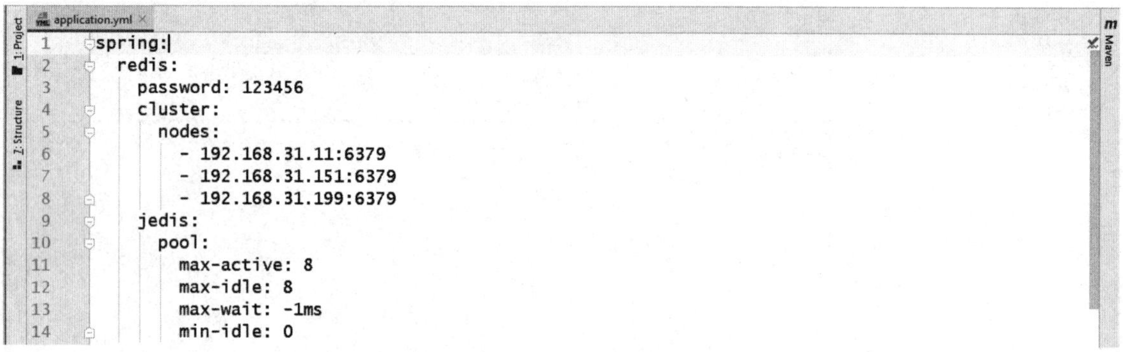

```yaml
spring:
  redis:
    password: 123456
    cluster:
      nodes:
        - 192.168.31.11:6379
        - 192.168.31.151:6379
        - 192.168.31.199:6379
    jedis:
      pool:
        max-active: 8
        max-idle: 8
        max-wait: -1ms
        min-idle: 0
```

图 13-20 添加 Jedis 依赖项

经过以上操作后，已经默默地把 Lettuce 替换为了 Jedis，Java 代码无需做任何更改，这是 RedisTemplate 方式带给我们的好处。

13.7 Redisson

Redisson 是具有内存数据网格等丰富功能的 Redis Java 客户端。它支持 Redis 哨兵设置，支持安卓平台，支持自动重新连接，支持发送命令失败自动重试，支持 SSL，支持许多流行的编解码器。它提供了反应式流 API、异步 API 和事务 API。

想要查看 Redisson 的源代码、文档等更多有用信息，请访问以下网址：

https://github.com/redisson/redisson

在 Spring Boot 应用程序中，我们习惯于用 starter 引入相应的技术或服务，有关 Redisson starter 的详细信息，请访问以下网址：

https://github.com/redisson/redisson/tree/master/redisson-spring-boot-starter

Redisson 中文文档，如下所示：

https://github.com/redisson/redisson/wiki/%E7%9B%AE%E5%BD%95

在 Spring Boot 应用程序中，想要用 Redisson 连接 Redis，请添加 Redisson 依赖项，如下所示：

```xml
<dependency>
    <groupId>org.redisson</groupId>
    <artifactId>redisson-spring-boot-starter</artifactId>
```

```
<version>3.13.1</version>
</dependency>
```

接着,在 application.yml 配置文件中添加配置项,如图 13-21 所示。

```yaml
spring:
  redis:
    host: 192.168.226.128
    port: 6379
    password: 123456
#   cluster:
#     nodes:
#       - 192.168.226.128:6379
#       - 192.168.226.133:6379
#       - 192.168.226.134:6379
#   sentinel:
#     master:
#     nodes:
#   redisson:
#     config: classpath:redisson.yaml
```

图 13-21 添加 Redisson 依赖项

从上图所示的配置中,我们可以看出:其中有属于 Spring Data 的公共配置也有 Redisson 特有的配置。接下来,我们使用 RedissonClient 接口操作 Redis 服务器,编写 controller 层的代码,如下所示:

```java
package com.gf.redis.controller;
import org.redisson.api.RedissonClient;
import org.springframework.http.ResponseEntity;
import org.springframework.web.bind.annotation.GetMapping;
import org.springframework.web.bind.annotation.PostMapping;
import org.springframework.web.bind.annotation.RequestMapping;
import org.springframework.web.bind.annotation.RestController;
import java.util.Arrays;
import java.util.Collections;
import java.util.List;
import static com.gf.common.GfResponseEntity.self;
@RestController
@RequestMapping("/redis")
public class RedisMenuController {
    private final RedissonClient redissonClient;
    public RedisMenuController(RedissonClient redissonClient) {
        this.redissonClient = redissonClient;
    }
    /**
     * 存储一个关联 key 的字符串
     */
    @PostMapping("/create")
    public ResponseEntity create(String key, String val) {
```

```java
        redissonClient.getBucket(key).set(val);
        return self().ok();
    }
    /**
     * 根据 key 获取对应的字符串
     */
    @GetMapping("/view")
    public ResponseEntity view(String key) {
        Object val = redissonClient.getBucket(key).get();
        return self().ok(Collections.singletonMap(key, val));
    }
    /**
     * 存储一个关联 key 的列表
     */
    @PostMapping("/list/create")
    public ResponseEntity create(String key, String[] vals) {
        redissonClient.getList(key).addAll(Arrays.asList(vals));
        return self().ok();
    }
    /**
     * 根据 key 获取对应的列表
     */
    @GetMapping("/list/view")
    public ResponseEntity list(String key) {
        List<Object> list = redissonClient.getList(key).readAll();
        return self().ok(list);
    }
}
```

13.8 本章小结

本章讲解了 Redis 单节点和集群的搭建，及更详细的配置内容，还讲解了在 Spring Boot 中如何使用 Redis。最后是关于 Redis UI 客户端、Jedis 和 Redisson 的内容。

第十四章 MongoDB

MongoDB 是为现代应用程序开发者和云时代构建的通用、基于文档的分布式数据库。它是高性能、可扩展、易部署、易使用的，存储数据非常方便。

14.1 在线安装

可以在 Linux、Windows、macOS 等系列的操作系统上安装使用 MongoDB。在这里，我们演示联网情况下在 CentOS 8 中安装 MongoDB 服务器，其中 MongoDB 的版本为 4.2.6。

在目录 /etc/yum.repos.d 中新建文件 mongodb-org-4.2.repo，添加的内容，如图 14-1 至图 14-3 所示。

```
[mongodb-org-4.2]
name=MongoDB Repository
baseurl=https://repo.mongodb.org/yum/redhat/$releasever/mongodb-org/4.2/x86_64/
gpgcheck=1
enabled=1
gpgkey=https://www.mongodb.org/static/pgp/server-4.2.asc
```

图 14-1 在线安装 MongoDB（1）

要安装 MongoDB 的最新稳定版本，请执行命令，如下所示：
sudo yum install -y mongodb-org

要安装 MongoDB 4.2.6 版本，请执行命令，如下所示：

$ sudo yum install -y mongodb-org-4.2.6 \
mongodb-org-server-4.2.6 \
mongodb-org-shell-4.2.6 mongodb-org-mongos-4.2.6 mongodb-org-tools-4.2.6

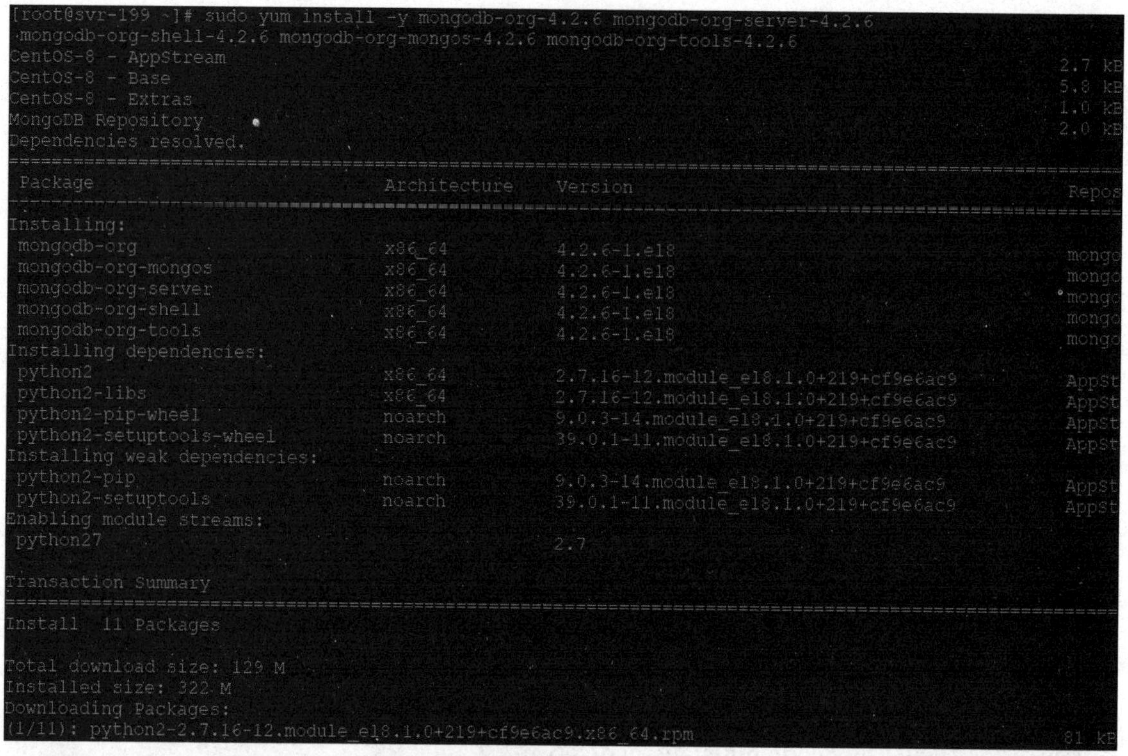

图 14-2 在线安装 MongoDB（2）

图 14-3 在线安装 MongoDB（3）

控制台的最后一行输出了"Complete!"，证明安装成功了。

安装 MongoDB 时，会创建一个用户 mongod（所属的组为 mongod）。这个用户用于运行 MongoDB。

14.2 启动 MongoDB

启动 MongoDB 服务器，请执行命令，如下所示：

$ sudo systemctl start mongod

上面的命令执行后，控制台不会产生任何输出。要想查看是否启动成功，执行命令：$ sudo systemctl status mongod。如图 14-4 所示。

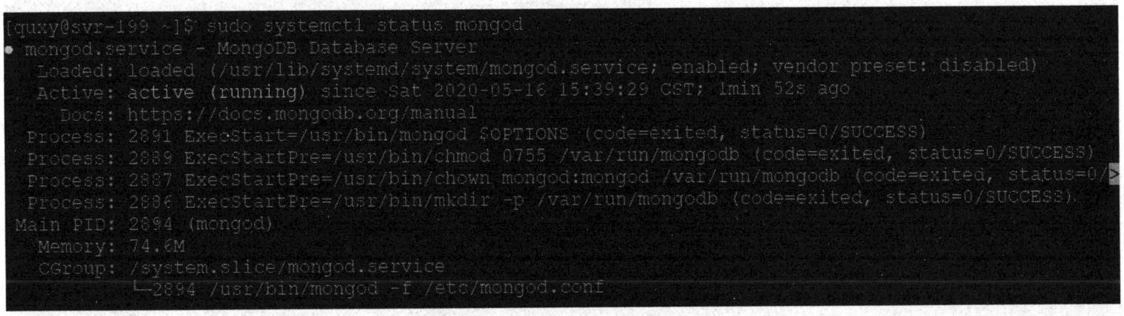

```
[quxy@svr-199 ~]$ sudo systemctl status mongod
● mongod.service - MongoDB Database Server
   Loaded: loaded (/usr/lib/systemd/system/mongod.service; enabled; vendor preset: disabled)
   Active: active (running) since Sat 2020-05-16 15:39:29 CST; 1min 52s ago
     Docs: https://docs.mongodb.org/manual
  Process: 2891 ExecStart=/usr/bin/mongod $OPTIONS (code=exited, status=0/SUCCESS)
  Process: 2889 ExecStartPre=/usr/bin/chmod 0755 /var/run/mongodb (code=exited, status=0/SUCCESS)
  Process: 2887 ExecStartPre=/usr/bin/chown mongod:mongod /var/run/mongodb (code=exited, status=0/>
  Process: 2886 ExecStartPre=/usr/bin/mkdir -p /var/run/mongodb (code=exited, status=0/SUCCESS)
 Main PID: 2894 (mongod)
   Memory: 74.6M
   CGroup: /system.slice/mongod.service
           └─2894 /usr/bin/mongod -f /etc/mongod.conf
```

图 14-4 启动 MongoDB

在图 14-4 中，我们可以看到“Active: active (running)”字样，证明 MongoDB 服务器已经启动正常运行。

14.3 Shell 客户端

接下来，我们通过命令行客户端使用 MongoDB，访问本机的 MongoDB 服务器，请执行命令：$ mongo。如图 14-5 所示。

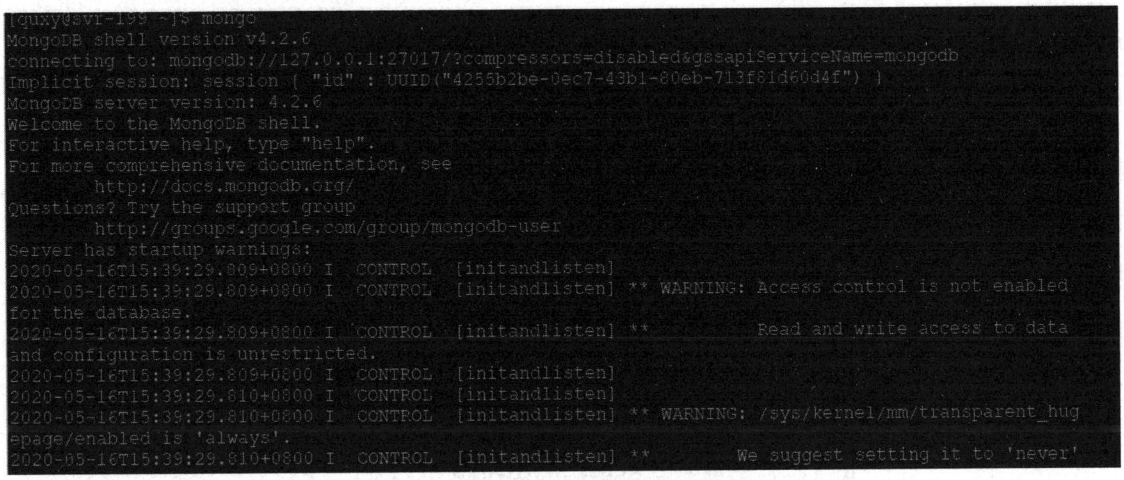

```
[quxy@svr-199 ~]$ mongo
MongoDB shell version v4.2.6
connecting to: mongodb://127.0.0.1:27017/?compressors=disabled&gssapiServiceName=mongodb
Implicit session: session { "id" : UUID("4255b2be-0ec7-43b1-80eb-713f81d60d4f") }
MongoDB server version: 4.2.6
Welcome to the MongoDB shell.
For interactive help, type "help".
For more comprehensive documentation, see
        http://docs.mongodb.org/
Questions? Try the support group
        http://groups.google.com/group/mongodb-user
Server has startup warnings:
2020-05-16T15:39:29.809+0800 I  CONTROL  [initandlisten]
2020-05-16T15:39:29.809+0800 I  CONTROL  [initandlisten] ** WARNING: Access control is not enabled
for the database.
2020-05-16T15:39:29.809+0800 I  CONTROL  [initandlisten] **          Read and write access to data
and configuration is unrestricted.
2020-05-16T15:39:29.809+0800 I  CONTROL  [initandlisten]
2020-05-16T15:39:29.810+0800 I  CONTROL  [initandlisten]
2020-05-16T15:39:29.810+0800 I  CONTROL  [initandlisten] ** WARNING: /sys/kernel/mm/transparent_hug
epage/enabled is 'always'.
2020-05-16T15:39:29.810+0800 I  CONTROL  [initandlisten] **        We suggest setting it to 'never'
```

图 14-5 访问 MongoDB

新建集合 gf，并向其中插入一条数据，然后查看该数据，如图 14-6 所示。

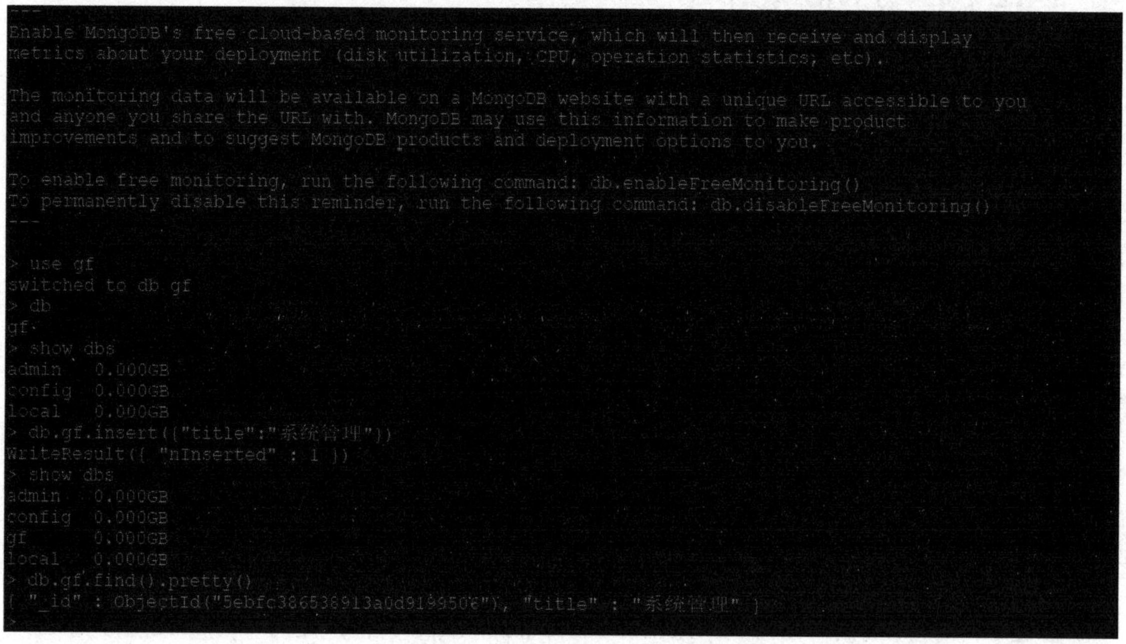

图 14-6 测试 MongoDB

14.4 本地测试用局域网访问

要想在从其他机器访问 MongoDB 服务器，执行修改配置文件命令：$ sudo vim /etc/mongod.conf。如图 14-7 所示。

图 14-7 本地测试用局域网内访问 MongoDB（1）

如图 14-7 所示，将 bindIp 的值修改为 0.0.0.0。本地测试在局域网中的其他机器上访问 MongoDB 服务器，请执行命令：$ bin\mongo --host 192.168.31.199 --port 27017。如图 14-8 所示。

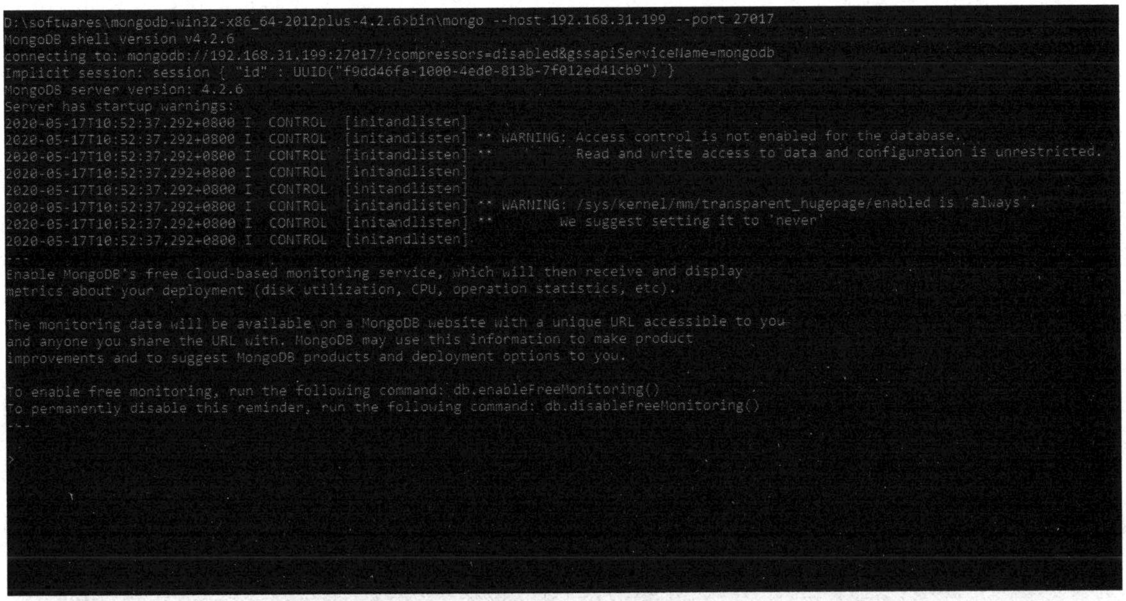

图 14-8 本地测试在局域网内访问 MongoDB（2）

图 14-8 中的本地测试在局域网访问，采用了无用户名、无密码的访问方式。

14.5 离线安装

我们使用 Win 10 操作系统来讲解离线安装。下载页面的网址：https://www.mongodb.com/download-center/community。如图 14-9 所示。

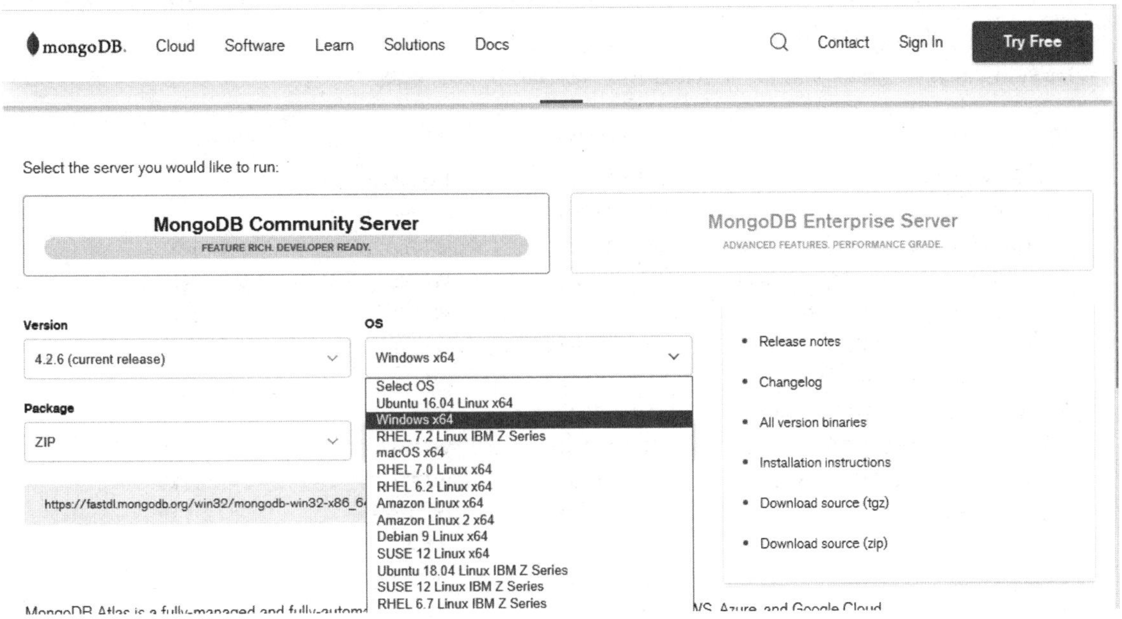

图 14-9 离线安装图 MongoDB（1）

MongoDB 下载地址，如下所示：

https://fastdl.mongodb.org/win32/mongodb-win32-x86_64-2012plus-4.2.6.zip

解压下载好的 MongoDB 压缩包，并提前创建好数据目录 ...\data\db，打开 CMD 窗口，进入到 MongoDB 主目录中，启动服务器，请执行命令：$ bin\mongod –dbpath D:\softwares\mongodb\data\db。如图 14-10 所示。

```
D:\softwares\mongodb-win32-x86_64-2012plus-4.2.6>bin\mongod --dbpath D:\softwares\mongodb\data\db
2020-05-17T09:31:15.314+0800 I  CONTROL  [main] Automatically disabling TLS 1.0, to force-enable TLS 1.0 specify --sslDisabledProtocols 'none'

2020-05-17T09:31:15.317+0800 W  ASIO     [main] No TransportLayer configured during NetworkInterface startup
2020-05-17T09:31:15.319+0800 I  CONTROL  [initandlisten] MongoDB starting : pid=9972 port=27017 dbpath=D:\softwares\mongodb\data\db 64-bit hos
t=DESKTOP-DGFGBUD
2020-05-17T09:31:15.319+0800 I  CONTROL  [initandlisten] targetMinOS: Windows 7/Windows Server 2008 R2
2020-05-17T09:31:15.320+0800 I  CONTROL  [initandlisten] db version v4.2.6
2020-05-17T09:31:15.320+0800 I  CONTROL  [initandlisten] git version: 20364840b8f1af16917e4c23c1b5f5efd8b352f8
2020-05-17T09:31:15.321+0800 I  CONTROL  [initandlisten] allocator: tcmalloc
2020-05-17T09:31:15.321+0800 I  CONTROL  [initandlisten] modules: none
2020-05-17T09:31:15.322+0800 I  CONTROL  [initandlisten] build environment:
2020-05-17T09:31:15.322+0800 I  CONTROL  [initandlisten]     distmod: 2012plus
2020-05-17T09:31:15.323+0800 I  CONTROL  [initandlisten]     distarch: x86_64
2020-05-17T09:31:15.323+0800 I  CONTROL  [initandlisten]     target_arch: x86_64
2020-05-17T09:31:15.324+0800 I  CONTROL  [initandlisten] options: { storage: { dbPath: "D:\softwares\mongodb\data\db" } }
2020-05-17T09:31:15.331+0800 I  STORAGE  [initandlisten] wiredtiger_open config: create,cache_size=3535M,cache_overflow=(file_max=0M),session_
max=33000,eviction=(threads_min=4,threads_max=4),config_base=false,statistics=(fast),log=(enabled=true,archive=true,path=journal,compressor=sn
appy),file_manager=(close_idle_time=100000,close_scan_interval=10,close_handle_minimum=250),statistics_log=(wait=0),verbose=[recovery_progress
,checkpoint_progress],
2020-05-17T09:31:16.391+0800 I  STORAGE  [initandlisten] WiredTiger message [1589679076:391300][9972:140729968582224], txn-recover: Set global
 recovery timestamp: (0, 0)
2020-05-17T09:31:16.626+0800 I  RECOVERY [initandlisten] WiredTiger recoveryTimestamp. Ts: Timestamp(0, 0)
2020-05-17T09:31:16.831+0800 I  STORAGE  [initandlisten] Timestamp monitor starting
2020-05-17T09:31:16.962+0800 I  CONTROL  [initandlisten]
2020-05-17T09:31:16.963+0800 I  CONTROL  [initandlisten] ** WARNING: Access control is not enabled for the database.
2020-05-17T09:31:16.966+0800 I  CONTROL  [initandlisten] **          Read and write access to data and configuration is unrestricted.
2020-05-17T09:31:16.967+0800 I  CONTROL  [initandlisten]
2020-05-17T09:31:16.969+0800 I  CONTROL  [initandlisten] ** WARNING: This server is bound to localhost.
2020-05-17T09:31:16.973+0800 I  CONTROL  [initandlisten] **          Remote systems will be unable to connect to this server.
2020-05-17T09:31:16.974+0800 I  CONTROL  [initandlisten] **          Start the server with --bind_ip <address> to specify which IP
2020-05-17T09:31:16.974+0800 I  CONTROL  [initandlisten] **          addresses it should serve responses from, or with --bind_ip_all to
2020-05-17T09:31:16.975+0800 I  CONTROL  [initandlisten] **          bind to all interfaces. If this behavior is desired, start the
2020-05-17T09:31:16.975+0800 I  CONTROL  [initandlisten] **          server with --bind_ip 127.0.0.1 to disable this warning.
2020-05-17T09:31:16.976+0800 I  CONTROL  [initandlisten]
2020-05-17T09:31:16.978+0800 I  STORAGE  [initandlisten] createCollection: admin.system.version with provided UUID: 688b6542-0cce-4726-817e-78
4902043c1d and options: { uuid: UUID("688b6542-0cce-4726-817e-784902043c1d") }
2020-05-17T09:31:17.130+0800 I  INDEX    [initandlisten] index build: done building index _id_ on ns admin.system.version
2020-05-17T09:31:17.134+0800 I  SHARDING [initandlisten] Marking collection admin.system.version as collection version: <unsharded>
```

图 14-10 离线安装图 MongoDB（2）

上图是 MongoDB 服务器启动时，控制台的输出信息。在数据目录中创建的各类初始化文件夹和文件，如图 14-11 所示。

› This PC › Seagate Expansion Drive (D:) › softwares › mongodb › data › db

Name	Date modified	Type	Size
diagnostic.data	2020/5/17 9:34	File folder	
journal	2020/5/17 9:31	File folder	
_mdb_catalog.wt	2020/5/17 9:32	WT File	20 KB
collection-0--5882321908792977596.wt	2020/5/17 9:32	WT File	20 KB
collection-2--5882321908792977596.wt	2020/5/17 9:32	WT File	20 KB
collection-4--5882321908792977596.wt	2020/5/17 9:31	WT File	4 KB
index-1--5882321908792977596.wt	2020/5/17 9:32	WT File	20 KB
index-3--5882321908792977596.wt	2020/5/17 9:32	WT File	20 KB
index-5--5882321908792977596.wt	2020/5/17 9:31	WT File	4 KB
index-6--5882321908792977596.wt	2020/5/17 9:33	WT File	12 KB
mongod.lock	2020/5/17 9:31	LOCK File	1 KB
sizeStorer.wt	2020/5/17 9:33	WT File	20 KB
storage.bson	2020/5/17 9:31	BSON File	1 KB
WiredTiger	2020/5/17 9:31	File	1 KB

图 14-11 离线安装图 MongoDB（3）

要与 MongoDB 服务器交互，请打开 CMD 命令行客户端，并执行命令：$ bin\mongo。如图 14-12 所示。

```
Microsoft Windows [Version 10.0.18363.778]
(c) 2019 Microsoft Corporation. All rights reserved.

D:\softwares\mongodb-win32-x86_64-2012plus-4.2.6>bin\mongo
MongoDB shell version v4.2.6
connecting to: mongodb://127.0.0.1:27017/?compressors=disabled&gssapiServiceName=mongodb
Implicit session: session { "id" : UUID("68b6f503-d171-422c-a1a8-7efaf6955feb") }
MongoDB server version: 4.2.6
Welcome to the MongoDB shell.
For interactive help, type "help".
For more comprehensive documentation, see
        http://docs.mongodb.org/
Questions? Try the support group
        http://groups.google.com/group/mongodb-user
Server has startup warnings:
2020-05-17T09:31:16.962+0800 I  CONTROL  [initandlisten]
2020-05-17T09:31:16.963+0800 I  CONTROL  [initandlisten] ** WARNING: Access control is not enabled for the database.
2020-05-17T09:31:16.966+0800 I  CONTROL  [initandlisten] **          Read and write access to data and configuration is unrestricted.
2020-05-17T09:31:16.967+0800 I  CONTROL  [initandlisten]
2020-05-17T09:31:16.969+0800 I  CONTROL  [initandlisten] ** WARNING: This server is bound to localhost.
2020-05-17T09:31:16.973+0800 I  CONTROL  [initandlisten] **          Remote systems will be unable to connect to this server.
2020-05-17T09:31:16.974+0800 I  CONTROL  [initandlisten] **          Start the server with --bind_ip <address> to specify which IP
2020-05-17T09:31:16.974+0800 I  CONTROL  [initandlisten] **          addresses it should serve responses from, or with --bind_ip_all to

2020-05-17T09:31:16.975+0800 I  CONTROL  [initandlisten] **          bind to all interfaces. If this behavior is desired, start the
2020-05-17T09:31:16.975+0800 I  CONTROL  [initandlisten] **          server with --bind_ip 127.0.0.1 to disable this warning.
2020-05-17T09:31:16.976+0800 I  CONTROL  [initandlisten]
---
Enable MongoDB's free cloud-based monitoring service, which will then receive and display
metrics about your deployment (disk utilization, CPU, operation statistics, etc).

The monitoring data will be available on a MongoDB website with a unique URL accessible to you
and anyone you share the URL with. MongoDB may use this information to make product
improvements and to suggest MongoDB products and deployment options to you.

To enable free monitoring, run the following command: db.enableFreeMonitoring()
To permanently disable this reminder, run the following command: db.disableFreeMonitoring()
---

> show dbs
```

图 14-12 访问 MongoDB

新建集合 gf，并向其中插入一条数据，然后查看该数据，如图 14-13 所示。

```
To enable free monitoring, run the following command: db.enableFreeMonitoring()
To permanently disable this reminder, run the following command: db.disableFreeMonitoring()
---

> show dbs
admin   0.000GB
config  0.000GB
local   0.000GB
> use gf
switched to db gf
> db
gf
> db.gf.insert({"title":"日志管理"})
WriteResult({ "nInserted" : 1 })
> show dbs
admin   0.000GB
config  0.000GB
gf      0.000GB
local   0.000GB
> db.gf.find().pretty()
{ "_id" : ObjectId("5ec09643ad24322fc2030dc3"), "title" : "日志管理" }
>
```

图 14-13 测试 MongoDB

14.6 Compass 客户端

作为 MongoDB 的图形用户界面，Compass 可以进行文档结构、查询、索引和文档验证等操作。帮助开发者更高效地进行 MongoDB 相关的开发工作。Compass 的下载地址为：https://www.mongodb.com/try/download/compass。下载页面如图 14-14 所示。

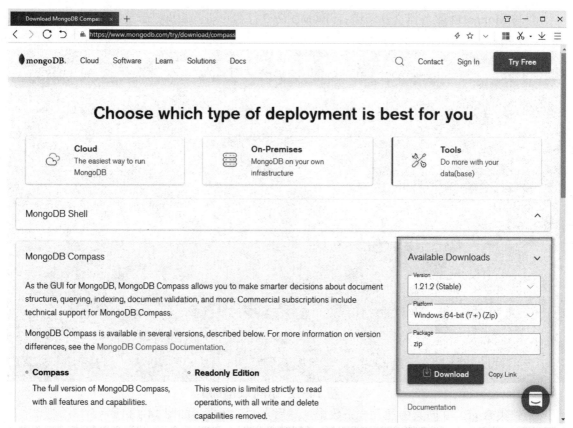

图 14-14 下载 Compass 客户端

在下载页面选择合适的选项，然后进行下载，下载地址，如下所示：
https://downloads.mongodb.com/compass/mongodb-compass-1.21.2-win32-x64.zip

下载完成后，解压文件，进入到 Compass 主目录，如图 14-15 所示。

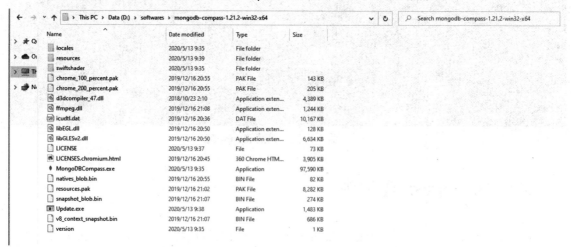

图 14-15 Compass 主目录

双击 MongoDBCompass.exe 文件，即可启动 Compass 客户端，启动后，进入系统欢迎界面，如图 14-16 所示。

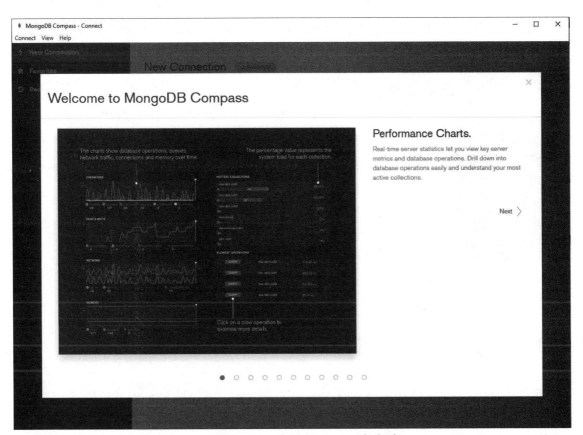

图 14-16 启动 Compass 客户端

在界面中填写连接字符串：$ mongodb://localhost/test，访问本地的 MongoDB 服务器。如图 14-17 所示。

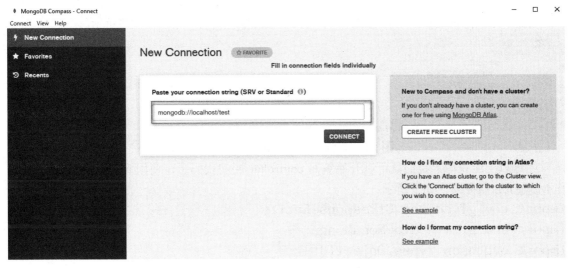

图 14-17 访问 MongDB

连接字符串的更完整的写法，如下所示：

mongodb://username:password@localhost:27017/test

填写好连接字符串后，单击按钮[CONNECT]，进入到操作界面，如图 14-18 所示。

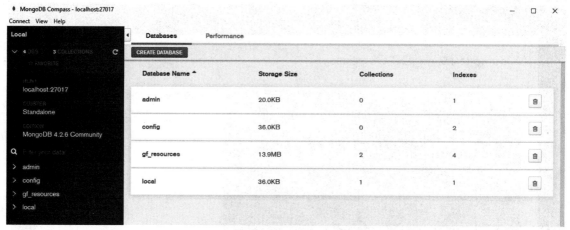

图 14-18 连接 MongDB

在上图所示的操作界面中，我们可以对 MongoDB 服务器中的数据等进行各种操作。

14.7 MongoDB 示例

想要在 Spring Boot 应用程序中使用 MongoDB 服务器，请添加 MongoDB 依赖项，如下所示：

```
<dependency>
    <groupId>org.springframework.boot</groupId>
    <artifactId>spring-boot-starter-data-mongodb</artifactId>
</dependency>
```

然后，在 application.yml 配置文件中，添加相应的配置项，如图 14-19 所示。

图 14-19 添加 application.yml 配置项

经过以上配置后，我们就可以开始编写 controller 层的代码了，使用 MongoTemplate 来操作 MongoDB 服务器，如下所示：

```
import com.gf.common.GfResponseEntity;
import com.gf.mvc.dto.MenuMongo;
import org.bson.types.ObjectId;
import org.springframework.data.mongodb.core.MongoTemplate;
import org.springframework.http.ResponseEntity;
import org.springframework.web.bind.annotation.*;
```

```java
import java.util.Collection;
import java.util.Collections;
import java.util.List;
/**
 * MongoDB 操作示例
 */
@RestController
@RequestMapping("/mvc/mongodb")
public class MongoDBController {
    private final MongoTemplate mongoTemplate;
    private final String COLLECTION_NAME = "menu";
    public MongoDBController(MongoTemplate mongoTemplate) {
        this.mongoTemplate = mongoTemplate;
    }
    /**
     * 新建一条数据,自动生成 objectId
     */
    @PostMapping("/create")
    public ResponseEntity create(@RequestBody MenuMongo params) {
        Collection<MenuMongo> result = mongoTemplate.insert(Collections.singletonList(params), COLLECTION_NAME);
        return GfResponseEntity.self().ok();
    }
    /**
     * 根据 objectId,查找对应的数据。
     */
    @GetMapping("/view/{id}")
    public ResponseEntity view(@PathVariable ObjectId id) {
        MenuMongo entity = mongoTemplate.findById(id, MenuMongo.class, COLLECTION_NAME);
        return GfResponseEntity.self().ok(entity);
    }
    /**
     * 查找出集合中的所有数据。
     */
    @PostMapping("/list")
    public ResponseEntity list() {
```

```java
        List<MenuMongo> list = mongoTemplate.findAll(MenuMongo.class, COLLECTION_NAME);
        return GfResponseEntity.self().ok(list);
    }
}
```

实体类 MenuMongo 的定义，如下所示：

```java
package com.gf.mvc.dto;
import com.fasterxml.jackson.annotation.JsonIgnore;
import lombok.Data;
import org.bson.types.ObjectId;
@Data
public class MenuMongo {
    /**
     * 向客户端响应数据时，忽略本字段，转而展现 objectId 字段。
     */
    @JsonIgnore
    private ObjectId id;
    private String name;
    private Long parentId;
    private int sorted;
    private String objectId;
    /**
     * 向客户端响应数据时，展现本字段。
     */
    public String getObjectId() {
        return id.toHexString();
    }
}
```

14.8 本章小结

本章首先讲解了在 CentOS 8 操作系统中使用 yum 命令安装 MongoDB，然后启动它，并用 Shell 客户端操作它。后面讲解了在 Windows 操作系统中的离线安装 MongoDB 和 Compass 客户端。最后，讲解了在 Spring Boot 应用程序中如何访问 MongoDB。

第十五章 Cassandra

Apache Cassandra 是具有可伸缩性、高可用性和高响性能的分布式数据库。线性可扩展性和在商用硬件或云基础设施上已证实的容错性使其成为任务关键型数据的完美平台。Cassandra 对跨多个数据中心的复制的支持是同类中最好的，它为用户提供了更低的延迟，并且可以在区域性停机中生存。

15.1 下载

访问 Cassandra 官网：https://cassandra.apache.org/download，进入下载页面，如图 15-1 所示。

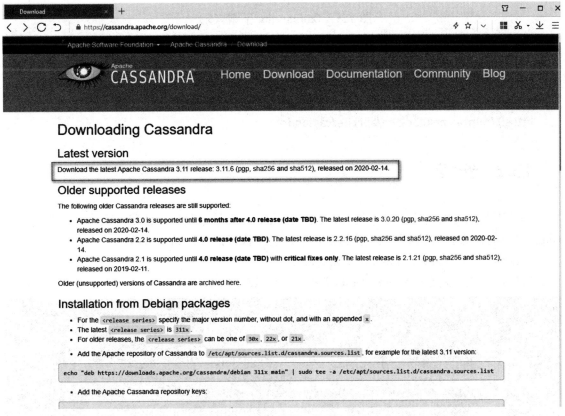

图 15-1 Cassandra 下载（1）

在上图的网页中，单击选框中的链接，进入镜像下载页面，如图 15-2 所示。

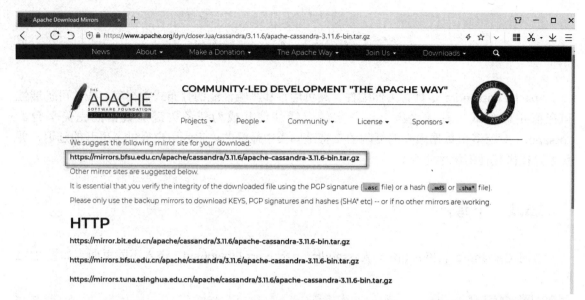

图 15-2 Cassandra 下载（2）

在图 15-2 的网页中，可以根据自己当前的网络情况来选择合适的镜像链接，本书选择的镜像链接，如下所示：

https://mirrors.bfsu.edu.cn/apache/cassandra/3.11.6/apache-cassandra-3.11.6-bin.tar.gz

若上面的网址不存在的话，则请自行选择合适的版本，访问以下网址：

https://mirrors.bfsu.edu.cn/apache/cassandra

15.2 安装

刚才下载的压缩包解压后，在 Linux 和 Windows 环境中都可以使用。在这里，我们举在 Linux 中使用的例子。下载到适当位置后，解压它，执行命令，如下所示：

$ tar xvf apache-cassandra-3.11.6-bin.tar.gz

解压后，进入到 Cassandra 的主目录中，如图 15-3 所示。

```
[quxy@svr-11 softwares]$ cd apache-cassandra-3.11.6/
[quxy@svr-11 apache-cassandra-3.11.6]$ ll
total 520
drwxr-xr-x. 2 quxy quxy   4096 May 29 16:52
-rw-r--r--. 1 quxy quxy   4832 Feb 11 06:57 CASSANDRA-14092.txt
-rw-r--r--. 1 quxy quxy 372667 Feb 11 06:57 CHANGES.txt
drwxr-xr-x. 3 quxy quxy   4096 May 29 16:52
drwxr-xr-x. 3 quxy quxy     33 May 29 16:52
drwxr-xr-x. 2 quxy quxy     30 May 29 16:52
drwxr-xr-x. 3 quxy quxy   4096 May 29 16:52
drwxr-xr-x. 4 quxy quxy   4096 May 29 16:52
-rw-r--r--. 1 quxy quxy  11609 Feb 11 06:57 LICENSE.txt
-rw-r--r--. 1 quxy quxy 115680 Feb 11 06:57 NEWS.txt
-rw-r--r--. 1 quxy quxy   2811 Feb 11 06:57 NOTICE.txt
drwxr-xr-x. 3 quxy quxy     94 May 29 16:52
drwxr-xr-x. 4 quxy quxy    135 May 29 16:52
[quxy@svr-11 apache-cassandra-3.11.6]$
```

图 15-3 安装 Cassandra（1）

可以无需修改 cassandra.yaml 配置文件中的默认配置项。在主目录中启动 Cassandra 服务器，请执行命令：$ bin/cassandra。如图 15-所示。

图 15-4 安装 Cassandra（2）

默认情况下，Cassandra 服务器是后台启动的。启动过程的最后，控制台输出的内容，如图 15-4 所示。

图 15-5 启动 Cassandra

启动完成后，要查看 Cassandra 服务器的运行情况，请执行命令：$ bin/nodetool status。如图 15-6 所示。

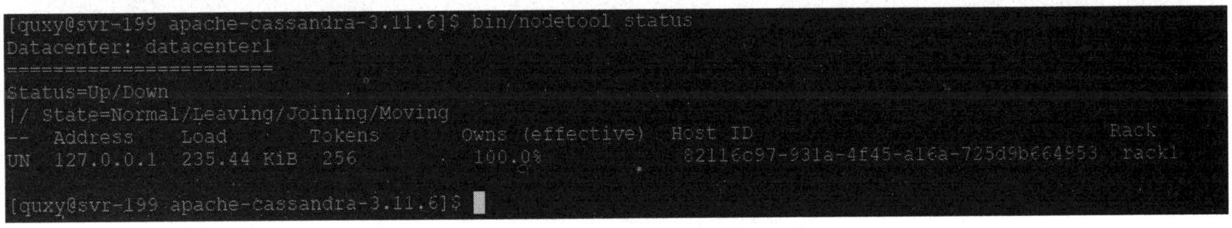

图 15-6 运行 Cassandra

在本机上连接 Cassandra 服务器，请执行命令：$ bin/cqlsh。如图 15-7 所示。

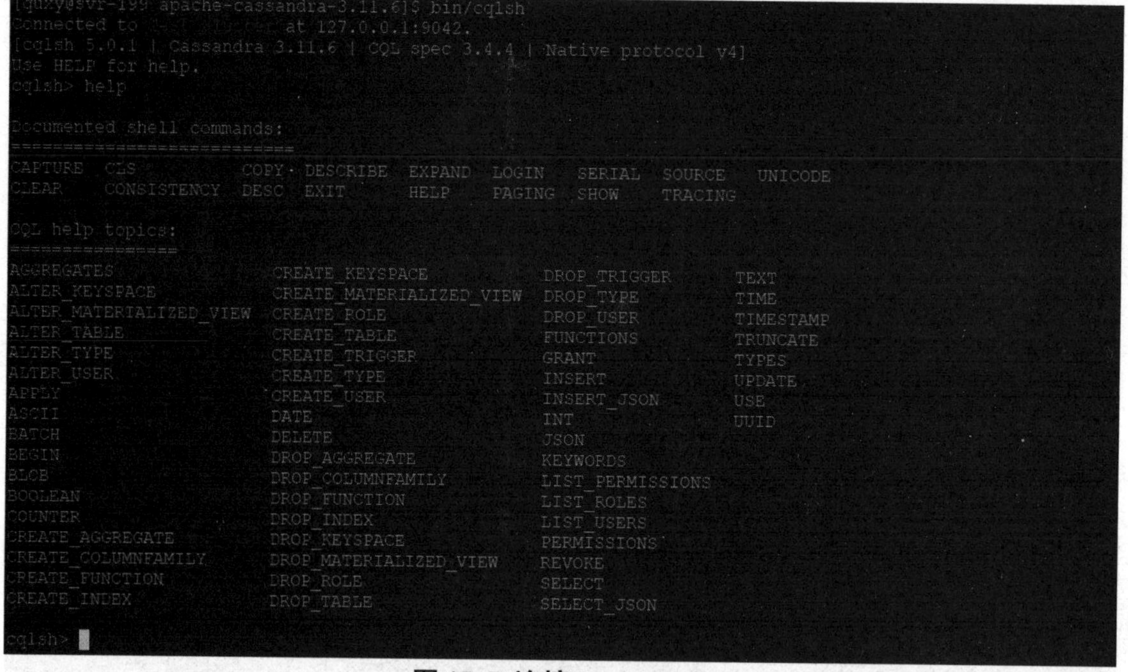

图 15-7 连接 Cassandra

在上图中列出了 Cassandra 命令行客户端可用的命令，例如：创建表和删除表等。创建键空间、新建表、插入数据和查看数据等操作示例，如图 15-8 所示。

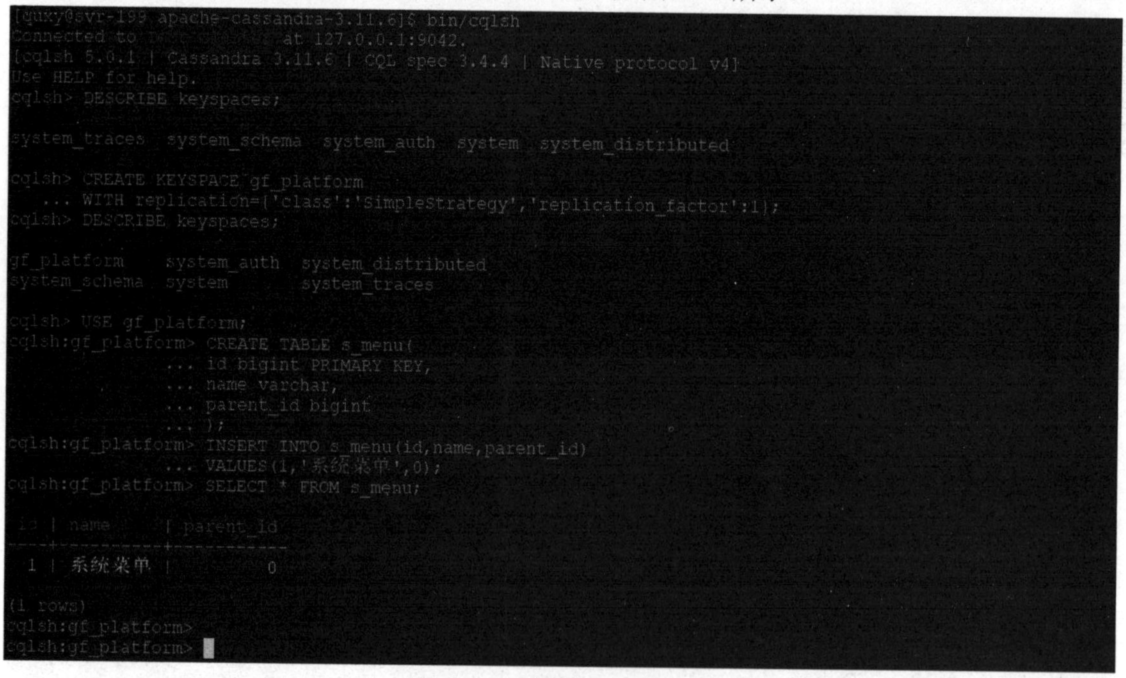

图 15-8 测试 Cassandra

15.3 集群搭建

在这里，我们创建一个拥有 3 节点的集群，具体规划，如表 15-1 所示。

表 15-1 集群列表

IP	端口
192.168.226.130	9042
192.168.226.131	9042
192.168.226.132	9042

首先，修改第一节点的 cassandra.yaml 配置文件中的各个配置项，如下所示：

cluster_name: 'gf-cassandra3-001'

start_rpc: true

listen_address: 192.168.226.130

rpc_address: 192.168.226.130

接下来，修改 seeds 配置项，如图 15-9 所示。

图 15-9 修改 seeds 配置项

把配置好的 Cassandra 复制到另外两台机器上，执行命令，如下所示：

$ scp -r ./apache-cassandra-3.11.6 quxy@192.168.226.131:/home/quxy/softwares

$ scp -r ./apache-cassandra-3.11.6 quxy@192.168.226.132:/home/quxy/softwares

分别使用 vim 编辑器打开 cassandra.yaml 配置文件，修改 listen_address 和 rpc_address 配置项，底行命令，如下所示：

0,$s/: 192\.168\.226\.130/: 192.168.226.131/g

0,$s/: 192\.168\.226\.130/: 192.168.226.132/g

进行完以上配置操作后，依次启动各个节点，执行命令，如下所示：

$ bin/cassandra

三个节点都启动成功之后，查看启动状态，执行命令：$ bin/nodetool status。如图 15-10 所示。

图 15-10 节点启动成功（1）

使用 Cassandra 命令行客户端连接本地测试用局域网中其他节点并进行操作：$ bin/cqlsh 192.168.226.132 9042。如图 15-11 所示。

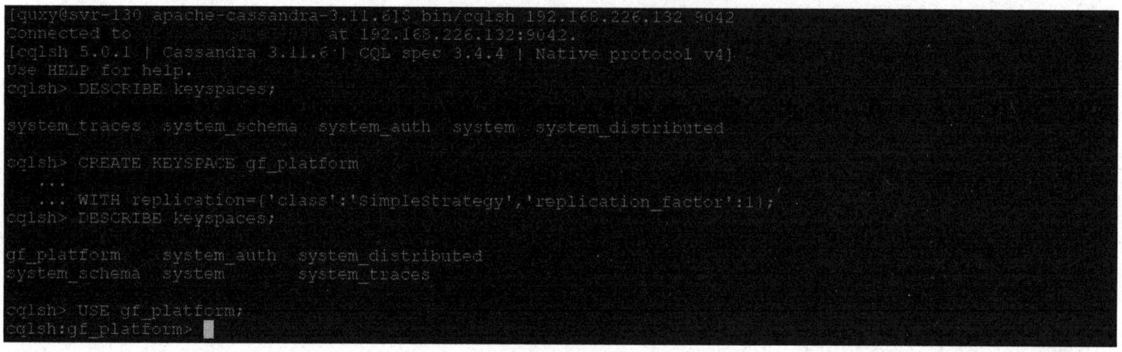

图 15-11 节点启动成功（2）

15.4 Cassandra 示例

借助于 Spring Data 对 Apache Cassandra 的封装，使开发者更容易开发 Apache Cassandra 应用程序。Spring Data 为 Apache Cassandra 提供了功能丰富的模板接口，以满足绝大多数情况下的开发需要。要在项目中使用 Cassandra 服务器，请添加 Cassandra 依赖项，如下所示：

```
<dependency>
    <groupId>org.springframework.boot</groupId>
    <artifactId>spring-boot-starter-data-cassandra</artifactId>
</dependency>
```

接着，在 application.yml 配置文件中添加配置项，如图 15-12 所示。

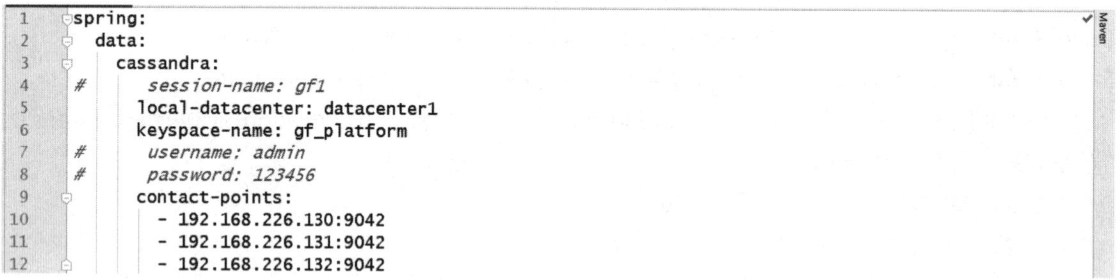

图 15-12 application.yml 配置

接下来，我们使用 CassandraTemplate 接口操作 Cassandra 服务器。编写 controller 层的代码，如下所示：

```
package com.gf.cassandra.controller;
import com.gf.cassandra.dto.MenuCreateDTO;
import com.gf.cassandra.entity.Menu;
import org.springframework.data.cassandra.core.CassandraTemplate;
import org.springframework.http.ResponseEntity;
```

```java
import org.springframework.web.bind.annotation.*;
import javax.validation.Valid;
import static com.gf.common.GfResponseEntity.self;
/**
 * 在这里省略了 service、dao 层，直接在 controller 层进行全部的编码工作。
 */
@RestController
@RequestMapping("/gf/cassandra")
public class CassandraController {
    private final CassandraTemplate cassandraTemplate;
    public CassandraController(CassandraTemplate cassandraTemplate) {
        this.cassandraTemplate = cassandraTemplate;
    }
    /**
     * 根据 ID，获取菜单信息
     */
    @GetMapping("/view")
    public ResponseEntity view(Long id) {
        Menu entity = cassandraTemplate.selectOneById(id, Menu.class);
        return self().ok(entity);
    }
    /**
     * 新建菜单：<br/>
     * 若数据库中已存在 ID 相同的数据，则默认直接覆盖掉。
     */
    @PostMapping("/create")
    public ResponseEntity create(@RequestBody @Valid MenuCreateDTO params) {
        Menu entity = new Menu();
        entity.setId(params.getId());
        entity.setName(params.getName());
        entity.setParentId(params.getParentId());
        cassandraTemplate.insert(entity);
        return self().ok();
    }
}
```

有关数据表的定义，上文已经介绍过，在此就不再说明了。定义实体类 Menu，如下所示：

```java
package com.gf.cassandra.entity;
import lombok.Data;
```

```java
import org.springframework.data.cassandra.core.mapping.Column;
import org.springframework.data.cassandra.core.mapping.PrimaryKey;
import org.springframework.data.cassandra.core.mapping.Table;
@Data
@Table("s_menu")
public class Menu {
    @PrimaryKey
    private Long id;
    private String name;
    @Column("parent_id")
    private Long parentId;
}
```

定义 MenuCreateDTO 类来接收 create 接口中传递过来的数据，如下所示：

```java
package com.gf.cassandra.dto;
import lombok.Data;
import javax.validation.constraints.Min;
import javax.validation.constraints.NotEmpty;
import javax.validation.constraints.NotNull;
@Data
public class MenuCreateDTO {
    @Min(1)
    private Long id;
    @NotEmpty
    private String name;
    @NotNull
    private Long parentId;
}
```

15.5 本章小结

本章讲解了 Cassandra 的简单介绍、下载安装和集群搭建。最后讲解了 Cassandra 在 Spring Boot 应用程序中的应用示例。通过这些内容的讲解希望大家能达到快速使用 Cassandra 的目的，更多有关 Cassandra 的内容，请访问其官网，请查看 Spring Data Cassandra 的内容。

第十六章 MyBatis Plus

MyBatis Plus（简称 MP）是持久层框架 MyBatis 的一个增强工具，在 MyBatis 的基础上只做增强不做改变，为简化开发、提高效率而生。相比于前面讲解的 Spring JDBC 和 Spring JPA，MyBatis Plus 真正做到了 SQL 与持久层代码的分离，并且极大地减少了重复的代码及相应 SQL。它以更灵活的方式将 SQL 编写入 XML 文件中，并且它还提供了针对单表的增删改查及代码自动生成等功能。

支持的国外数据库有 MySQL、MariaDB、Oracle、DB2、H2、HSQLDB、SQLite、PostgreSQL 和 SQL Server 等；支持的国内数据库有达梦数据库、虚谷数据库和人大金仓数据库等。

16.1 功能特性

MyBatis Plus 具有的功能特性，如下所示：
1) 无侵入：只做增强不做改变，引入它不会对现有工程产生影响；
2) 损耗小：启动即会自动注入基本 CURD，性能基本无损耗，直接面向对象操作；
3) 强大的 CRUD 操作：内置通用 Mapper、通用 Service，仅仅通过少量配置即可实现单表的大部分 CRUD 操作，更有强大的条件构造器，满足各类使用需求；
4) 支持主键自动生成：支持多达 4 种主键策略（内含分布式唯一 ID 生成器 Sequence），可自由配置，完美解决主键问题；
5) 内置代码生成器：采用代码或者 Maven 插件可快速生成 mapper、model、service 和 controller 层代码，支持模板引擎，更有超多自定义配置；
6) 内置分页插件：基于 MyBatis 物理分页，开发者无需关心具体操作，配置好插件之后，分页等同于普通 List 查询；
7) 内置性能分析插件：可输出 SQL 语句以及其执行时间，建议开发测试时启用该功能，能快速揪出慢查询；
8) 内置全局拦截插件：提供了全表 delete 和 update 操作的智能分析阻断，也可自定义拦截规则，预防误操作。

MyBatis Plus 的框架结构，如图 16-1 所示。

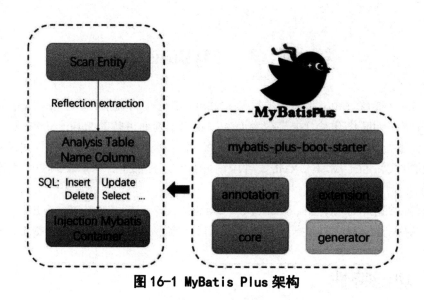

图 16-1 MyBatis Plus 架构

16.2 简单示例

要想在 Spring Boot 应用程序中引入 MyBatis Plus，只需添加相应依赖项，如下所示：

```
<dependency>
    <groupId>com.baomidou</groupId>
    <artifactId>mybatis-plus-boot-starter</artifactId>
    <version>3.3.1</version>
</dependency>
```

接下来，编写 Menu 实体类，如下所示：

```
import com.baomidou.mybatisplus.annotation.TableId;
import com.baomidou.mybatisplus.annotation.TableName;
@TableName("s_menu")
public class Menu {
    @TableId
    private Long id;
    private String name;
    private Long parentId;
    private int status;
    private int sorted;
    private String remark;
    // 省略 setter、getter ...
}
```

当数据表名称与实体类名称不同时，要记得添加 @TableName 注解来指定表名。接着，编写最为关键的 dao 层代码，如下所示：

```java
import com.baomidou.mybatisplus.core.mapper.BaseMapper;
import com.gf.mvc.entity.Menu;
//@Mapper
public interface MenuMapper extends BaseMapper<Menu> {}
```

这个接口看起来很简单,但是由于它继承了 BaseMapper,这让它变得很强大,它已具备增删改查的能力了。要想使这个接口真正的生效,我们还差最后一步,添加 @MapperScan 注解,如下所示:

```java
import org.mybatis.spring.annotation.MapperScan;
import org.springframework.boot.SpringApplication;
import org.springframework.boot.autoconfigure.SpringBootApplication;
@SpringBootApplication
@MapperScan({"com.gf.mvc.mapper"})
public class MvcApplication {
    public static void main(String[] args) {
        SpringApplication.run(MvcApplication.class, args);
    }
}
```

除了添加 @MapperScan 注解之外,我们还可以在 MenuMapper 接口上添加 @Mapper 注解来达到同样的效果。但是,这样需要在每个 XXXMapper 接口上添加 @Mapper 注解。若定义的 XXXMapper 接口较多的话就比较费事了。

所有的配置和定义工作已完成,下面就是开始使用了。我们在 service 层使用 MenuMapper 接口的实例对象(MyBatis-Plus 帮我们实现的),如下所示:

```java
import com.gf.mvc.dto.MenuDTO;
import com.gf.mvc.entity.Menu;
import com.gf.mvc.mapper.MenuMapper;
import com.gf.mvc.service.MenuService2;
import org.slf4j.Logger;
import org.slf4j.LoggerFactory;
import org.springframework.stereotype.Service;
import java.util.List;
@Service
public class MenuServiceImpl2 implements MenuService2 {
    private final Logger log = LoggerFactory.getLogger(getClass());
    private final MenuMapper menuMapper;
    public MenuServiceImpl2(MenuMapper menuMapper) {
        this.menuMapper = menuMapper;
    }
    @Override
```

```java
public boolean create(MenuDTO params) {
    Menu entity = new Menu();
    entity.setName(params.getName());
    entity.setParentId(params.getParentId());
    entity.setSorted(params.getSorted());
    int num = menuMapper.insert(entity);
    return num > 0;
}
@Override
public List<Menu> listAll() {
    return menuMapper.selectList(null);
}
@Override
public Menu view(Long id) {
    return menuMapper.selectById(id);
}
}
```

从以上代码片段中可以看出，MyBatis Plus 的配置和类型（类、接口）的定义及其调用都是非常简洁的，它是易于上手的。有关 MyBatis Plus 的更多详情，请访问以下网址：

https://mybatis.plus/guide

有关 MyBatis 的更多详情，请访问以下网址：

https://mybatis.org/mybatis-3/zh/index.html

16.3 分页

16.3.1 分页插件

MyBatis Plus 为我们提供了分页插件，主要是针对单表的分页操作。要想在应用中添加分页功能，只需注册 PaginationInterceptor bean，并进行相应的设置即可。当然，可以完全使用默认值，代码片段，如下所示：

```java
@Bean
public PaginationInterceptor paginationInterceptor() {
    PaginationInterceptor paginationInterceptor = new PaginationInterceptor();
    // 设置请求的页面大于最大页后操作， true 调回到首页，false 继续请求  默认 false
    // paginationInterceptor.setOverflow(false);
```

```
// 设置最大单页限制数量, 默认 500 条, -1 不受限制
// paginationInterceptor.setLimit(500);
// 开启 count 的 join 优化,只针对部分 left join
    paginationInterceptor.setCountSqlParser(new
JsqlParserCountOptimize(true));
    return paginationInterceptor;
}
```

注册完成分页插件 bean 之后，就可以使用 BaseMapper<T> 接口提供的分页方法，有如下两个接口方法：

1) selectPage(…)：接收两个参数，返回包含实体类列表的对象；
2) selectMapsPage(…)：和上面一样，接收两个参数，返回包含 map 数据列表的对象。

在 service 层调用分页方法，代码片段，如下所示：

```
/**
 * 根据页号（num）、每页数据大小（size）等输入参数返回相应的分页数据。
 */
@Override
public GfPage<Menu> page2(MenuPageDTO params) {
    // 指定页号（num）、每页数据大小（size）
    Page<Menu> page = new Page<>(params.getNum(), params.getSize());
    // 设置查询的列, 及查询条件。当然, 你还可以根据需要添加更多的查询条件。
    QueryWrapper<Menu> queryWrapper = new QueryWrapper<>();
    queryWrapper.select("id", "name")
            .ge("parent_id", params.getParentId());
    IPage<Menu> pageData = menuMapper.selectPage(page, queryWrapper);
    // 转换为符合要求的数据格式。
    GfPage<Menu> gfPage = GfPage.toConvert(pageData);
    return gfPage;
}
```

controller 层的调用代码，这里就不再展示了。其中，入参类型 MenuPageDTO 类的定义代码，如下所示：

```
package com.gf.mvc.dto;
import lombok.Data;
@Data
public class MenuPageDTO {
    private int size = 10;
    private int num = 1;
    private long parentId;
}
```

这里，我们使用了 @Data 注解，及其 IDE 相关的插件。因此，本类中就可以无需编写 setter 和 getter 方法了。GfPage 类的定义代码，如下所示：

```java
package com.gf.mvc.common;
import com.baomidou.mybatisplus.core.metadata.IPage;
import lombok.Data;
import java.util.List;
@Data
public class GfPage<T> {
    private long total;
    private long num;
    private List<T> list;
    public static <T> GfPage<T> toConvert(IPage<T> pageData) {
        GfPage<T> gfPage = new GfPage<>();
        // 总数据条数
        gfPage.setTotal(pageData.getTotal());
        // 数据列表
        gfPage.setList(pageData.getRecords());
        // 当前页数
        gfPage.setNum(pageData.getCurrent());
        return gfPage;
    }
}
```

16.3.2 自定义分页

分页插件是用于单表的操作的。对于多表的分页查询，我们应该如何应对呢？这时，我们可以在相应的 XML 配置文件中编写 SQL 语句。示例中只列举了对单表的操作（多表操作同理可得），MenuMapper.xml 配置文件（src/main/resources/mapper/），如下所示：

```xml
<?xml version="1.0" encoding="UTF-8" ?>
<!DOCTYPE mapper PUBLIC "-//mybatis.org//DTD Mapper 3.0//EN"
"http://mybatis.org/dtd/mybatis-3-mapper.dtd">
<mapper namespace="com.gf.mvc.mapper.MenuMapper">
    <sql id="where3">
        <where>
            <if test="parentId != null">
                m.parent_id=#{parentId}
            </if>
        </where>
```

```xml
        </sql>
        <!-- 查询符合条件的数据的总数 -->
        <select id="count3" resultType="_long" parameterType="com.gf.mvc.dto.MenuPageDTO">
            SELECT count(0) FROM s_menu m
            <include refid="where3"/>
        </select>
        <!-- 查询符合条件的数据的列表 -->
        <select id="page3" resultType="com.gf.mvc.entity.Menu" parameterType="com.gf.mvc.dto.MenuPageDTO">
            SELECT m.id,m.name FROM s_menu m
            <include refid="where3"/>
            limit #{offset},#{size}
        </select>
</mapper>
```

以上片段显得有些繁杂，这是因为添加了判断条件和提取了重复 SQL 片段的缘故。这样做可以使 SQL 语句更加的灵活，以少量的 SQL 片段就可以应对较多的实际 SQL 需求。另外，也可以直接将完整的 SQL 语句写在这里。

定义完成 SQL 语句配置之后，还需要在 MenuMapper 接口中添加相应的方法声明，并且方法名与 XML 配置文件中的 select 元素的 id 值要相同。代码片段，如下所示：

```java
import com.baomidou.mybatisplus.core.mapper.BaseMapper;
import com.gf.mvc.dto.MenuPageDTO;
import com.gf.mvc.entity.Menu;
import java.util.List;
public interface MenuMapper extends BaseMapper<Menu> {
    long count3(MenuPageDTO params);
    List<Menu> page3(MenuPageDTO params);
}
```

相应的 service 层的代码，如下所示：

```java
/**
 * 自定义分页
 */
@Override
public GfPage<Menu> page3(MenuPageDTO params) {
    long total = menuMapper.count3(params);
    List<Menu> list = total > 0 ? menuMapper.page3(params) : new ArrayList<>();
    return new GfPage(total, params.getNum(), list);
}
```

16.4 单表操作

单表的增删改查,已由 MyBatis Plus 实现,无需再编写额外的 SQL 语句。我们所要做的是:创建数据表、编写实体类,以及编写 service 和 controller 层的代码。如果 service 层使用 IService 接口的话,本层的代码也无需编写。以下代码片段给出了典型的单表操作的示例。

在这里,无需 MenuMapper3.xml 配置文件。定义一个空的 dao 层,如下所示:

```
package com.gf.mvc.mapper;
import com.baomidou.mybatisplus.core.mapper.BaseMapper;
import com.gf.mvc.entity.Menu;
public interface MenuMapper3 extends BaseMapper<Menu> {}
```

定义空的 service 层:MenuService3 接口和 MenuServiceImpl3 实现类,如下所示:

```
package com.gf.mvc.service;
import com.baomidou.mybatisplus.extension.service.IService;
import com.gf.mvc.entity.Menu;
public interface MenuService3 extends IService<Menu> {}
```

```
package com.gf.mvc.service.impl;
import com.baomidou.mybatisplus.extension.service.impl.ServiceImpl;
import com.gf.mvc.entity.Menu;
import com.gf.mvc.mapper.MenuMapper3;
import com.gf.mvc.service.MenuService3;
import org.springframework.stereotype.Service;
@Service
public class MenuServiceImpl3 extends ServiceImpl<MenuMapper3, Menu>
implements MenuService3 {}
```

定义 controller 层,如下所示:

```
package com.gf.mvc.controller;
import static com.gf.mvc.common.GfResponseEntity.*;
import com.baomidou.mybatisplus.core.conditions.query.QueryWrapper;
import com.baomidou.mybatisplus.core.metadata.IPage;
import com.baomidou.mybatisplus.extension.plugins.pagination.Page;
import com.gf.mvc.common.GfPage;
import com.gf.mvc.dto.MenuCreateDTO;
import com.gf.mvc.dto.MenuPageDTO;
import com.gf.mvc.dto.MenuUpdateDTO;
import com.gf.mvc.entity.Menu;
import com.gf.mvc.service.MenuService3;
```

```java
import org.springframework.http.ResponseEntity;
import org.springframework.web.bind.annotation.*;
import javax.validation.Valid;
import java.io.Serializable;
@RestController
@RequestMapping("/mvc/menu3")
public class MenuController3 {
    private final MenuService3 service;
    public MenuController3(MenuService3 service) {
        this.service = service;
    }
    @PostMapping("/create")
    public ResponseEntity create(@RequestBody @Valid MenuCreateDTO params) {
        Menu entity = new Menu();
        entity.setName(params.getName());
        entity.setParentId(params.getParentId());
        entity.setSorted(params.getSorted());
        service.save(entity);
        return self().ok();
    }
    @GetMapping("/del/{id}")
    public ResponseEntity del(@PathVariable Serializable id) {
        service.removeById(id);
        return self().ok();
    }
    @PostMapping("/update")
    public ResponseEntity update(@RequestBody @Valid MenuUpdateDTO params) {
        Menu entity = new Menu();
        entity.setId(params.getId());
        entity.setName(params.getName());
        entity.setParentId(params.getParentId());
        entity.setSorted(params.getSorted());
        service.updateById(entity);
        return self().ok();
    }
    @GetMapping("/view/{id}")
    public ResponseEntity view(@PathVariable Serializable id) {
```

```java
        Menu entity = service.getById(id);
        return self().ok(entity);
    }
    @PostMapping("/page")
    public ResponseEntity page(@RequestBody MenuPageDTO params) {
        Page<Menu> page = new Page<>(params.getNum(), params.getSize());
        QueryWrapper<Menu> queryWrapper = new QueryWrapper<>();
        queryWrapper.ge("parent_id", params.getParentId());
        IPage<Menu> pageData = service.page(page, queryWrapper);
        GfPage<Menu> gfPage = GfPage.toConvert(pageData);
        return self().ok(gfPage);
    }
}
```

16.5 多表操作

想要使用 LEFT JOIN 和 RIGHT JOIN 等多表操作，可以直接在 MenuMapper3.xml 配置文件中编写 SQL 语句，并设置相应的参数，如下所示：

```xml
<?xml version="1.0" encoding="UTF-8" ?>
<!DOCTYPE mapper PUBLIC "-//mybatis.org//DTD Mapper 3.0//EN"
"http://mybatis.org/dtd/mybatis-3-mapper.dtd">
<mapper namespace="com.gf.mvc.mapper.MenuMapper3">
    <select id="list3" resultType="com.gf.mvc.po.MenuPO" parameterType="com.gf.mvc.dto.MenuPageDTO">
        SELECT m.id,m.name,m.parent_id,m2.name parent_name FROM s_menu m
        LEFT JOIN s_menu m2 on m.parent_id=m2.id
        <where>
            <if test="parentId != null">
                m.parent_id=#{parentId}
            </if>
        </where>
    </select>
</mapper>
```

16.6 代码自动生成

MyBatis Plus 为我们提供了代码自动生成的功能，并且提供各种配置参数以满足不同的需求。默认是基于命令行窗口的交互生成模式，也可以改造为基于网页的（可参考网络上的开源项目）。根据项目的不同，大都需要调整默认的一些配置，使之符合项目要求。

AutoGenerator 是 MyBatis Plus 的代码生成器，通过 AutoGenerator 可以快速生成 Entity、Mapper、Mapper XML、Service 和 Controller 等各个模块的代码，极大的提升了开发效率。有关官方文档的更多信息，请访问以下网址：

https://mybatis.plus/guide/generator.html

要想使用代码自动生成功能，需要添加相关的依赖项，如下所示：

```xml
<dependency>
    <groupId>com.baomidou</groupId>
    <artifactId>mybatis-plus-generator</artifactId>
    <version>3.3.1.tmp</version>
</dependency>
<dependency>
    <groupId>org.freemarker</groupId>
    <artifactId>freemarker</artifactId>
    <version>2.3.30</version>
</dependency>
```

然后，配置数据库的相关参数，并根据需要调整其他相关设置，示例代码，如下所示：

```java
package com.gf.generator;
import com.baomidou.mybatisplus.core.exceptions.MybatisPlusException;
import com.baomidou.mybatisplus.core.toolkit.StringPool;
import com.baomidou.mybatisplus.core.toolkit.StringUtils;
import com.baomidou.mybatisplus.generator.AutoGenerator;
import com.baomidou.mybatisplus.generator.InjectionConfig;
import com.baomidou.mybatisplus.generator.config.*;
import com.baomidou.mybatisplus.generator.config.po.TableInfo;
import com.baomidou.mybatisplus.generator.config.rules.NamingStrategy;
import com.baomidou.mybatisplus.generator.engine.FreemarkerTemplateEngine;
import java.sql.SQLException;
import java.util.*;
public class CodeGenerator {
    /**
```

```java
 * 读取控制台内容
 */
public static String scanner(String tip) {
    Scanner scanner = new Scanner(System.in);
    StringBuilder help = new StringBuilder();
    help.append("请输入" + tip + ": ");
    System.out.println(help.toString());
    if (scanner.hasNext()) {
        String ipt = scanner.next();
        if (StringUtils.isNotEmpty(ipt)) {
            return ipt;
        }
    }
    throw new MybatisPlusException("请输入正确的" + tip + "! ");
}

public static void main(String[] args) {
    String subProjectName = "gf-generator";
    // 代码生成器
    AutoGenerator mpg = new AutoGenerator();
    // 全局配置
    GlobalConfig gc = new GlobalConfig();
    String projectPath = System.getProperty("user.dir");
    gc.setOutputDir(projectPath + "/" + subProjectName + "/src/main/java");
    gc.setAuthor("gf");
    gc.setOpen(false);
    // gc.setSwagger2(true); 实体属性 Swagger2 注解
    mpg.setGlobalConfig(gc);
    // 数据源配置
    DataSourceConfig dsc = new DataSourceConfig();
    dsc.setUrl("jdbc:mysql://localhost:3306/gf-platform2?serverTimezone=Asia/Shanghai&useUnicode=true&useSSL=false&characterEncoding=utf8");
    // dsc.setSchemaName("public");
    //dsc.setDriverName("com.mysql.jdbc.Driver");
    dsc.setDriverName("com.mysql.cj.jdbc.Driver");
    dsc.setUsername("root");
    dsc.setPassword("123456");
```

```java
mpg.setDataSource(dsc);
// 包配置
PackageConfig pc = new PackageConfig();
String modelName=scanner("模块名");
pc.setModuleName(modelName);
pc.setParent("com.gf");
mpg.setPackageInfo(pc);
// 自定义配置
InjectionConfig cfg = new InjectionConfig() {
    @Override
    public void initMap() {
        Map<String, Object> map = new HashMap<>();
        map.put("modelName", modelName);
        this.setMap(map);
    }
};
// 如果模板引擎是 freemarker
String templatePath = "/templates/mapper.xml.ftl";
// 如果模板引擎是 velocity
// String templatePath = "/templates/mapper.xml.vm";
// 自定义输出配置
List<FileOutConfig> focList = new ArrayList<>();
// 自定义配置会被优先输出
focList.add(new FileOutConfig(templatePath) {
    @Override
    public String outputFile(TableInfo tableInfo) {
        // 自定义输出文件名，如果你 Entity 设置了前后缀、此处注意 xml 的名称会跟着发生变化！！
        return projectPath + "/" + subProjectName +
"/src/main/resources/mapper/" + pc.getModuleName()
            + "/" + tableInfo.getEntityName() + "Mapper" +
StringPool.DOT_XML;
    }
});
/*
cfg.setFileCreate(new IFileCreate() {
    @Override
    public boolean isCreate(ConfigBuilder configBuilder,
FileType fileType, String filePath) {
```

```java
            // 判断自定义文件夹是否需要创建
            checkDir("调用默认方法创建的目录，自定义目录用");
            if (fileType == FileType.MAPPER) {
                // 已经生成 mapper 文件判断存在,不想重新生成返回 false
                return !new File(filePath).exists();
            }
            // 允许生成模板文件
            return true;
        }
    });
    */
    cfg.setFileOutConfigList(focList);
    mpg.setCfg(cfg);
    // 配置模板
    TemplateConfig templateConfig = new TemplateConfig();
    // 配置自定义输出模板
    //指定自定义模板路径，注意不要带上.ftl/.vm，会根据使用的模板引擎自动识别
    // templateConfig.setEntity("templates/entity2.java");
    // templateConfig.setService();
    // templateConfig.setController();
    templateConfig.setXml(null);
    mpg.setTemplate(templateConfig);
    // 策略配置
    StrategyConfig strategy = new StrategyConfig();
    strategy.setNaming(NamingStrategy.underline_to_camel);
    strategy.setColumnNaming(NamingStrategy.underline_to_camel);
    // strategy.setSuperEntityClass("你自己的父类实体,没有就不用设置!");
    strategy.setEntityLombokModel(true);
    strategy.setRestControllerStyle(true);
    // 公共父类
    // strategy.setSuperControllerClass("你自己的父类控制器,没有就不用设置!");
    // 写于父类中的公共字段
    //     strategy.setSuperEntityColumns("id");
    strategy.setInclude(scanner("表名，多个英文逗号分割").split(","));
    strategy.setControllerMappingHyphenStyle(true);
```

```
        // strategy.setTablePrefix(pc.getModuleName() + "_");
        strategy.setTablePrefix(scanner("表名前缀（例如：s_menu 的前缀为 s）"));
        mpg.setStrategy(strategy);
        mpg.setTemplateEngine(new FreemarkerTemplateEngine());
        mpg.execute();
    }
}
```

可以使用默认模板，也可以使用自己定义的模板。在这里，自定义了 controller.java.ftl 模板，其中实现了对单表的增删改查接口，如下所示：

```
package ${package.Controller};

import static com.gf.common.GfResponseEntity.*;
import com.baomidou.mybatisplus.core.conditions.query.QueryWrapper;
import com.baomidou.mybatisplus.core.metadata.IPage;
import com.baomidou.mybatisplus.extension.plugins.pagination.Page;
import com.gf.common.GfPage;
import com.gf.${cfg.modelName}.dto.${entity}CreateDTO;
import com.gf.${cfg.modelName}.dto.${entity}PageDTO;
import com.gf.${cfg.modelName}.dto.${entity}UpdateDTO;
import ${package.Entity}.${entity};
import ${package.Service}.${table.serviceName};
import org.springframework.http.ResponseEntity;
import org.springframework.web.bind.annotation.*;
<#if restControllerStyle>
import org.springframework.web.bind.annotation.RestController;
<#else>
import org.springframework.stereotype.Controller;
</#if>
<#if superControllerClassPackage??>
import ${superControllerClassPackage};
</#if>
import java.io.Serializable;

/**
 * ${table.comment!} 前端控制器
 * @author ${author}
 * @since ${date}
```

```
*/
<#if restControllerStyle>
@RestController
<#else>
@Controller
</#if>
@RequestMapping("<#if
package.ModuleName??>/${package.ModuleName}</#if>/<#if
controllerMappingHyphenStyle??>${controllerMappingHyphen}<#else>${t
able.entityPath}</#if>")
<#if kotlin>
class ${table.controllerName}<#if superControllerClass??> :
${superControllerClass}()</#if>
<#else>
<#if superControllerClass??>
public class ${table.controllerName} extends ${superControllerClass}
{
<#else>
public class ${table.controllerName} {
</#if>
    private final ${table.serviceName} service;

    public ${table.controllerName}(${table.serviceName} service) {
        this.service = service;
    }

    @PostMapping("/create")
    public ResponseEntity create(@RequestBody ${entity}CreateDTO
params) {
        ${entity} entity = new ${entity}();
        // TODO: 设置字段值
        service.save(entity);
        return self().ok();
    }

    @GetMapping("/del/{id}")
    public ResponseEntity del(@PathVariable Serializable id) {
        service.removeById(id);
        return self().ok();
```

```
    }

    @PostMapping("/update")
    public ResponseEntity update(@RequestBody ${entity}UpdateDTO params) {
        ${entity} entity = new ${entity}();
        entity.setId(params.getId());
        // TODO: 设置字段值
        service.updateById(entity);
        return self().ok();
    }

    @GetMapping("/view/{id}")
    public ResponseEntity view(@PathVariable Serializable id) {
        ${entity} entity = service.getById(id);
        return self().ok(entity);
    }

    @PostMapping("/page")
    public ResponseEntity page(@RequestBody ${entity}PageDTO params) {
        Page<${entity}> page = new Page<>(params.getPage().getNum(), params.getPage().getSize());
        QueryWrapper<${entity}> queryWrapper = new QueryWrapper<>();
        // TODO: 设置查询条件。
        IPage<${entity}> pageData = service.page(page, queryWrapper);
        GfPage<${entity}> gfPage = GfPage.toConvert(pageData);
        return self().ok(gfPage);
    }
}</#if>
```

有关其他模板文件的内容，请访问以下网址：
https://github.com/baomidou/mybatis-plus/tree/3.0/mybatis-plus-generator/src/main/resources/templates

完成以上配置后，运行 CodeGenerator 类中的主方法以开始生成代码，如图 16-2 所示。

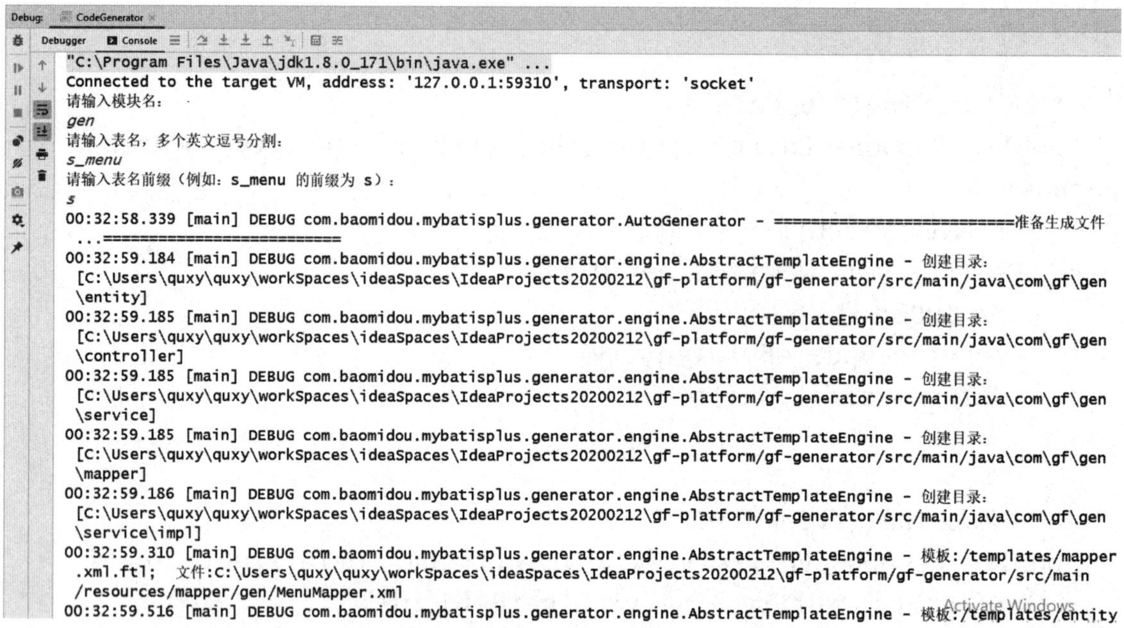

图 16-2 运行 CodeGenerator（1）

执行完成以上操作后，生成的各个模块的文件，如图 16-3 所示。

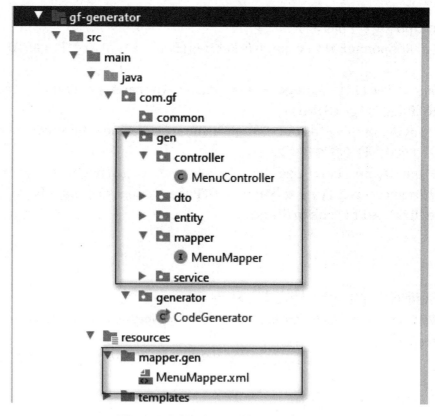

图 16-3 运行 CodeGenerator（2）

16.7 本章小结

本章讲解了 MyBatis Plus 各个部分的知识点,包括:分页和代码自动生成等,并列举了一些示例。想要了解更多的 MyBatis Plus 内容,还请查看其官方文档。

第十七章 Spring 安全

Spring Security 是一个安全框架，它提供身份验证、授权和针对常见攻击的保护。凭借对命令式和反应式应用程序的一流支持，它是保护基于 Spring 的应用程序的事实上的标准。它还提供与其他库的集成，以简化其使用。

17.1 简单示例

想要在 Spring Boot 应用程序中使用 Spring Security，只需添加依赖项，如下所示：

```xml
<dependency>
    <groupId>org.springframework.boot</groupId>
    <artifactId>spring-boot-starter-security</artifactId>
</dependency>
```

以上依赖项的依赖关系，如图 17-1 所示。

```
[INFO] \- org.springframework.boot:spring-boot-starter-security:jar:2.3.1.RELEASE:compile
[INFO]    +- org.springframework:spring-aop:jar:5.2.7.RELEASE:compile
[INFO]    +- org.springframework.security:spring-security-config:jar:5.3.3.RELEASE:compile
[INFO]    |  \- org.springframework.security:spring-security-core:jar:5.3.3.RELEASE:compile
[INFO]    \- org.springframework.security:spring-security-web:jar:5.3.3.RELEASE:compile
[INFO] ------------------------------------------------------------------------
[INFO] BUILD SUCCESS
[INFO] ------------------------------------------------------------------------
[INFO] Total time:  01:17 min
[INFO] Finished at: 2020-06-29T14:08:09+08:00
[INFO] ------------------------------------------------------------------------
```

图 17-1 Spring Security 依赖项

接下来，添加 controller 层代码，如下所示：

```java
import com.gf.common.GfResponseEntity;
import org.springframework.http.ResponseEntity;
import org.springframework.security.core.userdetails.UserDetails;
import org.springframework.security.core.userdetails.UserDetailsService;
import org.springframework.web.bind.annotation.GetMapping;
import org.springframework.web.bind.annotation.PathVariable;
import org.springframework.web.bind.annotation.RequestMapping;
import org.springframework.web.bind.annotation.RestController;
@RestController
@RequestMapping("/gf/security")
public class UserController {
    /**
     * 本 bean 是由 Spring Security 提供的
     */
    private final UserDetailsService userDetailsService;
    public UserController(UserDetailsService userDetailsService) {
        this.userDetailsService = userDetailsService;
    }
    /**
     * 根据用户名获取用户的详细信息
```

```
    */
@GetMapping("/username/{user}")
public ResponseEntity userInfo(@PathVariable String user) {
    UserDetails userDetails = userDetailsService.loadUserByUsername(user);
    return GfResponseEntity.self().ok(userDetails);
    }
}
```

编写好代码之后，我们启动当前的 web 应用，控制台输出一个用户密码，如图 17-2 所示。

```
  .   ____          _            __ _ _
 /\\ / ___'_ __ _ _(_)_ __  __ _ \ \ \ \
( ( )\___ | '_ | '_| | '_ \/ _` | \ \ \ \
 \\/  ___)| |_)| | | | | || (_| |  ) ) ) )
  '  |____| .__|_| |_|_| |_\__, | / / / /
 =========|_|==============|___/=/_/_/_/
 :: Spring Boot ::        (v2.3.1.RELEASE)

2020-06-29 16:48:14.381  INFO 5140 --- [           main] com.gf.security.SecurityApplication
2020-06-29 16:48:14.381  INFO 5140 --- [           main] com.gf.security.SecurityApplication
2020-06-29 16:48:18.765  INFO 5140 --- [           main] o.s.b.w.embedded.tomcat.TomcatWebServer
2020-06-29 16:48:18.791  INFO 5140 --- [           main] o.apache.catalina.core.StandardService
2020-06-29 16:48:18.792  INFO 5140 --- [           main] org.apache.catalina.core.StandardEngine
2020-06-29 16:48:18.966  INFO 5140 --- [           main] o.a.c.c.C.[Tomcat].[localhost].[/]
2020-06-29 16:48:18.966  INFO 5140 --- [           main] w.s.c.ServletWebServerApplicationContext
2020-06-29 16:48:19.538  INFO 5140 --- [           main] o.s.s.concurrent.ThreadPoolTaskExecutor
2020-06-29 16:48:20.149  INFO 5140 --- [           main] .s.s.UserDetailsServiceAutoConfiguration

Using generated security password: aa93935a-7486-42bf-8216-05a796266ae2

2020-06-29 16:48:20.498  INFO 5140 --- [           main] o.s.s.web.DefaultSecurityFilterChain
2020-06-29 16:48:20.632  INFO 5140 --- [           main] o.s.b.w.embedded.tomcat.TomcatWebServer
2020-06-29 16:48:20.648  INFO 5140 --- [           main] com.gf.security.SecurityApplication
```

图 17-2 控制台

默认 UserDetailsService 只有一个用户，用户名为 user，每次启动应用时产生随机密码，如上图中控制台的输出。在当前只有默认配置的情况下，所有自定义的 URL 都会受到 Spring Security 的保护，也就是说不输入用户名和密码而直接访问 URL 会响应 401 状态码，如图 17-3 所示。

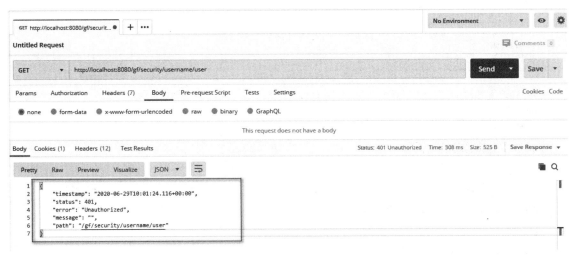

图 17-3 URL 受 Spring Security 的保护

默认情况下，添加 Spring Security 后，web 应用程序获得的基本功能，如下所示：

1) 带有内存存储的 UserDetailsService（或者，如果是 Spring WebFlux 应用程序，则为 ReactiveUserDetailsService）bean 和具有生成密码的单个用户；
2) 整个应用程序的基于表单的登录或 HTTP 基本安全性（取决于请求中的 Accept 头）；
3) 用于发布身份验证事件的 DefaultAuthenticationEventPublisher 实现类，可以通过实现 AuthenticationEventPublisher 接口来覆盖它。

在 application.yml 配置文件中修改用户名和密码，请添加配置项，如下所示：

spring.security.user.name: user
spring.security.user.password: 123456

添加以上配置项后，启动应用程序，之后在浏览器中访问以下 URL：http://localhost:8080//gf/security/username/user，会跳转到登录页面，如图 17-4 所示。

图 17-4 登录页面

在登录页面中，输入用户名和密码并登录成功之后才能正常获取相应 URL 的响应。另外，我们在 Postman 中也可以做同样的操作，如图 17-5 所示。

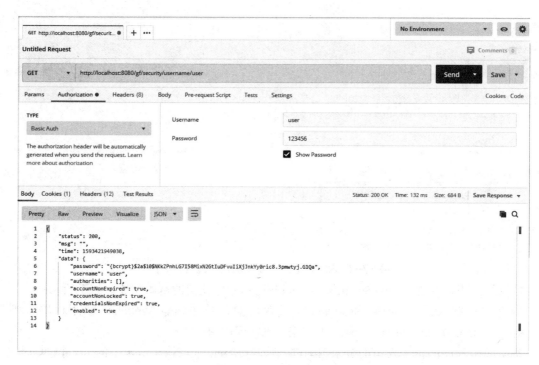

图 17-5 正确登录

17.2 @EnableGlobalMethodSecurity

通过本注解可以启用 Spring Security 全局方法安全性，类似于 <global-method-security> xml 支持。更高级的配置可以通过继承 GlobalMethodSecurityConfiguration 并重写受保护的方法来提供自定义实现。请注意，@EnableGlobalMethodSecurity 注解仍然必须包含在继承 GlobalMethodSecurityConfiguration 的类中，才能确定设置。在本注解中有三个属性，其功能说明，如表 17-1 所示。

表 17-1 注解属性列表

名称	说明	默认值
prePostEnabled()	确定是否应启用前置和后置注解	false
securedEnabled()	确定是否应启用安全注解	false
jsr250Enabled()	确定是否应启用 JSR-250 注解	false

17.2.1 前置和后置注解

@PreAuthorize 和 @PostAuthorize 这两个注解在类和方法上都是可以使用的，要使用它们，请在主类上添加注解，如下所示：

@EnableGlobalMethodSecurity(prePostEnabled= true)

各个注解所表示的意思，如下所示：
1) @PreAuthorize：在方法执行之前验证授权，支持 Spring EL 表达式；
2) @PostAuthorize：在方法执行后再进行权限验证，支持 Spring EL 表达式，本注解较少使用；
3) @PreFilter：对输入的集合类型参数进行过滤，支持 Spring EL 表达式；
4) @PostFilter：对方法返回的集合类型参数进行过滤，支持 Spring EL 表达式。

@PreAuthorize 注解的用法，示例代码，如下所示：

```
/**
 * 1、根据用户名获取用户的详细信息
 * 2、当前用户必须具有 user 和 admin 角色，才能执行本方法。
 */
@PreAuthorize("hasRole('user') AND hasRole('admin')")
@GetMapping("/username/{user}")
public ResponseEntity userInfo(@PathVariable String user) {
    UserDetails userDetails = userDetailsService.loadUserByUsername(user);
    return GfResponseEntity.self().ok(userDetails);
}
```

Spring Security 3.0 引入了使用 Spring EL 表达式作为授权机制的能力，此外还简单地使用了配置属性和以前见过的访问决策投票者。基于表达式的访问控制建立在相同的体系结构上，但允许将复杂的布尔逻辑封装在单个表达式中。表达式根对象的基类是 SecurityExpressionRoot。这里提供了一些常见的表达式，这些表达式在 web 和方法安全性中都是可用的，有关具体信息，如表 17-2 所示。

表 17-2 表达式属性描述

表达式	描述
hasRole(String)	如果当前主体具有指定的角色，则返回 true。默认情况下，如果提供的角色不以"ROLE_"开头，它将被添加。 例如：hasRole('admin') 这可以通过修改 DefaultWebSecurityExpressionHandler 类的方法 defaultRolePrefix 来定制。
hasAnyRole(String...)	如果当前主体具有任何提供的角色之一（以逗号分隔的字符串列表形式给出），则返回 true。 例如：hasAnyRole('admin', 'user') 这可以通过修改 DefaultWebSecurityExpressionHandler 类的方法 defaultRolePrefix 来定制。
hasAuthority(String)	如果当前主体具有指定的权限，则返回 true。 例如：hasAuthority('read')
hasAnyAuthority(String...)	如果当前主体具有任何提供的权限之一（以逗号分隔的字符串列表形式给出），则返回 true。 例如：hasAnyAuthority('read', 'write')

表达式	描述
principal	允许直接访问表示当前用户的主体对象。
authentication	允许直接访问从 SecurityContext 获取的当前 Authentication 对象。
permitAll	总是允许,也就是说始终返回 true。
denyAll	总是拒绝,也就是说始终返回 false。
isAnonymous()	如果当前主体是匿名用户,则返回 true。
isRememberMe()	如果当前主体是 remember-me 用户,则返回 true。
isAuthenticated()	如果用户不是匿名的,则返回 true。
isFullyAuthenticated()	如果用户不是匿名用户或 remember-me 用户,则返回 true。
hasPermission(Object, Object)	如果用户有权访问给定权限的提供的目标,则返回 true。 例如:hasPermission(domainObject, 'read')
hasPermission(Object, String, Object)	如果用户有权访问给定权限的提供的目标,则返回 true。 例如: hasPermission(1, 'com.example.domain.Message', 'read')

有关表达式的更详细的信息,请访问以下网址:

https://docs.spring.io/spring-security/site/docs/5.3.2.RELEASE/reference/html5/#el-common-built-in

17.2.2 @Secured 注解

要使用本注解,请在主类上添加注解,如下所示:
@EnableGlobalMethodSecurity(securedEnabled = true)
本注解可以用在类上,表示其中的所有方法都受到控制;也可以用在方法上,表示该方法受到控制。把前缀"ROLE_"+角色名称组合的字符串添加到本注解中,表示该角色可以访问相应方法,可以使用此方式指定多个角色。示例代码,如下所示:
@Secured({"ROLE_admin", "ROLE_user"})
@GetMapping("/username/{user}")
public ResponseEntity userInfo(@PathVariable String user) {
    // 本代码上文已经出现过,故在此省略。
}

@Secured 注解使用起来比较简单容易上手,但是,它是不支持 Spring EL 表达式的。

17.2.3 JSR-250 注解

@DenyAll、@RolesAllowed 和 @PermitAll 这三个注解在类和方法上都是可以使用的,要使用它们,请在主类上添加注解,如下所示:

@EnableGlobalMethodSecurity(jsr250Enabled = true)

各个注解所表示的意思，如下所示：
1) @DenyAll：拒绝所有访问；
2) @RolesAllowed：允许指定的角色访问，用法：@RolesAllowed({"user", "admin"});
3) @PermitAll：允许所有访问。

17.3 数据库支持

在上一节中，用户信息是存放在配置文件中的，并且只有一个用户，这显然不能满足实际的需求。在本节中，我们介绍基于数据库的 Spring 安全功能。使用 MariaDB 作为数据库服务器，创建数据库 gf_platform2，并向其中添加表 users 和 authorities。

新建数据表 users，如下所示：

```
CREATE TABLE `users` (
  `username` varchar(50) NOT NULL,
  `password` varchar(100) NOT NULL DEFAULT '',
  `enabled` tinyint(1) NOT NULL,
  PRIMARY KEY (`username`)
) ENGINE=InnoDB DEFAULT CHARSET=utf8mb4;
```

创建后向表中插入数据，如下所示：

```
INSERT INTO `users` VALUES
('user','{bcrypt}$2a$10$MTAbjiPAFadY9/uyQCIRq.X/w1DkXcSYC5q7YXVMtdUbvuEw.dTxC',1);
```

新建数据表 authorities，如下所示：

```
CREATE TABLE `authorities` (
  `id` bigint(20) NOT NULL AUTO_INCREMENT,
  `username` varchar(50) NOT NULL,
  `authority` varchar(50) NOT NULL,
  PRIMARY KEY (`id`),
  UNIQUE KEY `ix_auth_username` (`username`,`authority`)
) ENGINE=InnoDB AUTO_INCREMENT=3 DEFAULT CHARSET=utf8mb4;
```

创建后向表中插入数据，如下所示：

```
INSERT INTO `authorities` VALUES (1,'user','ROLE_user');
```

```sql
INSERT INTO `authorities` VALUES (2,'user','ROLE_admin');
```

执行完上面的 SQL 语句后，默认情况下，并不需要创建对应的实体类及其相关类，因为 Spring Security 已经编写好了相应的代码。接下来，添加数据库相关的依赖项，如下所示：

```xml
<dependency>
    <groupId>org.springframework.boot</groupId>
    <artifactId>spring-boot-starter-data-jpa</artifactId>
</dependency>
<dependency>
    <groupId>mysql</groupId>
    <artifactId>mysql-connector-java</artifactId>
</dependency>
```

请注意：如果比较喜欢使用 MyBatis Plus 的话，可以把 JPA 的依赖项替换为 MyBatis Plus 的依赖项。然后，在 application.yml 配置文件中添加数据连接配置，没什么特殊的，和通常的配置一样，具体内容，如图 17-6 所示。

```yaml
spring:
  datasource:
    url: jdbc:mysql://localhost:3306/gf_platform2?serverTimezone=Asia/Shanghai&useUnicode=true&characterEncoding=utf8
    username: root
    password: 123456
#    type: com.zaxxer.hikari.HikariDataSource  # 配置数据连接池
#  security:
#    user:
#      name: user
#      password: 123456
#      roles:
#        - user
#        - admin
```

图 17-6 application.yml 配置

下面这个 bean 是最为关键的，它指定了数据源为数据库，如下所示：

```java
/**
 * 为 Spring Security 提供来自数据库的用户信息。
 */
@Bean
public UserDetailsManager users(DataSource dataSource) {
    return new JdbcUserDetailsManager(dataSource);
}
```

controller 层的代码和上文中的一样，仍然可以使用 UserDetailsService。接口调用，如图 17-7 所示。

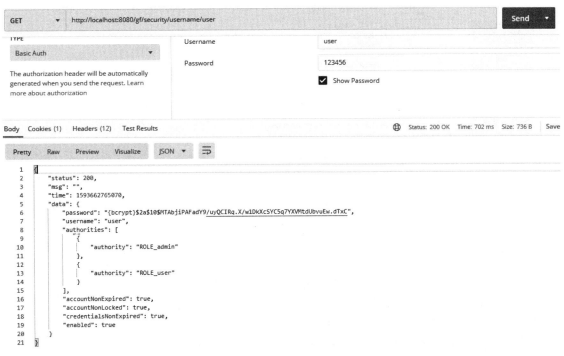

图 17-7 UserDetailsService 接口调用

17.4 "记住我"身份验证

记住我或持久登录身份验证是指网站能够在会话之间记住主体的身份。这通常通过向浏览器发送 cookie 来实现,在以后的会话中检测到相应的 cookie 并触发自动登录。Spring Security 为这些操作提供了必要的挂钩,并有两个具体的"记住我"实现。一个使用哈希来保护基于 cookie 的令牌的安全性,另一个使用数据库或其他持久存储机制来存储生成的令牌。

17.4.1 自定义登录页面

我们使用 Thymeleaf 模板技术来自定义登录页面,添加依赖项,如下所示:
```
<dependency>
    <groupId>org.springframework.boot</groupId>
    <artifactId>spring-boot-starter-thymeleaf</artifactId>
</dependency>
```
然后,在 src\main\resources\templates 目录中新建 login.html 模板文件,如图 17-8 所示。

```
<!DOCTYPE html>
<html xmlns="http://www.w3.org/1999/xhtml" xmlns:th="https://www
.thymeleaf.org">
<head>
    <meta charset="UTF-8">
    <title>登录页面</title>
</head>
<body>
<h1>登录页面</h1>
<div th:if="${param.error}">
    无效的用户名或密码！ </div>
<div th:if="${param.logout}">
    你已经退出登录！ </div>
<form th:action="@{/gf/login}" method="post">
    <div>
        <input type="text" name="username" placeholder="用户名"/>
    </div>
    <div>
        <input type="password" name="password" placeholder="密码"/>
    </div>
    <input type="submit" value="登录" />
</form>
</body>
</html>
```

图 17-8 login.html 模板文件

在 controller 层中指定与模板文件对应的 API，如下所示：

```
package com.gf.security.controller;
import org.springframework.stereotype.Controller;
import org.springframework.web.bind.annotation.GetMapping;
import org.springframework.web.bind.annotation.RequestMapping;
@Controller
@RequestMapping("/gf")
public class LoginController {
    /**
     * 访问本接口，返回登录页面
     */
    @GetMapping("/login")
    String login() {
        return "login";}
}
```

最后，添加 MySecurityConfig 配置类，以指定自定义的 API，如下所示：

```
package com.gf.security.config;
import org.springframework.context.annotation.Configuration;
import org.springframework.security.config.annotation.web.builders.HttpSecurity;
import org.springframework.security.config.annotation.web.configuration.WebSecurityConfigurerAdapter;
```

```
import static org.springframework.security.config.Customizer.withDefaults;
@Configuration
public class MySecurityConfig extends WebSecurityConfigurerAdapter {
    @Override
    protected void configure(HttpSecurity http) throws Exception {
        http.httpBasic(withDefaults())
                //指定登录页面 API
                .formLogin(form -> form.loginPage("/gf/login"))
                .authorizeRequests()
                // 允许匹配的 API 的非授权访问
                .mvcMatchers("/gf/login").permitAll()
                // 除了上面允许的,禁止所有其他 API 的非授权访问
                .anyRequest().authenticated();}
}
```

现在,启动应用程序,访问登录页面,如下所示:

http://localhost:8080/gf/login

17.4.2 基于哈希的令牌方式

remember-me 令牌仅在指定的时间段(默认 14 天)内有效,前提是用户名、密码和键不更改。值得注意的是,这种方式有一个潜在的安全问题,即在令牌过期之前,任何用户代理(浏览器等)都可以使用捕获的 remember-me 令牌。如果委托人知道已捕获令牌,则他们可以轻松更改密码,并立即使所有发行的 remember-me 令牌无效。

通过基于哈希的令牌方式实现持久登录身份验证,下面给出相应的示例。首先,新建 login.html 文件,其中的表单要包含复选框 remember-me 元素,如图 17-9 所示。

```
<!DOCTYPE html>
<html xmlns="http://www.w3.org/1999/xhtml" xmlns:th="https://www.thymeleaf.org">
<head>
    <meta charset="UTF-8">
    <title>登录页面</title>
</head>
<body>
<h1>登录页面</h1>
<div th:if="${param.error}">
    无效的用户名或密码! </div>
<div th:if="${param.logout}">
    你已经退出登录! </div>
<form th:action="@{/gf/login}" method="post">
    <div>
        <input type="text" name="username" placeholder="用户名"/>
    </div>
    <div>
        <input type="password" name="password" placeholder="密码"/>
    </div>
    <div><input type="checkbox" id="remember-me" name="remember-me" value="yes"/>
        <label for="remember-me">14天内免登录</label></div>
    <input type="submit" value="登录" />
</form>
</body>
</html>
```

图 17-9 login.html 文件

然后，添加 HashBasedTokenConfig 配置类进行具体的设置，如下所示：

```
package com.gf.security.config;
import org.springframework.beans.factory.annotation.Autowired;
import org.springframework.context.annotation.Bean;
import org.springframework.context.annotation.Configuration;
import org.springframework.security.config.annotation.web.builders.HttpSecurity;
import org.springframework.security.config.annotation.web.configuration.WebSecurityConfigurerAdapter;
import org.springframework.security.core.userdetails.UserDetailsService;
import org.springframework.security.web.authentication.rememberme.TokenBasedRememberMeServices;
import static org.springframework.security.config.Customizer.withDefaults;
@Configuration
public class HashBasedTokenConfig extends WebSecurityConfigurerAdapter {
    private static final String SECRET = "gf";
    @Autowired
    private UserDetailsService userDetailsService;
    @Bean("tokenBaseRememberMeServices")
    public TokenBasedRememberMeServices tokenBasedRememberMeServices() {
        TokenBasedRememberMeServices services =
                new TokenBasedRememberMeServices(SECRET, userDetailsService);
        // 以下都是默认值，也可以根据需要调整。
//        services.setAlwaysRemember(false);
        services.setCookieName(AbstractRememberMeServices.SPRING_SECURITY_REMEMBER_ME_COOKIE_KEY);
        services.setTokenValiditySeconds(AbstractRememberMeServices.TWO_WEEKS_S);
        return services;
    }
    @Override
```

```
protected void configure(HttpSecurity http) throws Exception {
    http.httpBasic(withDefaults())
            .formLogin(form -> form.loginPage("/gf/login"))
            .authorizeRequests()
            .mvcMatchers("/gf/login").permitAll()
            .anyRequest().authenticated().and().rememberMe()
            .rememberMeServices(tokenBasedRememberMeServices());
    }
}
```

完成以上操作后,启动应用程序,并在浏览器中访问 URL,如下所示:
http://localhost:8080/gf/security/username/user

由于该请求是未授权的,立即跳转到了登录页面,在其中输入用户名和密码,并勾选复选框,如图 17-10 所示。

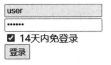

图 17-10 登录页面(1)

正确地填写好字段后,单击按钮[登录],向服务端发送请求。在身份验证成功后,服务端向浏览器响应 remember-me cookie,并跳转到之前的 URL 请求。最终,响应 JSON 数据,如图 17-11 所示。

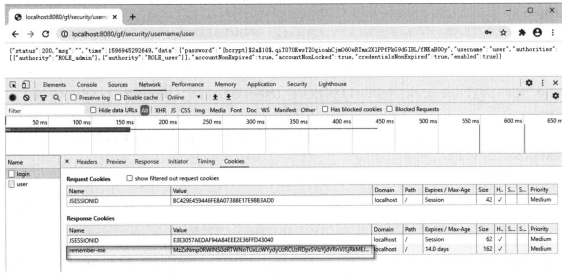

图 17-11 登录页面(2)

17.4.3 持久的令牌方式

使用数据库来存储生成的令牌,这比上一种方式更加安全。首先,新建 persistent_logins 数据表,如下所示:

```sql
CREATE TABLE `persistent_logins` (
  `username` varchar(64) NOT NULL,
  `series` varchar(64) NOT NULL,
  `token` varchar(64) NOT NULL,
  `last_used` timestamp NOT NULL DEFAULT current_timestamp() ON UPDATE current_timestamp(),
  PRIMARY KEY (`series`)
) ENGINE=InnoDB DEFAULT CHARSET=utf8mb4;
```

然后,添加 PersistentTokenConfig 配置类进行具体的设置,如下所示:

```java
package com.gf.security.config;
import org.springframework.beans.factory.annotation.Autowired;
import org.springframework.context.annotation.Bean;
import org.springframework.context.annotation.Configuration;
import org.springframework.jdbc.core.JdbcTemplate;
import org.springframework.security.config.annotation.web.builders.HttpSecurity;
import org.springframework.security.config.annotation.web.configuration.WebSecurityConfigurerAdapter;
import org.springframework.security.core.userdetails.UserDetailsService;
import org.springframework.security.web.authentication.rememberme.JdbcTokenRepositoryImpl;
import org.springframework.security.web.authentication.rememberme.PersistentTokenBasedRememberMeServices;
import org.springframework.security.web.authentication.rememberme.PersistentTokenRepository;
@Configuration
```

```java
public class PersistentTokenConfig extends
WebSecurityConfigurerAdapter {
    private static final String SECRET = "gf";
    @Autowired
    private UserDetailsService userDetailsService;
    @Autowired
    private JdbcTemplate jdbcTemplate;
    @Bean
    public PersistentTokenRepository persistentTokenRepository() {
        JdbcTokenRepositoryImpl repository = new JdbcTokenRepositoryImpl();
        repository.setJdbcTemplate(jdbcTemplate);
        return repository;
    }
    @Bean
    public PersistentTokenBasedRememberMeServices
persistentTokenBasedRememberMeServices() {
        return
                new PersistentTokenBasedRememberMeServices(SECRET,
userDetailsService, persistentTokenRepository());
    }
    @Override
    protected void configure(HttpSecurity http) throws Exception {
        http.formLogin(form -> form.loginPage("/gf/login"))
                .authorizeRequests()
                .mvcMatchers("/gf/login").permitAll()
                .anyRequest().authenticated().and().rememberMe()
                .rememberMeServices(persistentTokenBasedRememberMeServices());
    }
}
```

至此，持久的令牌方式已经配置完成，其他的内容和上一种方式是一样的。

17.5 OAuth 2

OAuth 2（Open Authorization）是行业标准的授权协议。OAuth 2 专注于客户端开发者的简单性，同时为 web 应用程序、桌面应用程序、移动电话和客厅设备提供特定的授权流程。

OAuth 是一个开放标准，它允许用户授权第三方应用访问特定的数据，这些数据是用户存储在另外的服务提供者上的。在这一过程中，不需要将用户名和密码提供给第三方应用或分享用户数据的全部。OAuth 2 是 OAuth 协议的延续版本，但不向后兼容 OAuth 1。

OAuth 授权类型，如下所示：

授权码（Authorization Code）

本类型由机密客户端和公共客户端用于将授权码交换为访问令牌。用户通过重定向 URL 返回客户端（web 应用等）后，该客户端将从该 URL 获取授权码，并使用它来请求访问令牌。建议所有客户端也在此流程中使用 PKCE 扩展，以提供更好的安全性。

客户端凭证（Client Credentials）

本类型被客户端用于获取用户上下文之外的访问令牌。客户端通常使用它来访问自己的资源，而不是访问用户的资源。

设备码（Device Code）

设备流中的无浏览器或受输入约束的设备使用本类型，以将先前获得的设备码交换为访问令牌。本类型值为 urn:ietf:params:oauth:grant-type:device_code。

刷新令牌（Refresh Token）

当访问令牌过期时，客户端使用本类型，以将刷新令牌交换为访问令牌。这允许客户端继续拥有有效的访问令牌，而无需与用户进一步交互。

传统：隐式流（Implicit Flow）

隐式流是一个简化的 OAuth 流，以前推荐用于本机应用程序和 JavaScript 应用程序，在这些应用程序中，访问令牌在没有额外授权码交换步骤的情况下立即返回。不建议使用隐式流（有些服务器完全禁止这种流），因为在 HTTP 重定向中返回访问令牌存在固有的风险，而不需要确认客户端已收到它。公共客户端（如本机应用程序和 JavaScript 应用程序）现在应该使用 PKCE 扩展的授权码流。

传统：密码授予（Password Grant）

密码授予类型是一种将用户凭据交换为访问令牌的方法。因为客户端应用程序必须收集用户密码并将其发送到授权服务器，所以不建议再使用此授权。此流程不提供多因素身份验证或委托帐户等机制，因此在实践中非常有限。最新的 OAuth 2.0 安全最佳当前实践完全不允许密码授予。

有关授权类型的更多信息，请访问以下网址：
https://oauth.net/2/grant-types

17.5.1 OAuth 客户端

OAuth 2.0 登录特性为应用程序（OAuth 客户端）提供了一种功能，通过使用 OAuth 2.0 提供商（如 GitHub）或 OpenID Connect 1.0 提供商（如 Google）上的现有帐户可以让用户登录到该应用程序（OAuth 客户端）。在 Spring Boot 应用程序中引入 OAuth 客户端，请添加依赖项，如下所示：

<dependency>

```xml
    <groupId>org.springframework.boot</groupId>
    <artifactId>spring-boot-starter-oauth2-client</artifactId>
</dependency>
```

接着添加 web 和 thymeleaf 依赖项，为下面的 GitHub 作为身份验证提供者的示例做准备，如下所示：

```xml
<dependency>
    <groupId>org.springframework.boot</groupId>
    <artifactId>spring-boot-starter-web</artifactId>
</dependency>
<dependency>
    <groupId>org.springframework.boot</groupId>
    <artifactId>spring-boot-starter-thymeleaf</artifactId>
</dependency>
```

下面是使用 GitHub 作为身份验证提供者的示例，讲解了 OAuth 客户端的使用。GitHub 是框架默认配置的授权服务提供商，因此，可以省略一些配置项。默认情况下，OAuth 授权类型是授权码。

第一步，访问 https://github.com/settings/applications/new，在 GitHub 上注册 OAuth 客户端，在注册页面填写内容，如图 17-12 所示。

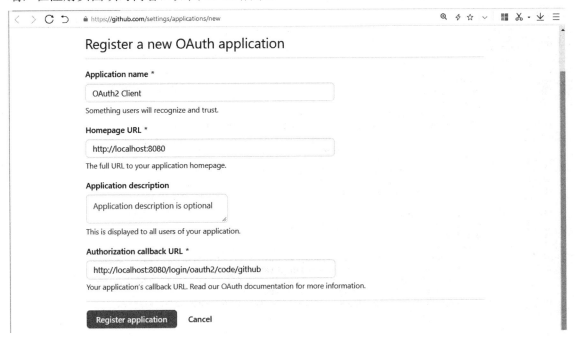

图 17-12 注册 OAuth 客户端

上图中的回调 URL 是框架默认提供的 API。注册完成之后，就可以得到 Client ID 和 Client Secret 的值，如图 17-13 所示。

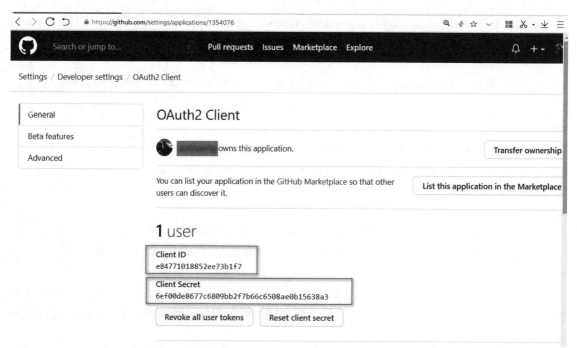

图 17-13 获得 Client ID 和 Client Secret 值

在 application.yml 配置文件中,把得到的值分别设置给 client-id 和 client-secret,如图 17-14 所示。

图 17-14 在 application.yml 配置文件赋值

编写 controller 层,示例代码,如下所示:

```java
package com.gf.security.oauth2.controller;
import org.springframework.http.ResponseEntity;
import org.springframework.web.bind.annotation.GetMapping;
import org.springframework.web.bind.annotation.RequestMapping;
import org.springframework.web.bind.annotation.RestController;
import java.util.Collections;
import static com.gf.common.GfResponseEntity.self;
@RestController
@RequestMapping("/")
public class HelloController {
    @GetMapping("/hello")
    public ResponseEntity hello() {
        return self().ok(Collections.singletonMap("city", "郑州"));
    }
}
```

上面的普通 API 是需要授权才能访问的。来自官方的 controller 层代码，如下所示：

```java
package com.gf.security.oauth2.controller;
import org.springframework.security.core.annotation.AuthenticationPrincipal;
import org.springframework.security.oauth2.client.OAuth2AuthorizedClient;
import org.springframework.security.oauth2.client.annotation.RegisteredOAuth2AuthorizedClient;
import org.springframework.security.oauth2.core.user.OAuth2User;
import org.springframework.stereotype.Controller;
import org.springframework.ui.Model;
import org.springframework.web.bind.annotation.GetMapping;
/**
 * @author Joe Grandja
 * @author Rob Winch
 */
@Controller
public class OAuth2LoginController {
    @GetMapping("/")
    public String index(Model model,
            @RegisteredOAuth2AuthorizedClient OAuth2AuthorizedClient authorizedClient,
```

```
                @AuthenticationPrincipal OAuth2User oauth2User) {
        model.addAttribute("userName", oauth2User.getName());
        model.addAttribute("clientName",
authorizedClient.getClientRegistration().getClientName());
        model.addAttribute("userAttributes",
oauth2User.getAttributes());
        return "index";
    }
}
```

上面的示例代码和下面 index.html 文件都来自官方示例，想要查看完整的官方示例，请访问以下网址：

https://github.com/spring-projects/spring-security/tree/5.3.4.RELEASE/samples/boot/oauth2login

编写模板 index.html 文件，用于展现授权成功后的用户信息，如图 17-15 所示。

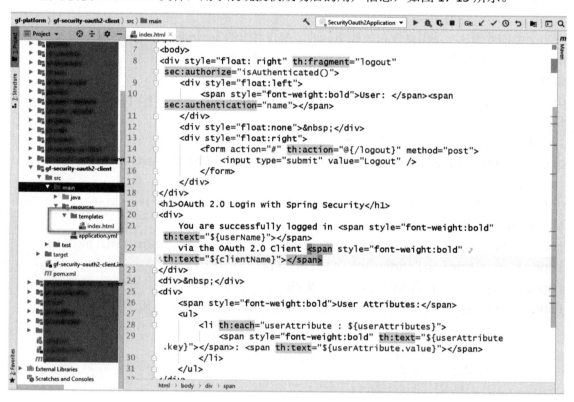

图 17-15 模板 index.html 文件

以上配置和编码工作完成后，启动应用程序，在浏览器中访问 URL，如下所示：

http://localhost:8080

由于是未授权访问 URL，会重定向到登录选择页面（若只有一个第三方登录，则会自动跳转到第三方页面），从页面中选择一个第三方登录，如图 17-16 所示。

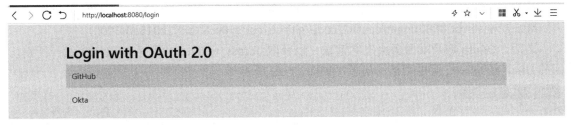

图 17-16 登录 Oauth（1）

默认情况下，上面的页面是框架提供的，当然，我们也可以自定义它。这里，我们选择 GitHub 登录，跳转到 GitHub 登录页面（若已经在浏览器中登录过 GitHub，则会直接跳转到授权页面），如图 17-17 所示。

图 17-17 登录 Oauth（2）

登录成功之后，进入授权页面，如图 17-18 所示。

图 17-18 登录 Oauth（3）

授权成功之后，会重定向到我们注册 OAuth 客户端时的回调 URL 并带有参数，如下所示：
http://localhost:8080/login/oauth2/code/github?code=0cc81b2512018a8f600f

之后，OAuth 客户端服务会发送请求，以获取 access_token。再之后，重定向到一开始访问的 URL，这次 URL 可以正常访问了，如图 17-19 所示。

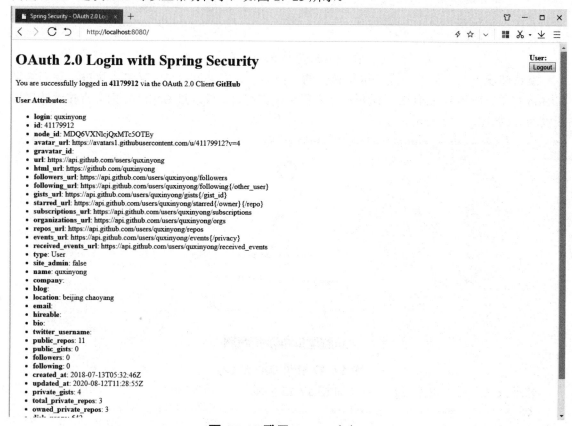

图 17-19 登录 Oauth（4）

17.5.2 资源服务器

在 Spring Boot 项目中引入 OAuth 2 资源服务器，添加依赖项，如下所示：

```
<dependency>
    <groupId>org.springframework.boot</groupId>
    <artifactId>spring-boot-starter-oauth2-resource-server</artifactId>
</dependency>
```

执行 Maven 命令，查看 spring-boot-starter-oauth2-resource-server 的依赖关系，如图 17-20 所示。

```
[INFO] com.gf:gf-security-oauth2-res-server:jar:0.0.1-SNAPSHOT
[INFO] \- org.springframework.boot:spring-boot-starter-oauth2-resource-server:jar:2.3.1.RELEASE:compile
[INFO]    +- org.springframework.boot:spring-boot-starter:jar:2.3.1.RELEASE:compile
[INFO]    |  +- org.springframework.boot:spring-boot:jar:2.3.1.RELEASE:compile
[INFO]    |  +- org.springframework.boot:spring-boot-autoconfigure:jar:2.3.1.RELEASE:compile
[INFO]    |  +- org.springframework.boot:spring-boot-starter-logging:jar:2.3.1.RELEASE:compile
[INFO]    |  |  +- ch.qos.logback:logback-classic:jar:1.2.3:compile
[INFO]    |  |  |  +- ch.qos.logback:logback-core:jar:1.2.3:compile
[INFO]    |  |  |  \- org.slf4j:slf4j-api:jar:1.7.30:compile
[INFO]    |  |  +- org.apache.logging.log4j:log4j-to-slf4j:jar:2.13.3:compile
[INFO]    |  |  |  \- org.apache.logging.log4j:log4j-api:jar:2.13.3:compile
[INFO]    |  |  \- org.slf4j:jul-to-slf4j:jar:1.7.30:compile
[INFO]    |  +- jakarta.annotation:jakarta.annotation-api:jar:1.3.5:compile
[INFO]    |  +- org.springframework:spring-core:jar:5.2.7.RELEASE:compile
[INFO]    |  |  \- org.springframework:spring-jcl:jar:5.2.7.RELEASE:compile
[INFO]    |  \- org.yaml:snakeyaml:jar:1.26:compile
[INFO]    +- org.springframework.security:spring-security-config:jar:5.3.3.RELEASE:compile
[INFO]    |  +- org.springframework:spring-aop:jar:5.2.7.RELEASE:compile
[INFO]    |  +- org.springframework:spring-beans:jar:5.2.7.RELEASE:compile
[INFO]    |  \- org.springframework:spring-context:jar:5.2.7.RELEASE:compile
```

图 17-20 OAuth 2 资源服务器

Spring Security 和 Spring Security OAuth2（过时的）配置资源服务器的示例，如下所示：

https://github.com/spring-projects/spring-security/wiki/OAuth-2.0-Migration-Guide#examples-matrix-1

在配置 OAuth 2 资源服务器的时候，请根据自己项目的实际需要，参照以上示例进行配置。

17.5.3 授权服务器

之前，Spring Security OAuth 提供了 spring-security-oauth2-autoconfigure 模块来搭建 OAuth 2 授权服务器，但是这种方式目前已经过时。使用这种方式所需的依赖项，如下所示：

```
<dependency>
    <groupId>org.springframework.security.oauth.boot</groupId>
    <artifactId>spring-security-oauth2-autoconfigure</artifactId>
</dependency>
```

有关 Spring Security OAuth 过时的更多信息，请访问以下网址：

https://spring.io/projects/spring-security-oauth#overview

目前，Spring Authorization Server 提供授权服务器的支持，它还是一个成长中的项目，刚刚发布了 0.0.1 版本（2020 年 8 月 20 日）。有关 Spring 授权服务器的通告，如下所示：

https://spring.io/blog/2020/04/15/announcing-the-spring-authorization-server

Spring Authorization Server 是一个由 Spring Security 团队领导的社区驱动的项目，专注于向 Spring 社区提供授权服务器支持。本项目将作为一个独立项目在 Spring 的实验项目中开始，以便它能够更快地发展。本项目的最终目标是替换 Spring Security OAuth 提供的授权服务器支持。在社区帮助下，本项目将以与原始 Spring Security OAuth 项目相同的方式发展。有关 Spring 授权服务器的源代码及其使用文档，如下所示：

https://github.com/spring-projects-experimental/spring-authorization-server

Spring 授权服务器的版本发布页面，如下所示：

https://github.com/spring-projects-experimental/spring-authorization-server/releases

17.6　Apache Shiro

　　Apache Shiro 是一个功能强大且易于使用的 Java 安全框架，它可执行身份验证、授权、加密和会话管理。借助 Shiro 易于理解的 API，可以快速轻松地保护任何应用程序，从最小的移动应用程序到最大的 web 和企业应用程序。Apache Shiro 的官网，如下所示：

http://shiro.apache.org

　　Apache Shiro 中的一些概念，如下所示：

1) Subject：应用程序用户的特定于安全的用户"视图"。它可以是人、第三方进程和连接到应用程序的服务器，甚至是 cron 作业。基本上，它是与应用程序通信的任何东西或人。
2) Principals：标识属性的主体（例如，用户名等）。
3) Credentials：用于验证身份的保密数据（例如：密码、生物识别数据和 x509 证书等）。
4) Realms：与后端数据源通信的组件。

　　下面给出了一个登陆验证的示例。想要在 Spring Boot 项目中引入 Shiro，请添加依赖项，如下所示：

```xml
<dependency>
    <groupId>org.apache.shiro</groupId>
    <artifactId>shiro-spring-boot-web-starter</artifactId>
    <version>1.5.3</version>
</dependency>
```

　　查看 shiro-spring-boot-web-starter 的依赖关系，其中包含了 spring-boot-starter-web 依赖项，如图 17-21 所示。

```
[INFO] com.gf:gf-security-shiro:jar:0.0.1-SNAPSHOT
[INFO] +- org.apache.shiro:shiro-spring-boot-web-starter:jar:1.5.3:compile
[INFO] |  +- org.apache.shiro:shiro-spring-boot-starter:jar:1.5.3:compile
[INFO] |  |  \- org.apache.shiro:shiro-spring:jar:1.5.3:compile
[INFO] |  |     +- org.apache.shiro:shiro-core:jar:1.5.3:compile
[INFO] |  |     |  +- org.apache.shiro:shiro-lang:jar:1.5.3:compile
[INFO] |  |     |  +- org.apache.shiro:shiro-cache:jar:1.5.3:compile
[INFO] |  |     |  +- org.apache.shiro:shiro-crypto-hash:jar:1.5.3:compile
[INFO] |  |     |  |  \- org.apache.shiro:shiro-crypto-core:jar:1.5.3:compile
[INFO] |  |     |  +- org.apache.shiro:shiro-crypto-cipher:jar:1.5.3:compile
[INFO] |  |     |  +- org.apache.shiro:shiro-config-core:jar:1.5.3:compile
[INFO] |  |     |  +- org.apache.shiro:shiro-config-ogdl:jar:1.5.3:compile
[INFO] |  |     |  |  \- commons-beanutils:commons-beanutils:jar:1.9.4:compile
[INFO] |  |     |  |     \- commons-collections:commons-collections:jar:3.2.2:compile
[INFO] |  |     |  \- org.apache.shiro:shiro-event:jar:1.5.3:compile
[INFO] |  |     \- org.apache.shiro:shiro-web:jar:1.5.3:compile
[INFO] |  |        \- org.owasp.encoder:encoder:jar:1.2.2:compile
[INFO] |  +- org.springframework.boot:spring-boot-autoconfigure:jar:2.3.1.RELEASE:compile
[INFO] |  |  \- org.springframework.boot:spring-boot:jar:2.3.1.RELEASE:compile
[INFO] |  +- org.springframework.boot:spring-boot-starter:jar:2.3.1.RELEASE:compile
[INFO] |  |  +- org.springframework.boot:spring-boot-starter-logging:jar:2.3.1.RELEASE:compile
[INFO] |  |  |  +- ch.qos.logback:logback-classic:jar:1.2.3:compile
[INFO] |  |  |  |  \- ch.qos.logback:logback-core:jar:1.2.3:compile
[INFO] |  |  |  +- org.apache.logging.log4j:log4j-to-slf4j:jar:2.13.3:compile
[INFO] |  |  |  |  \- org.apache.logging.log4j:log4j-api:jar:2.13.3:compile
[INFO] |  |  |  \- org.slf4j:jul-to-slf4j:jar:1.7.30:compile
[INFO] |  |  +- jakarta.annotation:jakarta.annotation-api:jar:1.3.5:compile
[INFO] |  |  \- org.yaml:snakeyaml:jar:1.26:compile
[INFO] |  \- org.springframework.boot:spring-boot-starter-web:jar:2.3.1.RELEASE:compile
[INFO] |     +- org.springframework.boot:spring-boot-starter-json:jar:2.3.1.RELEASE:compile
```

图 17-21　shiro-spring-boot-web-starter 的依赖关系

在官网页面中，可以查看 shiro-spring-boot-web-starter 最新的版本信息，如图 17-22 所示。

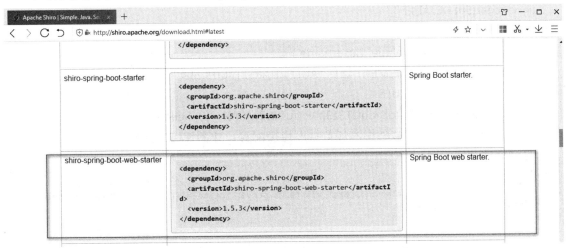

图 17-22 shiro-spring-boot-web-starter 版本信息

使用数据库来存储用户数据，添加相应依赖项，如下所示：
```
<dependency>
    <groupId>org.springframework.boot</groupId>
    <artifactId>spring-boot-starter-data-jpa</artifactId>
</dependency>
<dependency>
    <groupId>mysql</groupId>
    <artifactId>mysql-connector-java</artifactId>
</dependency>
```
一般情况下，用户登录都需要验证码来规避恶意行为，添加相应依赖项，如下所示：
```
<dependency>
    <groupId>com.github.whvcse</groupId>
    <artifactId>easy-captcha</artifactId>
    <version>1.6.2</version>
</dependency>
```
对于通过 API 传入的参数进行校验，添加相应依赖项，如下所示：
```
<dependency>
    <groupId>org.springframework.boot</groupId>
    <artifactId>spring-boot-starter-validation</artifactId>
</dependency>
```
添加数据表 shiro_user 来存放用户数据，其中用户的密码为：123456，如下所示：
```
CREATE TABLE `shiro_user` (
  `Id` bigint(11) NOT NULL AUTO_INCREMENT,
  `username` varchar(200) NOT NULL DEFAULT '',
  `password` varchar(200) NOT NULL DEFAULT '',
```

```
`status` int(11) DEFAULT 0,
`salt` varchar(50) DEFAULT '',
PRIMARY KEY (`Id`),
UNIQUE KEY `username_unique` (`username`)
) ENGINE=InnoDB AUTO_INCREMENT=2 DEFAULT CHARSET=utf8mb4 COMMENT='用户表';
INSERT INTO `shiro_user` VALUES (1,'user','9rh0n3wdKyl+eo79QoDRJqQT3z9atvcLfuJ6NZMCPeE=',0,'hello');
```

在 application.yml 配置文件中添加数据库连接配置，如图 17-23 所示。

```yaml
spring:
  datasource:
    url: jdbc:mysql://localhost:3306/gf_platform2?serverTimezone=Asia/Shanghai&useUnicode=true&characterEncoding=utf8
    username: root
    password: 123456
```

图 17-23 数据库连接配置

编写自定义的 UserRealm，示例代码，如下所示：

```java
package com.gf.security.shiro.config;
import com.gf.security.shiro.entity.User;
import com.gf.security.shiro.repository.UserRepository;
import org.apache.shiro.SecurityUtils;
import org.apache.shiro.authc.*;
import org.apache.shiro.authz.AuthorizationInfo;
import org.apache.shiro.authz.SimpleAuthorizationInfo;
import org.apache.shiro.realm.AuthorizingRealm;
import org.apache.shiro.subject.PrincipalCollection;
import org.apache.shiro.util.ByteSource;
import org.springframework.data.domain.Example;
import java.util.HashSet;
import java.util.Optional;
import java.util.Set;
public class UserRealm extends AuthorizingRealm {
    private UserRepository userRepository;
    public UserRealm(UserRepository userRepository) {
        this.userRepository = userRepository;
    }
    @Override
    protected AuthorizationInfo doGetAuthorizationInfo(PrincipalCollection principalCollection) {
        User user = (User) SecurityUtils.getSubject().getPrincipal();
```

```java
        SimpleAuthorizationInfo authInfo = new SimpleAuthorizationInfo();
        // 角色
        Set<String> roles = new HashSet<>();
        // 测试用权限
        authInfo.setRoles(roles);
//        authorizationInfo.setStringPermissions(permissions);
        return authInfo;
    }
    @Override
    protected AuthenticationInfo doGetAuthenticationInfo(AuthenticationToken authenticationToken) throws AuthenticationException {
        String username = (String) authenticationToken.getPrincipal();
        Optional<User> optionalUser = userRepository.findOne(Example.of(new User(username)));
        User user = optionalUser.orElse(null);
        // 账号不存在
        if (user == null) {
            throw new UnknownAccountException();
        }
        // 账号被锁定
        if (user.getStatus() != 0) {
            throw new LockedAccountException();
        }
        String salt = user.getSalt();
        return new SimpleAuthenticationInfo(user, user.getPassword(),
            ByteSource.Util.bytes(salt), getName());
    }
}
```

编写 ShiroConfig 配置类,来指定不受权限约束的 API 和密码比较规则,如下所示:

```java
package com.gf.security.shiro.config;
import com.gf.security.shiro.repository.UserRepository;
import org.apache.shiro.authc.credential.HashedCredentialsMatcher;
import org.apache.shiro.crypto.hash.Sha256Hash;
import org.apache.shiro.mgt.SessionsSecurityManager;
import org.apache.shiro.spring.web.config.DefaultShiroFilterChainDefinition;
```

```java
import org.apache.shiro.spring.web.config.ShiroFilterChainDefinition;
import org.apache.shiro.web.mgt.DefaultWebSecurityManager;
import org.springframework.beans.factory.annotation.Autowired;
import org.springframework.context.annotation.Bean;
import org.springframework.context.annotation.Configuration;
import java.util.HashMap;
import java.util.Map;
@Configuration
public class ShiroConfig {
    @Autowired
    private UserRepository userRepository;
    @Bean
    public UserRealm userRealm() {
        UserRealm userRealm = new UserRealm(userRepository);
        userRealm.setCredentialsMatcher(credentialsMatcher());
        return userRealm;
    }
    @Bean
    public ShiroFilterChainDefinition shiroFilterChainDefinition() {
        DefaultShiroFilterChainDefinition definition = new DefaultShiroFilterChainDefinition();
        Map<String, String> paths = new HashMap<>();
        paths.put("/captcha", "anon");
        paths.put("/logout", "anon");
        paths.put("/login", "anon");
        paths.put("/**", "authc");
        definition.addPathDefinitions(paths);
        return definition;
    }
    @Bean
    public HashedCredentialsMatcher credentialsMatcher() {
        HashedCredentialsMatcher matcher = new HashedCredentialsMatcher();
        matcher.setHashAlgorithmName(Sha256Hash.ALGORITHM_NAME);
        // 数据库存储的密码字段使用 HEX 还是 BASE64 方式加密
        matcher.setStoredCredentialsHexEncoded(false);
        // 迭代次数
        matcher.setHashIterations(1024);
```

```
        return matcher;
    }
    @Bean
    public SessionsSecurityManager securityManager() {
        return new DefaultWebSecurityManager(userRealm());
    }
}
```

编写 controller 层，定义登入、登出和验证码 API，如下所示：

```
package com.gf.security.shiro.controller;
import com.gf.security.shiro.dto.LoginUserDTO;
import com.wf.captcha.utils.CaptchaUtil;
import org.apache.shiro.SecurityUtils;
import org.apache.shiro.authc.*;
import org.apache.shiro.subject.Subject;
import org.springframework.http.HttpStatus;
import org.springframework.http.ResponseEntity;
import org.springframework.web.bind.annotation.*;
import javax.servlet.http.HttpServletRequest;
import javax.servlet.http.HttpServletResponse;
import javax.validation.Valid;
import static com.gf.common.GfResponseEntity.self;
import static org.springframework.http.ResponseEntity.status;
@RestController
@RequestMapping("/")
public class LoginController {
    /**
     * 登录 API
     */
    @PostMapping("/login")
    public ResponseEntity doLogin(@Valid @RequestBody LoginUserDTO params, HttpServletRequest req) {
        if (!CaptchaUtil.ver(params.getCaptcha(), req)) {
            //CaptchaUtil.clear(request);
            return self().msg("验证码不正确").build(null, status(HttpStatus.UNAUTHORIZED));
        }
        try {
            UsernamePasswordToken token = new UsernamePasswordToken(params.getUsername(), params.getPassword());
```

```java
            token.setRememberMe(true);
            Subject curUser = SecurityUtils.getSubject();
            curUser.login(token);
            // 生成令牌
//          Map<String, String> map = Collections.singletonMap("access_token", TokenUtils.generateToken());
            return self().msg("登录成功").ok();
        } catch (IncorrectCredentialsException ice) {
            return self().msg("密码错误").build(null, status(HttpStatus.UNAUTHORIZED));
        } catch (UnknownAccountException uae) {
            return self().msg("账号不存在").build(null, status(HttpStatus.UNAUTHORIZED));
        } catch (LockedAccountException e) {
            return self().msg("账号被锁定").build(null, status(HttpStatus.UNAUTHORIZED));
        } catch (ExcessiveAttemptsException eae) {
            return self().msg("操作频繁，请稍后再试").build(null, status(HttpStatus.UNAUTHORIZED));
        }
    }
    /**
     * 退出登录
     */
    @GetMapping("/logout")
    public ResponseEntity logout() {
        Subject curSubject = SecurityUtils.getSubject();
        curSubject.logout();
        return self().msg("已经退出系统！").ok();
    }
    /**
     * 获取验证码
     */
    @GetMapping("/captcha")
    public void captcha(HttpServletRequest request, HttpServletResponse response) throws Exception {
        CaptchaUtil.out(request, response);
    }
```

}

编写LoginUserDTO类来接收入参,并添加了字段校验,其中的数字401表示HTTP状态码,如下所示:

```
package com.gf.security.shiro.dto;
import lombok.Data;
import javax.validation.constraints.NotEmpty;
@Data
public class LoginUserDTO {
    @NotEmpty(message = "账号或密码不能为空#401")
    private String username;
    @NotEmpty(message = "账号或密码不能为空#401")
    private String password;
    @NotEmpty(message = "验证码不能为空#401")
    private String captcha;
}
```

编写User实体类,用于与数据表shiro_user建立映射,如下所示:

```
package com.gf.security.shiro.entity;
import lombok.Data;
import javax.persistence.Entity;
import javax.persistence.Id;
import javax.persistence.Table;
import java.io.Serializable;
@Entity
@Data
@Table(name = "shiro_user")
public class User implements Serializable {
    public User() {}
    public User(String username) {
        this.username = username;
    }
    @Id
    private Long id;
    private String username;
    private String password;
    private int status;
    private String salt;
}
```

编写UserRepository接口,并继承JpaRepository,以利用现成的增删改查功能,如下所示:

```
package com.gf.security.shiro.repository;
```

```
import com.gf.security.shiro.entity.User;
import org.springframework.data.jpa.repository.JpaRepository;
public interface UserRepository extends JpaRepository<User, Long>{}
```

编写 ExceptionControllerAdvice 类，本类在章节"Spring MVC：验证"中已经介绍过了，在此略有调整，本类用于入参字段验证失败后的响应提示信息，如图 17-24、图 17-25 所示。

```java
@ResponseBody
@ExceptionHandler(MethodArgumentNotValidException.class)
public ResponseEntity<?> validationBodyException(HttpServletRequest request,
                                                  MethodArgumentNotValidException ex) {
    StringBuilder strBuilder = new StringBuilder();
    BindingResult result = ex.getBindingResult();
    int statusCode = 0;
    if (result.hasErrors()) {
        List<ObjectError> errors = result.getAllErrors();
        statusCode = statusCode(errors);
        errors.forEach(p -> {
            FieldError fieldError = (FieldError) p;
            String msg = cutText(fieldError.getDefaultMessage());
            log.warn("请求参数验证: object{" + fieldError.getObjectName() + "},field{" +
                    fieldError.getField() + "},errorMessage{" + msg + "}");
            strBuilder.append(", ");
            strBuilder.append(fieldError.getField());
            strBuilder.append(": ");
            strBuilder.append(msg);
        });
    }
    if (strBuilder.length() > 0) {
        strBuilder.deleteCharAt(0);
    }
    HttpStatus status = statusCode <= 0 ? getStatus(request) : HttpStatus.valueOf(statusCode);
    String path = request.getRequestURI();
    return new ResponseEntity<>(new CustomErrorType(status.value(), strBuilder.toString(), path),
            status);
}
```

图 17-24 验证失败后响应提示（1）

```java
/**
 * 解析出 HTTP 状态码 statusCode
 */
public int statusCode(List<ObjectError> errors) {
    int statusCode = 0;
    if (errors == null || errors.isEmpty()) {
        return statusCode;
    }
    String msg = errors.get(0).getDefaultMessage();

    if (!StringUtils.hasText(msg)) {
        return statusCode;
    }

    String[] arr = msg.split("#");
    if (arr.length > 1) {
        statusCode = Integer.parseInt(arr[arr.length - 1]);
    }
    return statusCode;
}

/**
 * 去掉尾部的文本
 */
public String cutText(String text) {
    String[] arr = Optional.of(text).orElse("").split("#");
    return arr[0];
}
```

图 17-25 验证失败后响应提示（2）

17.7 本章小结

本章讲解了 Spring Security、OAuth 2 和 Apache Shiro，它们都是重要的安全框架或协议，它们之间相互融合一起构成功能完备的安全系统。除了本章所讲解的内容，还有许多安全领域的技术在此并未涉及，希望大家多多学习，扩宽知识面。

第十八章 缓 存

Spring 框架支持向应用程序透明地添加缓存。缓存抽象的核心是将缓存应用于方法，从而根据缓存中可用的信息减少执行次数。缓存逻辑是透明应用的，不会对调用者造成任何干扰。Spring Boot 自动配置缓存基础设施，只要通过 @EnableCaching 注解启用缓存支持即可。对于基于注解的缓存，Spring 的缓存抽象提供了一组 Java 注解：

1) @Cacheable：触发缓存填充；
2) @CacheEvict：触发缓存清除；
3) @CachePut：在不干扰方法执行的情况下更新缓存；
4) @Caching：重新组合要应用于方法的多个缓存操作；
5) @CacheConfig：在类级别共享一些与缓存相关的公共设置。

有关 Spring 框架的缓存模块的内容，请访问以下网址：
https://docs.spring.io/spring/docs/5.2.7.RELEASE/spring-framework-reference/integration.html#cache

有关 Spring Boot 的缓存模块的内容，请访问以下网址：
https://docs.spring.io/spring-boot/docs/2.3.1.RELEASE/reference/html/spring-boot-features.html#boot-features-caching

18.1 @Cacheable

@EnableCaching 和 @Cacheable 注解是从 Spring 框架 3.1 开始提供的，通过这两个注解就可以轻松地实现数据的缓存功能。我们通过下面的示例来介绍这两个注解的使用方式。

首先，新建区域数据表 s_area，并向其中插入数据，相应的 SQL 语句，如下所示：

```sql
CREATE TABLE `s_area` (
  `id` int(11) NOT NULL AUTO_INCREMENT,
  `name` varchar(50) DEFAULT '' COMMENT '区域名称',
  `code` varchar(20) DEFAULT '' COMMENT '区域编码',
  `parent_id` int(11) DEFAULT NULL COMMENT '上级 ID',
  `province_name` varchar(50) DEFAULT '' COMMENT '省份名称',
  `province_code` varchar(20) DEFAULT '' COMMENT '省份编码',
  `city_name` varchar(50) DEFAULT '' COMMENT '市/地区名称',
  `city_code` varchar(20) DEFAULT '' COMMENT '市/地区编码',
  `county_name` varchar(50) DEFAULT '' COMMENT '区县名称',
  `county_code` varchar(20) DEFAULT '' COMMENT '区县编码',
  `township_name` varchar(50) DEFAULT '' COMMENT '乡镇名称',
  `township_code` varchar(20) DEFAULT '' COMMENT '乡镇编码',
  `level` smallint(6) DEFAULT NULL COMMENT '级别',
  PRIMARY KEY (`id`)
```

) ENGINE=InnoDB AUTO_INCREMENT=6 DEFAULT CHARSET=utf8mb4;

INSERT INTO `s_area` VALUES (1,'北京','000001',0,'','','','','','','',0);
INSERT INTO `s_area` VALUES (2,'天津','000002',0,'','','','','','','',0);
INSERT INTO `s_area` VALUES (3,'广东省','000003',0,'','','','','','','',0);
INSERT INTO `s_area` VALUES (4,'河北省','000020',0,'','','','','','','',0);
INSERT INTO `s_area` VALUES (5,'河南省','000030',0,'','','','','','','',0);

然后，在 application.yml 配置文件中添加配置项，如图 18-1 所示。

```yaml
spring:
  datasource:
    url: jdbc:mysql://localhost:3306/gf_platform2?serverTimezone=Asia/Shanghai&useUnicode=true&characterEncoding=utf8
    username: root
    password: 123456
    type: com.zaxxer.hikari.HikariDataSource   # 配置数据连接池，这是默认的

# 在控制台中打印 SQL 语句。
mybatis-plus:
  configuration:
    log-impl: org.apache.ibatis.logging.stdout.StdOutImpl
```

图 18-1 application.yml 配置文件

在主类中添加 @EnableCaching 注解，如下所示：

```java
package com.gf.caching;
import org.mybatis.spring.annotation.MapperScan;
import org.springframework.boot.SpringApplication;
import org.springframework.boot.autoconfigure.SpringBootApplication;
import org.springframework.cache.annotation.EnableCaching;
@EnableCaching
@SpringBootApplication
@MapperScan({"com.gf.*.mapper"})
public class CachingApplication {
    public static void main(String[] args) {
        SpringApplication.run(CachingApplication.class, args);
    }
}
```

controller 层、service 层和 dao 层等的代码都可以利用 MyBatis Plus 的自动生成功能基于数据库表生成相应的代码。有关详情，请查看"MyBatis Plus：代码自动生成"。以下内容中只列出一些关键代码块。有关 service 层的代码，如下所示：

```java
package com.gf.caching.service.impl;
import com.baomidou.mybatisplus.core.conditions.query.QueryWrapper;
import com.baomidou.mybatisplus.extension.service.impl.ServiceImpl;
import com.gf.caching.entity.Area;
```

```java
import com.gf.caching.mapper.AreaMapper;
import com.gf.caching.service.IAreaService;
import org.springframework.cache.annotation.Cacheable;
import org.springframework.stereotype.Service;
import java.util.List;
@Service
public class AreaServiceImpl extends ServiceImpl<AreaMapper, Area>
implements IAreaService {
    @Cacheable("area_sublist")
    @Override
    public List<Area> sublist(Integer parentId) {
        QueryWrapper queryWrapper = new QueryWrapper();
        queryWrapper.select("id,name,code,parent_id")
            .eq("parent_id", parentId);
        return list(queryWrapper);
    }
}
```

有关 controller 层的代码，如下所示：

```java
package com.gf.caching.controller;
import com.gf.caching.entity.Area;
import com.gf.caching.service.IAreaService;
import org.springframework.http.ResponseEntity;
import org.springframework.web.bind.annotation.*;
import java.util.List;
import static com.gf.common.GfResponseEntity.self;
@RestController
@RequestMapping("/caching/area")
public class AreaController {
    private final IAreaService service;
    public AreaController(IAreaService service) {
        this.service = service;
    }
    /**
     * 根据上级 ID，查询所有下级数据
     */
    @GetMapping("/sublist/{parentId}")
    public ResponseEntity sublist(@PathVariable Integer parentId) {
        List<Area> list = service.sublist(parentId);
        return self().ok(list);
```

        }
}

经过以上配置和编写代码后，启动应用程序，并访问 URL，如下所示：
http://localhost:8080/caching/area/sublist/0
响应内容，如图 18-2 所示。

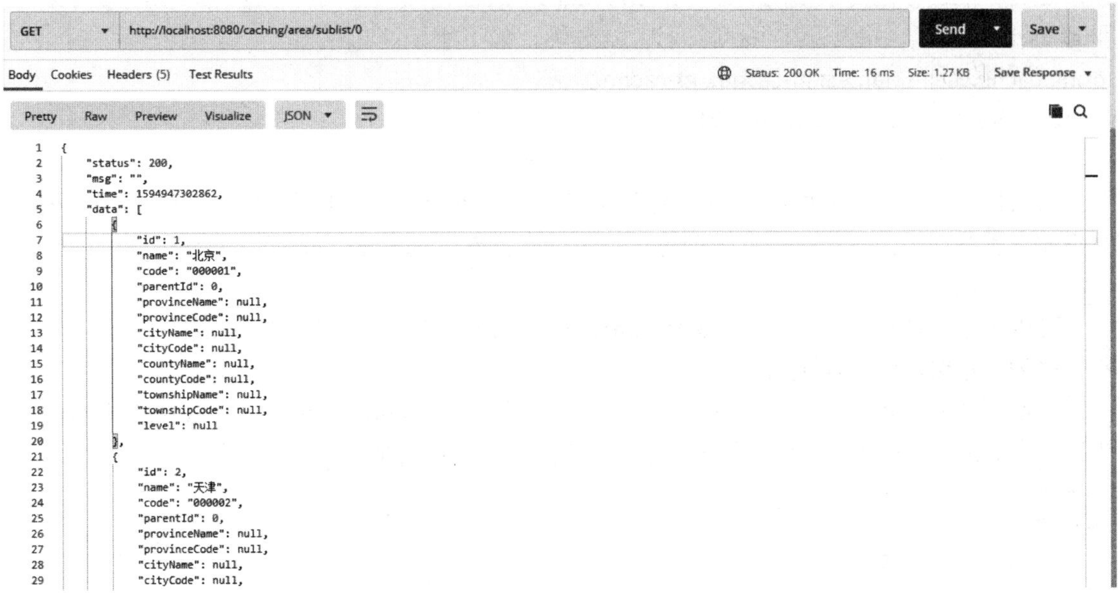

图 18-2 响应内容

控制台输出，如图 18-3 所示。

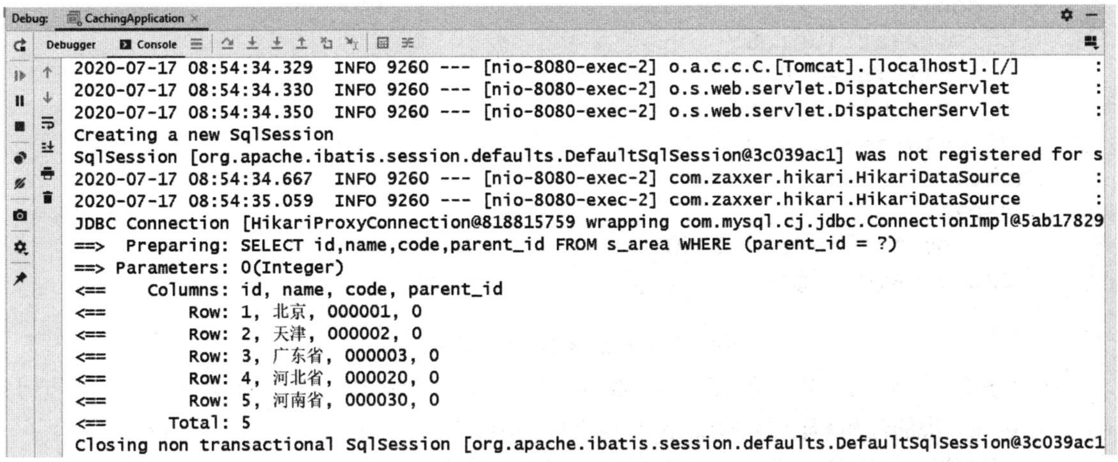

图 18-3 控制台输出

第二次执行相同的请求时，就不再访问数据库了，而是直接返回缓存中的数据。另外，如果需要在某些环境中完全禁用缓存，请添加配置项，如下所示：

spring.cache.type=none

18.2 Redis 缓存

当应用程序部署多个节点时，默认的 @Cacheable 方式的缓存数据也会存在多份。要解决该问题，可以为应用程序配置 Redis 缓存。可以通过设置 spring.cache-names 属性在启动时创建指定名称的缓存，可以使用 spring.cache.redis.* 属性修改缓存的各项配置。创建 gf-cache1 和 gf-cache2 缓存，并设置其生存时间为 10 分钟，如下所示：
spring.cache.cache-names: gf-cache1,gf-cache2
spring.cache.redis.time-to-live: 600000

默认情况下，会给键添加一个前缀，这样可以防止两个单独的缓存使用相同的键。在 Redis 中不能有重叠的键，也不能返回无效值。如果创建自定义的 RedisCacheManager，我们强烈建议启用此设置。

可以通过添加自定义的 RedisCacheConfiguration bean 来完全控制默认配置。如果要自定义默认序列化策略，这可能很有用。

要使用 Redis 缓存，请在 pom.xml 配置文件中添加 spring-boot-starter-data-redis 依赖项。然后，在 application.yml 配置文件中添加配置项，如图 18-4 所示。

```yaml
---
spring:
  profiles: dev
  redis:
    url: redis://:123456@192.168.226.128:6379
    jedis:
      pool:
        max-active: 8
        max-idle: 8
        max-wait: -1ms
        min-idle: 0
  cache:
    redis:
      time-to-live: 600000  #单位：秒，10分钟。
    cache-names: gf-cache1,gf-cache2  #缓存名称
```

图 18-4 Redis 缓存生存时间

从上图中，我们可以看出所有 Redis 缓存的生存时间都为 10 分钟。配置完成之后，把上一节中 service 层的注解修改为 @Cacheable("gf-cache1")。这样就可以开始使用 Redis 缓存了。

如果需要对 Redis 缓存做更进一步的配置调整，Spring Boot 也提供相应的方法，可以注册一个 RedisCacheManagerBuilderCustomizer bean，如下所示：

```
package com.gf.caching.config;
import org.springframework.boot.autoconfigure.cache.RedisCacheManagerBuilderCustomizer;
```

```java
import org.springframework.context.annotation.Bean;
import org.springframework.context.annotation.Configuration;
import org.springframework.data.redis.cache.RedisCacheConfiguration;
import java.time.Duration;
@Configuration
public class WebConfig {
    @Bean
    public RedisCacheManagerBuilderCustomizer myRedisCacheManagerBuilderCustomizer() {
        return (builder) ->
                builder.withCacheConfiguration("gf-cache1",
                        RedisCacheConfiguration.defaultCacheConfig()
                                .entryTtl(Duration.ofMinutes(30)))
                        .withCacheConfiguration("gf-cache2",
                                RedisCacheConfiguration.defaultCacheConfig()
                                        .entryTtl(Duration.ofMinutes(1)));
    }
}
```

18.3 本章小结

本章讲解了如何在 Spring Boot 应用程序中使用缓存，以此来提高 API 的响应速度，减少数据库的压力。当然，在实际的项目开发中，可用的缓存技术是比较多的，一定要结合具体的业务场景，选择适合的。

第十九章 作业调度

本章讲解作业调度框架的内容，以 Quartz 为例进行讲解。Quartz 是一个功能丰富、开源的作业调度库，可以集成到几乎任何 Java 应用程序中——从最小的独立应用程序到最大的电子商务系统。Quartz 可以创建简单或复杂的任务计划，用于执行数十个、数百个甚至数万个作业；这些作业的任务被定义为标准 Java 组件，可以执行几乎任何可编程执行的任务。Quartz 调度器包含许多企业级功能特性，例如：对 JTA 事务和集群的支持。

19.1 @Scheduled

从 3.0 开始，Spring 框架提供了一个简单的作业调度模块，它使用起来非常的简单，只需 @EnableScheduling 注解和 @Scheduled 注解。示例代码，如下所示：

```java
package com.gf.quartz;
import org.springframework.boot.SpringApplication;
import org.springframework.boot.autoconfigure.SpringBootApplication;
import org.springframework.scheduling.annotation.EnableScheduling;
@EnableScheduling
@SpringBootApplication
public class QuartzApplication {
    public static void main(String[] args) {
        SpringApplication.run(QuartzApplication.class, args);
    }
}
```

```java
package com.gf.quartz.job;
import org.springframework.scheduling.annotation.Scheduled;
import org.springframework.stereotype.Component;
import java.util.Date;
@Component
public class MyScheduledJob {
    /**
     * 每 10 秒钟执行一次
     */
    @Scheduled(cron = "0/10 * * * * ?")
    public void job1() {
        System.out.println("Scheduled" + new Date());
    }
}
```

Spring 中 @EnableScheduling 和 @Scheduled 标注的定时任务默认是单线程执行的，有可能任务之间发生阻塞现象，导致有些任务不能如期执行。使用 @Async 和 @EnableAsync 注解可以让任务异步地执行。但是还需要注意：如果任务执行时间大于间隔时间，可能出现任务交叉执行的现象。

19.2 Quartz

Spring Boot 为使用 Quartz 调度器提供了多种便利，包括 spring-boot-starter-quartz 依赖项。如果 Quartz 可用，则会自动配置 Scheduler 的实例（通过 SchedulerFactoryBean 抽象类）。

以下类型的 bean 会自动拾取并与 Scheduler 关联：
1) Jobdetail：它可以定义特定作业，可以使用 JobBuilder API 构建 JobDetail 实例；
2) Calendar：日历；
3) Trigger：定义何时触发特定作业。

Quartz 官网，如下所示：
https://www.quartz-scheduler.org
想要查看 Quartz 源代码，请访问以下网址：
https://github.com/quartz-scheduler/quartz
有关 Quartz 的 cron 表达式的详细信息，请自行通过搜索引擎检索，或查看官方教程文档，如下所示：
https://www.quartz-scheduler.org/documentation/quartz-2.3.0/tutorials/tutorial-lesson-06.html

19.3 Quartz 示例

要想在应用中引入 Quartz，只需添加以下依赖项：
```
<dependency>
    <groupId>org.springframework.boot</groupId>
    <artifactId>spring-boot-starter-quartz</artifactId>
</dependency>
```
在生成环境中，需要把 Quartz 的所有表放到外置的独立数据库中，添加数据库相关的依赖项，如下所示：
```
<dependency>
    <groupId>org.springframework.boot</groupId>
    <artifactId>spring-boot-starter-jdbc</artifactId>
</dependency>
<dependency>
    <groupId>mysql</groupId>
```

```xml
<artifactId>mysql-connector-java</artifactId>
</dependency>
```

在 application.yml 配置文件中，添加配置项，如图 19-1 所示。

```yaml
spring:
  datasource:
    url: jdbc:mysql://localhost:3306/gf_task?serverTimezone=Asia/Shanghai&useUnicode=true&characterEncoding=utf8
    username: root
    password: 123456
  quartz:
    overwrite-existing-jobs: true
    job-store-type: jdbc
    jdbc:
      initialize-schema: never # 可用的值：always、embedded（默认值）和 never
    properties:
      org:
        quartz:
          scheduler:
            instanceId: AUTO
          jobStore:
#            class: org.quartz.impl.jdbcjobstore.JobStoreTX
#            driverDelegateClass: org.quartz.impl.jdbcjobstore.StdJDBCDelegate
            useProperties: false
            tablePrefix: qrtz_   #数据库表前缀
            misfireThreshold: 60000
            clusterCheckinInterval: 5000
            isClustered: true
          threadPool:
            class: org.quartz.simpl.SimpleThreadPool
            threadCount: 10
            threadPriority: 5
            threadsInheritContextClassLoaderOfInitializingThread: true
```

图 19-1 application.yml 配置文件

第一次启动应用程序时，可以让 Quartz 自动创建数据表，设置配置项，如下所示：

spring.quartz.jdbc.initialize-schema: alway

定义 QuartzJobBean 抽象类的实现类，其中包含具体的业务代码，如下所示：

```java
package com.gf.quartz.job;
import lombok.extern.slf4j.Slf4j;
import org.quartz.JobExecutionContext;
import org.quartz.JobExecutionException;
import org.springframework.scheduling.quartz.QuartzJobBean;
@Slf4j
public class MyJob extends QuartzJobBean {
    @Override
    protected void executeInternal(JobExecutionContext context)
            throws JobExecutionException {
        log.info("Job 组/名称：" + context.getJobDetail().getKey() + ", 触发器：" + context.getTrigger());
    }
}
```

定义 JobDetail bean 以关联 MyJob 类，定义 Trigger bean 以指定业务代码何时执行，如下所示：

```
package com.gf.quartz.common;
import com.gf.quartz.job.MyJob;
import org.quartz.*;
import org.springframework.context.annotation.Bean;
import org.springframework.context.annotation.Configuration;
@Configuration
public class QuartzConfig {
    private final String jobName = "myJob";
    private final String jobGroupName = "group01";
    private final String jobTime = "0/10 * * * * ?";
    @Bean
    public JobDetail getJobDetail() {
        return JobBuilder.newJob(MyJob.class)
                .withIdentity(jobName, jobGroupName)
                .storeDurably().build();
    }
    @Bean
    public Trigger getTrigger() {
        return TriggerBuilder.newTrigger()
                .withIdentity(jobName + "T", jobGroupName + "T")
                .withSchedule(CronScheduleBuilder
                        .cronSchedule(jobTime))
                .forJob(jobName, jobGroupName).startNow().build();
}}
```

除了以上这种 Quartz 的使用方式之外，可以使用 Scheduler 来封装一个 service 层类，其中包括 Job 的增删改查方法等，这样就可以统一管理对 Job 的各种各样的操作。然后，定义一个 CommandLineRunner 接口的实现类，它负责向 service 层类的实例添加 Job 等。通过这些处理后，我们只需要关注 Job 的定义，而不需要关注 Jobdetail 和 Trigger。有了 service 层类，可以进一步定义 controller 层类，以提供相应的 web API。

19.4 其他

ElasticJob

ElasticJob 是基于 Zookepper 和 Quartz 开发并开源的一个 Java 分布式调度解决方案，它解决了 Quartz 不支持分布式的弊端。它由 2 个相互独立的子项目组成，如下所示：
1) ElasticJob-Lite 定位为轻量级无中心化解决方案，使用 jar 的形式提供分布式任务的协调服务；

2) ElasticJob-Cloud 使用 Mesos 的解决方案，额外提供资源治理、应用分发以及进程隔离等服务。

ElasticJob 的各个产品使用统一的作业 API，开发者仅需要一次开发，即可多处部署。ElasticJob 的官网，如下所示：

https://shardingsphere.apache.org/elasticjob

想要查看 ElasticJob 的源代码，请访问以下网址：

https://github.com/apache/shardingsphere-elasticjob

想要查看 ElasticJob 的中文文档，请访问以下网址：

https://shardingsphere.apache.org/elasticjob/legacy/lite-2.x/00-overview/

XXL-JOB

XXL-JOB 是一个开箱即用的开源的分布式任务调度平台，其核心设计目标是开发迅速、学习简单、轻量级和易扩展。想要查看 XXL-JOB 的源代码，请访问以下网址：

https://github.com/xuxueli/xxl-job

想要查看 XXL-JOB 的中文文档，请访问以下网址：

https://www.xuxueli.com/xxl-job

19.5 本章小结

本章讲解了作业调度的相关知识点，以 Spring 框架的 @Scheduled 和 Quartz 为示例来讲解的。希望本章中的示例内容能让大家轻松学习作业调度的知识点，并在自己的项目中能很好地应用这些知识点。

第二十章 其他技术

20.1 验证码

验证码是一种区分用户是计算机还是人的公共全自动程序。它可以防止恶意注册/登录、刷票和论坛灌水。验证码的分类有：静态图片验证码、gif 动画验证码、滑动验证码、拼图验证码、点选验证码、手机短信验证码、手机语音验证码和视频验证码等等。

20.1.1 纯前端验证码

先来介绍一个 GitHub 上的纯前端验证码，如下所示：
https://github.com/Hibear/verify
它提供了普通验证码、滑动验证码、点选验证码，如如图 20-1 至 20-3 所示。

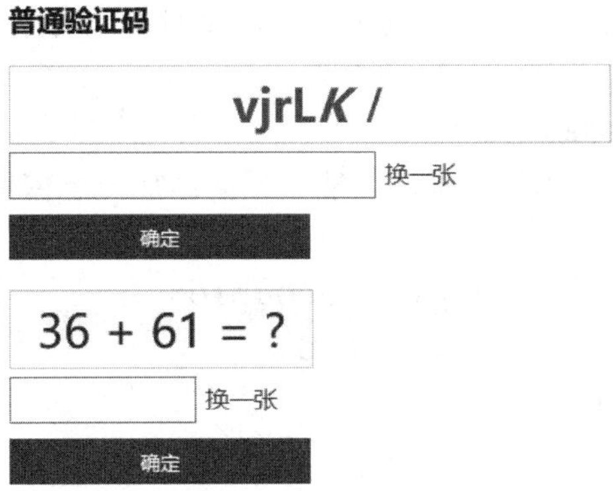

图 20-1 普通验证码

图 20-2 滑动验证码

图 20-3 点选验证码

如需在项目中使用纯前端验证码模块，可登录 GitHub 查看源代码及其发布的版本。

20.1.2 Kaptcha

Kaptcha 是谷歌公司开源的可高度配置的验证码工具。它有众多可配置的参数，可以满足不同的需求。在 GitHub 有其源代码，如下所示：

https://github.com/penggle/kaptcha

想要在 Spring Boot 项目中添加验证码功能，只需添加一个依赖项，如下所示：

```xml
<dependency>
    <groupId>com.github.penggle</groupId>
    <artifactId>kaptcha</artifactId>
    <version>2.3.2</version>
</dependency>
```

添加依赖项后，编写了一个最简单的测试代码，如下所示：

```java
package com.gf.other;
import com.google.code.kaptcha.impl.DefaultKaptcha;
import com.google.code.kaptcha.util.Config;
import org.junit.jupiter.api.Test;
import javax.imageio.ImageIO;
import java.awt.image.BufferedImage;
import java.io.File;
import java.io.IOException;
import java.util.Properties;
public class KaptchaTest {
    @Test
    public void test1() throws IOException {
        // 添加配置项：只使用默认值
```

```
        Config config = new Config(new Properties());
        DefaultKaptcha defaultKaptcha = new DefaultKaptcha();
        defaultKaptcha.setConfig(config);
        // 生成字符串
        String text = defaultKaptcha.createText();
        // 把生成的字符串按照默认样式绘制成验证码图片
        BufferedImage bi = defaultKaptcha.createImage(text);
        String pathName = "./qrcode/yanzhengma/code" +
System.currentTimeMillis() + ".jpg";
        File outFile = new File(pathName);
        ImageIO.write(bi, "jpg", outFile);
    }
}
```

有关 Kaptcha 的详细配置，可以查看类 com.google.code.kaptcha.Constants，这个类中包含的配置项是最全的。每个配置项的说明，如表 20-1 所示。

表 20-1 com.google.code.kaptcha.Constants 类中配置项属性

配置项	说明	默认值
*.border	图片边框，合法值：yes、no	yes
*.border.color	边框颜色，合法值： r,g,b (and optional alpha) 或者 white、black、blue。	black
*.image.width	图片宽	200px
*.image.height	图片高	50px
*.producer.impl	图片实现类	DefaultKaptcha
*.textproducer.impl	文本实现类	DefaultTextCreator
*.textproducer.char.string	验证码字符从此集合中获取	abcde2345678gfynmnpwx
*.textproducer.char.length	验证码长度	5
*.textproducer.font.names	字体	Arial、Courier
*.textproducer.font.size	字体大小	40px
*.textproducer.font.color	字体颜色，合法值：r,g,b 或者 white,black,blue.	black
*.textproducer.char.space	文字间隔	2
*.noise.impl	干扰实现类	DefaultNoise
*.noise.color	干扰颜色，合法值：r,g,b 或者 white、black、blue。	black
*.obscurificator.impl	水纹：WaterRipple 鱼眼：FishEyeGimpy 阴影：ShadowGimpy	WaterRipple

续表

配置项	说明	默认值
*.background.impl	背景实现类	DefaultBackground
*.background.clear.from	背景颜色渐变，开始颜色	light grey
*.background.clear.to	背景颜色渐变，结束颜色	white
*.word.impl	文字渲染器	DefaultWordRenderer
*.session.key	session key	KAPTCHA_SESSION_KEY
*.session.date	session date	KAPTCHA_SESSION_DATE

备注：* 表示为 kaptcha。

在 Spring Boot 项目中，使用 starter 方式接入第三方技术，这是最为便捷的。为此，在码云上找到一个 Kaptcha 的 starter，如下所示：

https://gitee.com/baomidou/kaptcha-spring-boot-starter

要引入该 starter，只需添加依赖项，如下所示：

```xml
<dependency>
    <groupId>com.baomidou</groupId>
    <artifactId>kaptcha-spring-boot-starter</artifactId>
    <version>1.1.0</version>
</dependency>
```

该 starter 在码云上有快速接入的步骤文档，其中包括获取验证码、校验验证码的示例接口。获取验证码时，验证码和时间戳会保存在对应的会话中，以便用于之后的验证。默认情况下，Kaptcha 接口的实现是 GoogleKaptcha 类，它是基于 servlet 构建的，因此，该 starter 不适用于 WebFlux 应用。要想让该 starter 适用于 WebFlux 应用，请自定义 Kaptcha 接口的实现 bean。

20.2 二维码

二维码（QR Code，Quick Response Code）是近几年在移动设备上非常流行的一种编码方式，它用途广泛：账号登录、手机支付、信息获取、优惠促销等。与条形码相比，它的优点有：信息容量大、容错能力强等。

本节所讲解的二维码功能是借助 GitHub 上的开源项目 zxing 来展开的，有关源码及更详细的文档信息，请访问网址，如下所示：

https://github.com/zxing/zxing

要想在项目中引入二维码功能，请添加相应的依赖项，如下所示：

```xml
<dependency>
    <groupId>com.google.zxing</groupId>
    <artifactId>core</artifactId>
    <version>3.4.0</version>
</dependency>
```

```xml
<dependency>
    <groupId>com.google.zxing</groupId>
    <artifactId>javase</artifactId>
    <version>3.4.0</version>
</dependency>
```

生成二维码的示例代码，如下所示：

```java
/**
 * 生成二维码
 */
private static BufferedImage generateQRCode(String content) throws Exception {
    Hashtable hints = new Hashtable();
    hints.put(EncodeHintType.ERROR_CORRECTION, ErrorCorrectionLevel.H);
    hints.put(EncodeHintType.CHARACTER_SET, CHARSET);
    hints.put(EncodeHintType.MARGIN, 1);
    BitMatrix bitMatrix = new MultiFormatWriter().encode(content, BarcodeFormat.QR_CODE, QRCODE_SIZE, QRCODE_SIZE,
            hints);
    int width = bitMatrix.getWidth();
    int height = bitMatrix.getHeight();
    BufferedImage image = new BufferedImage(width, height, BufferedImage.TYPE_INT_RGB);
    for (int x = 0; x < width; x++) {
        for (int y = 0; y < height; y++) {
            image.setRGB(x, y, bitMatrix.get(x, y) ? 0xFF000000 : 0xFFFFFFFF);
        }
    }
    return image;
}
```

识别二维码的示例代码，如下所示：

```java
/**
 * 识别二维码
 */
public static String extractQRCode(String imgPath) {
    String content = null;
    BufferedImage image;
    try {
```

```
        image = ImageIO.read(new File(imgPath));
        LuminanceSource source = new
BufferedImageLuminanceSource(image);
        Binarizer binarizer = new HybridBinarizer(source);
        BinaryBitmap binaryBitmap = new BinaryBitmap(binarizer);
        Map<DecodeHintType, Object> hits = new HashMap<>();
        hits.put(DecodeHintType.CHARACTER_SET,
StandardCharsets.UTF_8);
        Result result = new MultiFormatReader().decode(binaryBitmap,
hits);
        content = result.getText();
    } catch (IOException e) {
        e.printStackTrace();
    } catch (NotFoundException e) {
        e.printStackTrace();
    }
    return content;
}
```

以上示例代码只是二维码的简单处理，有关在二维码中间放置图片等内容，还请多多查看 GitHub 上的文档及搜索引擎中的信息。

20.3 Swagger

Swagger 是遵守 OpenAPI 规范的一套开源的工具集。它可以用于大规模地设计和文档化 API，并且提供了测试 API 的界面。Swagger 为服务端接口文档的格式提供了 OpenAPI 规范的实现。开发者编写描述接口的 YAML 或 JSON 格式的 Swagger 接口文档，然后 Swagger UI 将该文档解析成最终向用户展现的接口文档。有关 Swagger 各个项目的源代码，请访问以下网址：

https://github.com/swagger-api

20.3.1 Swagger 子项目

Swagger Codegen

Swagger Codegen 可以通过为任何 API 生成服务器存根和客户端 SDK 来简化构建过程，这些 API 是用 OpenAPI（以前称为 Swagger）规范定义的。有了它，团队可以更好地关注 API 的实现和使用。

可以把 Swagger Codegen 下载到自己电脑上来使用，也可以在 Swagger Hub 中使用它。其中，在 Swagger Hub 中使用 Swagger Codegen 时，可以使用更多的功能，还可以获得更多的支持。有关 Swagger Codegen 的更多详细信息，请访问以下网址：

https://swagger.io/tools/swagger-codegen

Swagger UI

Swagger UI 允许任何人在没有任何实现逻辑的情况下，可视化 API 资源并与之交互。这些 API 资源是根据 OpenAPI（以前称为 Swagger）规范自动生成的，可视化文档使后端实现和客户端消费变得容易。有关 Swagger UI 的更多详细信息，请访问以下网址：

https://swagger.io/tools/swagger-ui

Swagger UI 界面，如如图 20-4 所示。

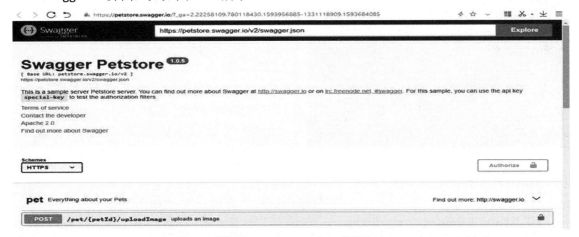

图 20-4 Swagger UI 界面

Swagger Editor

Swagger Editor 是第一个完全专用于 API（基于 OpenAPI 的）的开源编辑器，可以在其上设计、描述和文档化 API。Swagger Editor 是开始使用 OpenAPI 规范（以前称为 Swagger）的简单方法，支持 Swagger 2.0 和 OpenAPI 3.0。有关 Swagger Editor 的更多详细信息，请访问以下网址：

https://swagger.io/tools/swagger-editor

Swagger Editor 界面，如如图 20-5 所示。

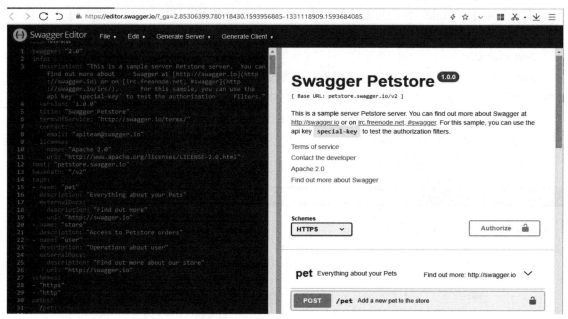

图 20-5 Swagger Editor 界面

Swagger Inspector

在 Swagger Inspector 的界面中，可以轻松测试和记录 API，界面如图 20-6 所示。

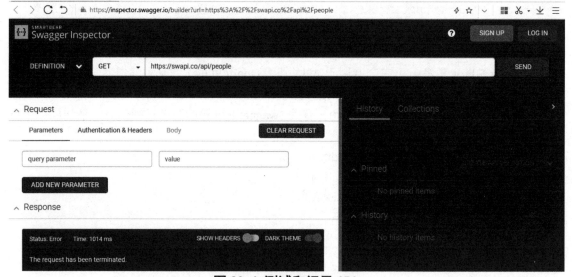

图 20-6 测试和记录 API

20.3.2 SpringFox-Swagger

Springfox-Swagger 为使用 Spring 构建的 API 自动化地生成 JSON 格式的 Swagger 接口文档。有了 Springfox-Swagger 的支持，从代码到 Swagger 接口文档，再到界面展现，整个过程完全实现了自动化。从此，后端读者再也不用手动编写接口文档了，再也无需担心给前端读者的接口文档与实际情况不一致了，并且想要更新 Swagger 接口文档只需启动应用即可。有关 Springfox 的源代码等更多详细信息，请访问以下网址：

https://github.com/springfox/springfox

要想在 Spring Boot 应用程序中引入 Springfox-Swagger，只需添加相应依赖项，如下所示：

```xml
<dependency>
    <groupId>io.springfox</groupId>
    <artifactId>springfox-swagger2</artifactId>
    <version>2.9.2</version>
</dependency>
<dependency>
    <groupId>io.springfox</groupId>
    <artifactId>springfox-swagger-ui</artifactId>
    <version>2.9.2</version>
</dependency>
```

添加依赖项后，我们启动应用程序，启动成功后访问 URL：http://localhost:8081/gf/v2/api-docs。JSONx 响应内容如图 20-7 所示。

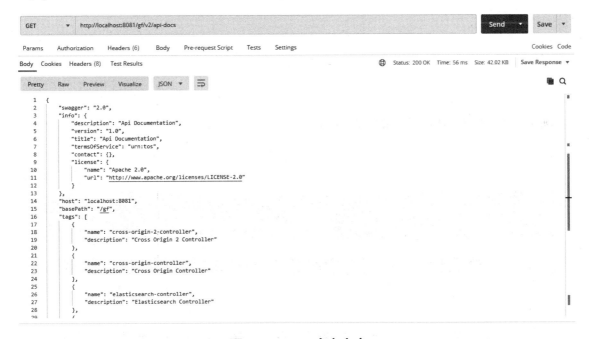

图 20-7 JSON 响应内容

从上图中可以看出，响应内容是一个比较大的 JSON 字符串，其中，包含了 controller 层的接口信息。接着，访问地址：http://localhost:8081/gf/swagger-ui.html，打开一个 API 信息界面，如图 20-8 所示。

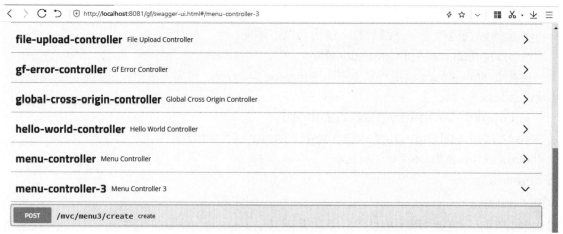

图 20-8 API 信息

上图中所展现的 API 信息，是前一个 URL 的响应 JSON 的可视化。我们单击其中一个 API 进行测试，如图 20-9 所示。

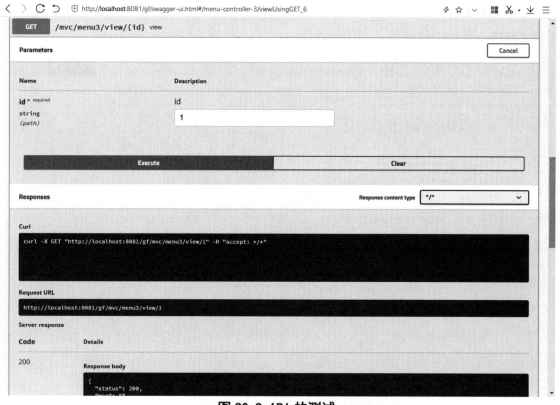

图 20-9 API 的测试

填写相应的参数，然后调用该 API，稍等片刻后，从服务端响应了 JSON 数据。在这个页面，我们完成了一个简单 API 的测试。

在整个过程中，我们没有对现有的 API 代码做任何的修改，通过 Springfox-Swagger 自动完成了 API 文档的生成及其展现，并可以允许在页面中进行测试。若想要更多地控制展现的内容，请查看上文给出的网页内容。

20.4 Excel 处理

20.4.1 Apache POI

Apache POI 是处理微软办公文档（Excel 和 Word 等）的 Java API，它是免费开源的。它的官方网址，如下所示：

http://poi.apache.org

Apache POI 的源代码，如下所示：

https://github.com/apache/poi

Apache POI 提供了许多组件，它们分别处理不同的文件类型，各个模块的功能说明，如表 20-2 所示。

表 20-2 Apache POI 组件列表

组件	说明
POIFS	OLE2 文件系统
HPSF	OLE2 属性集合
HSSF	提供读写 Excel XLS 格式文件的功能
HSLF	提供读写 PowerPoint PPT 格式文件的功能
HWPF	提供读写 Word DOC 格式文件的功能
HDGF	提供读 Visio VSD 格式文件的功能
HPBF	提供读 Publisher PUB 格式文件的功能
HSMF	提供读 Outlook MSG 格式文件的功能
DDF	Escher 公共图
HWMF	WMF 图
OpenXML4J	提供读写 OOXML 格式文件的功能
XSSF	提供读写 Excel XLSX 格式文件的功能
XSLF	提供读写 PowerPoint PPTX 格式文件的功能
XWPF	提供读写 Word DOCX 格式文件的功能
XDGF	提供读写 Visio VSDX 格式文件的功能
Common SL	用来处理 PowerPoint PPT 和 PPTX 格式的文件
Common SS	用来处理 Excel XLS 和 XLSX 格式的文件

如需在 Java 项目中使用 Apache POI，应添加相应依赖项，如下所示：

```
<dependency>
    <groupId>org.apache.poi</groupId>
    <artifactId>poi</artifactId>
    <version>4.1.2</version>
</dependency>
<dependency>
```

```xml
        <groupId>org.apache.poi</groupId>
        <artifactId>poi-ooxml</artifactId>
        <version>4.1.2</version>
</dependency>
```

添加依赖项后，编写测试代码来读写 Excel，如下所示：

```java
package com.gf.other;
import org.apache.poi.ss.usermodel.Row;
import org.apache.poi.xssf.usermodel.XSSFRow;
import org.apache.poi.xssf.usermodel.XSSFSheet;
import org.apache.poi.xssf.usermodel.XSSFWorkbook;
import org.junit.jupiter.api.Test;
import java.io.FileNotFoundException;
import java.io.FileOutputStream;
import java.io.IOException;
import java.util.Arrays;
import java.util.List;
public class ExcelPoiTest {
    /**
     * 从 Excel 文件中读取数据
     */
    @Test
    public void test1() throws IOException {
        String filePath = "./input/xlsx/test.xlsx";
        XSSFWorkbook workbook = new XSSFWorkbook(filePath);
        XSSFSheet sheet = workbook.getSheetAt(0);
        for (Row row : sheet) {
            if (row.getRowNum() == 0) {// 表头信息
                System.out.print(row.getCell(0).getStringCellValue() + "\t");
System.out.println(row.getCell(1).getStringCellValue());
            } else {// 表中的数据
                int num = (int) row.getCell(0).getNumericCellValue();
                System.out.print(num + "\t");
System.out.println(row.getCell(1).getStringCellValue());
            }
        }
    }
    /**
     * 把数据写入到 Excel 文件中
```

```java
*/
@Test
public void test2() {
    String outputFilePath = "./output/xlsx/test" + System.currentTimeMillis() + ".xlsx";
    try (XSSFWorkbook workbook = new XSSFWorkbook();
         FileOutputStream os = new FileOutputStream(outputFilePath);) {
        XSSFSheet sheet = workbook.createSheet();
        // 添加标题
        XSSFRow rowHead = sheet.createRow(0);
        rowHead.createCell(0).setCellValue("序号");
        rowHead.createCell(1).setCellValue("城市");
        // 添加列表数据
        List<String> list = Arrays.asList(new String[]{"石家庄", "青岛", "郑州", "天津", "北京"});
        for (int i = 0; i < list.size(); i++) {
            String city = list.get(i);
            XSSFRow row = sheet.createRow(i + 1);
            row.createCell(0).setCellValue(i + 1);
            row.createCell(1).setCellValue(city);
        }
        workbook.write(os);
    } catch (FileNotFoundException e) {
        e.printStackTrace();
    } catch (IOException e) {
        e.printStackTrace();
    }
}
```

20.4.2 Jxls

Jxls 是基于 Apache POI 开发的对 Excel 进行模板化操作的便捷的工具包。与 Apache POI 相比，Jxls 使用模板代替了大量用于描述样式、格式的 Java 代码，这些代码往往难以调试。要想查看 Jxls 的源代码，请访问以下网址：

https://bitbucket.org/leonate/jxls

想要在 Java 项目中使用 Jxls，只需添加相应依赖项，如下所示：

<dependency>

```xml
    <groupId>org.jxls</groupId>
    <artifactId>jxls</artifactId>
    <version>2.8.1</version>
</dependency>
<dependency>
    <groupId>org.jxls</groupId>
    <artifactId>jxls-poi</artifactId>
    <version>2.8.1</version>
</dependency>
<dependency>
    <groupId>org.jxls</groupId>
    <artifactId>jxls-jexcel</artifactId>
    <version>1.0.9</version>
</dependency>
```

之后，添加 template-list.xlsx 模板文件，如图 20-10 所示。

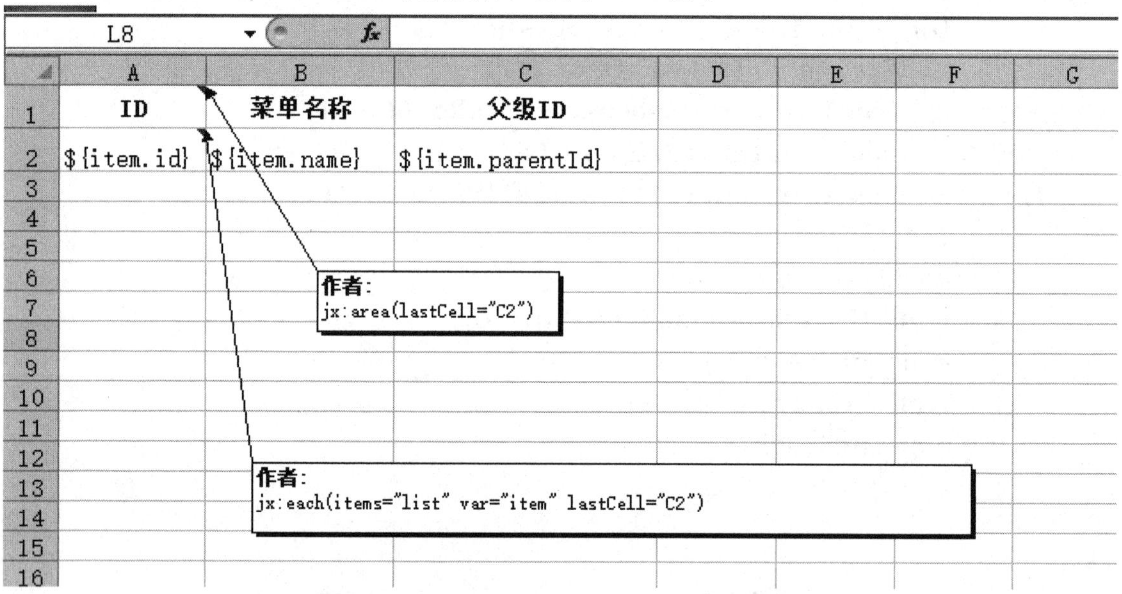

	A	B	C
1	ID	菜单名称	父级ID
2	${item.id}	${item.name}	${item.parentId}

作者:
jx:area(lastCell="C2")

作者:
jx:each(items="list" var="item" lastCell="C2")

图 20-10 template-list.xlsx 模板文件

之后，编写测试代码利用模板把数据导出到 Excel 文件中，如下所示：

```java
import com.gf.other.vo.MenuVO;
import org.junit.jupiter.api.Test;
import org.jxls.common.Context;
import org.jxls.util.JxlsHelper;
import org.springframework.util.ResourceUtils;
import java.io.*;
import java.util.ArrayList;
import java.util.List;
```

```java
public class JxlsTest {
    /**
     * 简单的入门示例:
     * 使用Excel批注的方式把list中的列表数据导出到Excel中。
     */
    @Test
    public void test1() throws FileNotFoundException {
        String resPath = "classpath:templates/xls/template-list.xlsx";
        File templateFile = ResourceUtils.getFile(resPath);
        String pathName = "./output/xls/list" + System.currentTimeMillis() + ".xlsx";
        try (InputStream is = new FileInputStream(templateFile);
             OutputStream os = new FileOutputStream(new File(pathName))) {
            // 准备数据
            List<MenuVO> list = new ArrayList<>();
            list.add(new MenuVO(1L, "菜单管理", 0L));
            list.add(new MenuVO(2L, "日志管理", 0L));
            list.add(new MenuVO(3L, "SQL监控", 0L));
            // 构建数据上下文
            Context context = new Context();
            context.putVar("list", list);
            // 执行导出操作
            JxlsHelper.getInstance().processTemplate(is, os, context);
        } catch (FileNotFoundException e) {
            e.printStackTrace();
        } catch (IOException e) {
            e.printStackTrace();
        }
    }
}
```

执行测试代码后导出的 Excel 表格，如图 20-11 所示。

图 20-11 导出 Excel 表格

20.4.3 EasyExcel

EasyExcel 是阿里巴巴公司开源的基于 Apache POI 开发的一个简单、省内存的读写 Excel 的项目。它在尽可能节约内存的情况下支持读写百兆的 Excel。有关 EasyExcel 的源代码，如下所示：

https://github.com/alibaba/easyexcel

想要查看 EasyExcel 的文档，请访问以下网址：

https://www.yuque.com/easyexcel/doc/easyexcel

要想在项目中引入 EasyExcel，只需添加相应依赖项，如下所示：

```
<dependency>
    <groupId>com.alibaba</groupId>
    <artifactId>easyexcel</artifactId>
    <version>2.2.6</version>
</dependency>
```

EasyExcel 是基于 Apache POI 构建的，具体的依赖关系，如图 20-12 所示。

```
+- com.alibaba:easyexcel:jar:2.2.6:compile
|  +- org.apache.poi:poi:jar:3.17:compile
|  |  +- commons-codec:commons-codec:jar:1.14:compile
|  |  \- org.apache.commons:commons-collections4:jar:4.1:compile
|  +- org.apache.poi:poi-ooxml:jar:3.17:compile
|  |  \- com.github.virtuald:curvesapi:jar:1.04:compile
|  +- org.apache.poi:poi-ooxml-schemas:jar:3.17:compile
|  |  \- org.apache.xmlbeans:xmlbeans:jar:2.6.0:compile
|  |     \- stax:stax-api:jar:1.0.1:compile
|  +- cglib:cglib:jar:3.1:compile
```

图 20-12 EasyExcel 依赖关系

添加完成依赖项后，编写 EasyMenuVO 实体类，如下所示：

```
import com.alibaba.excel.annotation.ExcelIgnore;
import com.alibaba.excel.annotation.ExcelProperty;
```

```java
import lombok.Data;
@Data
public class EasyMenuVO {
    @ExcelProperty("ID（easy）")
    private Long id;
    @ExcelProperty("名称")
    private String name;
    @ExcelProperty("父级ID")
    private Long parentId;
    /**
     * 忽略本字段
     */
    @ExcelIgnore
    private String remark;
    public EasyMenuVO(Long id, String name, Long parentId) {
        this.id = id;
        this.name = name;
        this.parentId = parentId;
    }
}
```

编写测试代码把数据导出到指定的Excel文件中，如下所示：

```java
import com.alibaba.excel.EasyExcel;
import com.gf.other.vo.EasyMenuVO;
import org.junit.jupiter.api.Test;
import java.util.ArrayList;
import java.util.List;
public class EasyExcelTest {
    /**
     * 无需模板的导出操作
     */
    @Test
    public void test1() {
        String pathName = "./output/xls/list-easyexcel" + System.currentTimeMillis() + ".xlsx";
        // 准备数据
        List<EasyMenuVO> list = new ArrayList<>();
        list.add(new EasyMenuVO(1L, "菜单管理", 0L));
        list.add(new EasyMenuVO(2L, "日志管理", 0L));
        list.add(new EasyMenuVO(3L, "SQL监控", 0L));
```

```java
// 执行导出操作
EasyExcel.write(pathName, EasyMenuVO.class).sheet().doWrite(list);
    }
}
```

导出成功后，打开相应的 Excel 文件，如图 20-13 所示。

图 20-13 导出并打开 Excel 文件

20.4.4 ExcelJS

ExcelJS 是一个用 JavaScript 语言编写的开源的处理 Excel 的软件项目。它可以读取、操作电子表格数据和样式，也可以把数据和电子表格样式写入到 XLSX 和 JSON 文件。想要安装它，请执行命令，如下所示：

npm install exceljs

有关 ExcelJS 的源代码及其文档，请访问以下网址：

https://github.com/exceljs/exceljs

有关 ExcelJS 中文文档的详细信息，请访问以下网址：

https://github.com/exceljs/exceljs/blob/master/README_zh.md

20.4.5 SheetJS

SheetJS 是一个 JavaScript 语言编写的开源的，可以解析和编辑每种 Excel 文件格式的软件项目。它可以用于 web 浏览器和服务器环境，并且具有统一 JS 表示的跨格式功能兼容性，以及 ES3/ES5 浏览器与 IE6 的兼容性。SheetJS 分为社区版和专业版，其中社区版是免费开源的。

想要全面地了解 SheetJS，请访问其官网，如下所示：
https://sheetjs.com
有关 SheetJS 的源代码及其文档，请访问以下网址：
https://github.com/SheetJS/sheetjs
有关 SheetJS 社区版的在线示例，请访问以下网址：
https://sheetjs.com/demo

20.4.6 MyExcel

MyExcel 是一个免费开源的，具有导入、导出和加密 Excel 等功能的工具包，它是基于 Apache POI 开发的。有关 MyExcel 的源代码，如下所示：
https://github.com/liaochong/myexcel

MyExcel 具有的优势，如下所示：
1) 可生成任意复杂表格；
2) 支持 xls、xlsx 和 csv 格式的文件生成；
3) 支持公式导出：支持在 Excel 模板中设置公式，降低服务端的计算量；
4) 支持低内存 SXSSF 模式：可利用极低的内存生成 xlsx 格式的文件；
5) 支持生产者消费者模式导出：无需一次性获取所有数据，分批获取数据配合 SXSSF 模式实现真正意义上海量数据导出；
6) 支持多种模板引擎：已内置 Freemarker、Groovy、Beetl 和 Thymeleaf 等常用模板引擎 Excel 构建器，推荐使用 Beetl 模板引擎；
7) 提供默认 Excel 构建器，直接输出简单 Excel：无需编写任何 html，已内置默认模板，可直接根据 POJO 数据列表输出；
8) 支持一次生成多 sheet：以数据表作为 sheet 单元，支持一份 Excel 文档中多 sheet 导出；
9) 支持 Excel 容量设定：支持设定 Excel 容量，到达容量后自动新建 Excel，可构建成 ZIP 压缩包导出。

想要在 Java 项目中使用 MyExcel，请添加依赖项，如下所示：

```xml
<dependency>
    <groupId>com.github.liaochong</groupId>
    <artifactId>myexcel</artifactId>
    <version>3.9.2</version>
</dependency>
```

执行 Maven 命令，查看 MyExcel 的依赖关系，如图 20-14 所示。

```
[INFO] --- maven-dependency-plugin:3.1.2:tree (default-cli) @ gf-other-myexcel ---
[INFO] com.gf:gf-other-myexcel:jar:0.0.1-SNAPSHOT
[INFO] \- com.github.liaochong:myexcel:jar:3.9.2:compile
[INFO]    +- org.apache.poi:poi-ooxml:jar:4.1.1:compile
[INFO]    |  +- org.apache.poi:poi:jar:4.1.1:compile
[INFO]    |  |  +- commons-codec:commons-codec:jar:1.14:compile
[INFO]    |  |  +- org.apache.commons:commons-collections4:jar:4.4:compile
[INFO]    |  |  \- org.apache.commons:commons-math3:jar:3.6.1:compile
[INFO]    |  +- org.apache.poi:poi-ooxml-schemas:jar:4.1.1:compile
[INFO]    |  |  \- org.apache.xmlbeans:xmlbeans:jar:3.1.0:compile
[INFO]    |  +- org.apache.commons:commons-compress:jar:1.19:compile
[INFO]    |  \- com.github.virtuald:curvesapi:jar:1.06:compile
[INFO]    +- org.jsoup:jsoup:jar:1.13.1:compile
[INFO]    +- org.slf4j:slf4j-api:jar:1.7.30:compile
[INFO]    \- com.twelvemonkeys.imageio:imageio-jpeg:jar:3.5:compile
[INFO]       +- com.twelvemonkeys.imageio:imageio-core:jar:3.5:compile
[INFO]       +- com.twelvemonkeys.imageio:imageio-metadata:jar:3.5:compile
[INFO]       +- com.twelvemonkeys.common:common-lang:jar:3.5:compile
[INFO]       +- com.twelvemonkeys.common:common-io:jar:3.5:compile
[INFO]       \- com.twelvemonkeys.common:common-image:jar:3.5:compile
[INFO] ------------------------------------------------------------------------
```

图 20-14 MyExcel 依赖关系

在本节中就不再编写 MyExcel 的示例代码，有关 MyExcel 的各个功能的文档，请访问以下网址：https://github.com/liaochong/myexcel/wiki/%E9%A6%96%E9%A1%B5

20.5 本章小结

本章讲解了验证码、二维码、Swagger 和 Excel 处理，这些技术在日常的开发中还是比较常用的。有了这些技术的加入，才能开发出更好用的应用程序，才能产出更具时效性的 API 文档。

第三部分：高级篇

本篇主要讲解 Kafka、应用监控、Spider 和搜索引擎等相关应用开发技术，涵盖了中大型应用项目所需要的知识点，包括流式计算、应用监控、爬虫开发、全文搜索、集群构建等，为大家开发中大型项目奠定技术基础。

第二十一章 Apache Kafka

Kafka 是由 Apache 软件基金会开发的一个开源流处理平台，由 Scala 和 Java 编写。它是一种高吞吐量的分布式发布订阅消息系统，它可以处理消费者在网站中的所有动作流数据。

Kafka 用于构建实时数据管道和流式应用程序。它是水平可扩展的，容错能力强，速度快，并被成千上万的公司用于生产环境中。流平台有三个关键功能，如下所示：
1) 发布和订阅记录流，类似于消息队列或企业消息传递系统；
2) 以容错的持久方式存储记录流；
3) 在记录流出现时对其进行处理。

Kafka 通常用于两大类应用，如图 21-1 所示：
1) 建立实时流数据管道，在系统或应用程序之间可靠地获取数据；
2) 构建对数据流进行转换或响应的实时流应用程序。

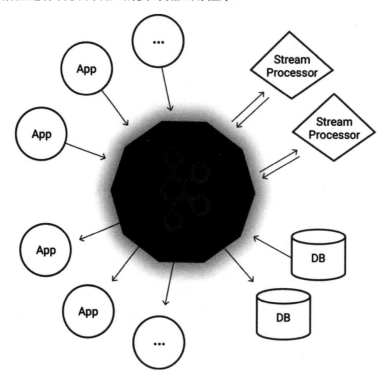

图 21-1 Kafka 应用范围

21.1 下载

访问 Kafka 官网：https://kafka.apache.org/downloads#2.5.0，进入下载页面，如图 21-2 所示。

图 21-2 下载 Kafka（1）

在上图的网页中，单击选框中的链接，进入镜像下载页面，如图 21-3 所示。

图 21-3 下载 Kafka（2）

在上图的网页中，可以根据自己当前的网络情况来选择合适的镜像链接，本书选择的镜像链接，如下所示：

https://mirrors.tuna.tsinghua.edu.cn/apache/kafka/2.5.0/kafka_2.13-2.5.0.tgz

21.2 单节点安装

Kafka 的软件包，在 Linux 和 Windows 环境中都可以使用。在这里，我们举在 Linux 中使用的例子。下载 Kafka 后，解压它，执行命令，如下所示：

$ tar xvf kafka_2.13-2.5.0.tgz

解压后，进入到 Kafka 的主目录中，如图 21-4 所示。

```
[quxy@svr-11 softwares]$ cd kafka_2.13-2.5.0/
[quxy@svr-11 kafka_2.13-2.5.0]$ ll
total 56
drwxr-xr-x. 3 quxy quxy  4096 Apr  8 09:18
drwxr-xr-x. 2 quxy quxy  4096 Apr  8 09:18
drwxr-xr-x. 2 quxy quxy  8192 May 27 11:18
-rw-r--r--. 1 quxy quxy 32216 Apr  8 09:13 LICENSE
-rw-r--r--. 1 quxy quxy   337 Apr  8 09:13 NOTICE
drwxr-xr-x. 2 quxy quxy    44 Apr  8 09:18
[quxy@svr-11 kafka_2.13-2.5.0]$
```

图 21-4 Kafka 目录

ZooKeeper 负责存储一些有关 Kafka 的 consumer 和 broker 的信息。因此，Kafka 是依赖于 ZooKeeper 的。Kafka 目录中默认包含了 ZooKeeper，可以无需修改 zookeeper.properties 配置文件中的默认配置项。先来启动 ZooKeeper 服务器，请执行命令：$ bin/zookeeper-server-start.sh config/zookeeper.properties。如图 21-5 所示。

```
[quxy@svr-11 kafka_2.13-2.5.0]$ bin/zookeeper-server-start.sh config/zookeeper.properties
[2020-05-27 11:25:41,484] INFO Reading configuration from: config/zookeeper.properties (org.apache.zookee
per.server.quorum.QuorumPeerConfig)
[2020-05-27 11:25:41,489] WARN config/zookeeper.properties is relative. Prepend ./ to indicate that you'r
e sure! (org.apache.zookeeper.server.quorum.QuorumPeerConfig)
[2020-05-27 11:25:41,504] INFO clientPortAddress is 0.0.0.0:2181 (org.apache.zookeeper.server.quorum.Quor
umPeerConfig)
[2020-05-27 11:25:41,504] INFO secureClientPort is not set (org.apache.zookeeper.server.quorum.QuorumPeer
Config)
[2020-05-27 11:25:41,551] INFO autopurge.snapRetainCount set to 3 (org.apache.zookeeper.server.DatadirCle
anupManager)
[2020-05-27 11:25:41,551] INFO autopurge.purgeInterval set to 0 (org.apache.zookeeper.server.DatadirClean
upManager)
[2020-05-27 11:25:41,551] INFO Purge task is not scheduled. (org.apache.zookeeper.server.DatadirCleanupMa
nager)
[2020-05-27 11:25:41,551] WARN Either no config or no quorum defined in config, running  in standalone mo
de (org.apache.zookeeper.server.quorum.QuorumPeerMain)
[2020-05-27 11:25:41,575] INFO Log4j found with jmx enabled. (org.apache.zookeeper.jmx.ManagedUtil)
[2020-05-27 11:25:41,617] INFO Reading configuration from: config/zookeeper.properties (org.apache.zookee
per.server.quorum.QuorumPeerConfig)
[2020-05-27 11:25:41,617] WARN config/zookeeper.properties is relative. Prepend ./ to indicate that you'r
e sure! (org.apache.zookeeper.server.quorum.QuorumPeerConfig)
[2020-05-27 11:25:41,618] INFO clientPortAddress is 0.0.0.0:2181 (org.apache.zookeeper.server.quorum.Quor
umPeerConfig)
[2020-05-27 11:25:41,618] INFO secureClientPort is not set (org.apache.zookeeper.server.quorum.QuorumPeer
Config)
```

图 21-5 启动 ZooKeeper

ZooKeeper 启动后，接着来启动 Kafka，也可以无需修改 server.properties 配置文件中的默认配置项，请执行命令：$ bin/kafka-server-start.sh config/server.properties，如图 21-6 所示。

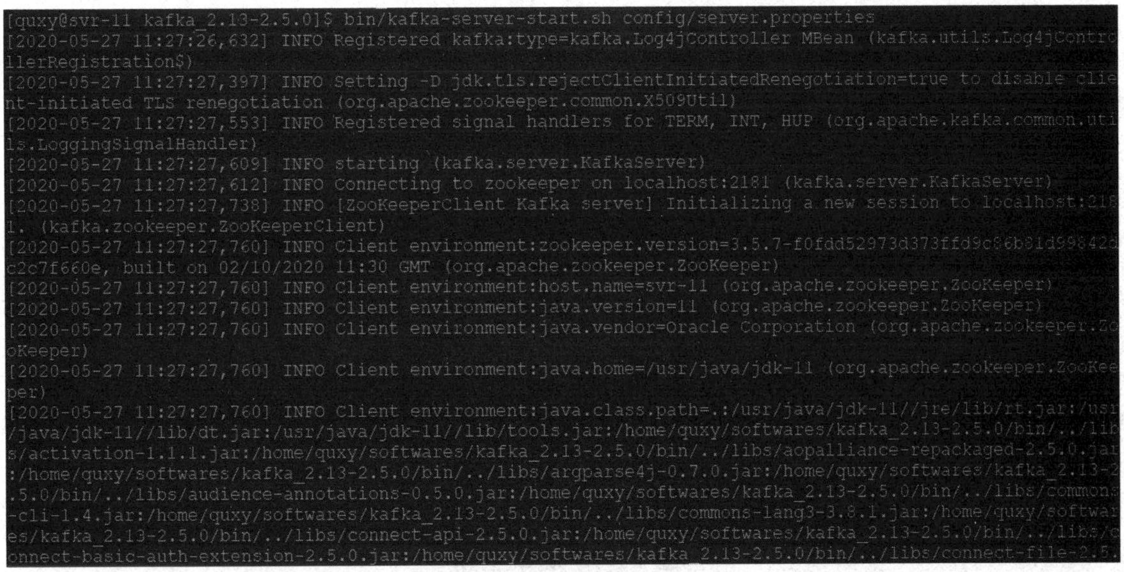

```
[quxy@svr-11 kafka_2.13-2.5.0]$ bin/kafka-server-start.sh config/server.properties
[2020-05-27 11:27:26,632] INFO Registered kafka:type=kafka.Log4jController MBean (kafka.utils.Log4jContro
llerRegistration$)
[2020-05-27 11:27:27,397] INFO Setting -D jdk.tls.rejectClientInitiatedRenegotiation=true to disable clie
nt-initiated TLS renegotiation (org.apache.zookeeper.common.X509Util)
[2020-05-27 11:27:27,553] INFO Registered signal handlers for TERM, INT, HUP (org.apache.kafka.common.uti
ls.LoggingSignalHandler)
[2020-05-27 11:27:27,609] INFO starting (kafka.server.KafkaServer)
[2020-05-27 11:27:27,612] INFO Connecting to zookeeper on localhost:2181 (kafka.server.KafkaServer)
[2020-05-27 11:27:27,738] INFO [ZooKeeperClient Kafka server] Initializing a new session to localhost:218
1. (kafka.zookeeper.ZooKeeperClient)
[2020-05-27 11:27:27,760] INFO Client environment:zookeeper.version=3.5.7-f0fdd52973d373ffd9c86b81d99842d
c2c7f660e, built on 02/10/2020 11:30 GMT (org.apache.zookeeper.ZooKeeper)
[2020-05-27 11:27:27,760] INFO Client environment:host.name=svr-11 (org.apache.zookeeper.ZooKeeper)
[2020-05-27 11:27:27,760] INFO Client environment:java.version=11 (org.apache.zookeeper.ZooKeeper)
[2020-05-27 11:27:27,760] INFO Client environment:java.vendor=Oracle Corporation (org.apache.zookeeper.Zo
oKeeper)
[2020-05-27 11:27:27,760] INFO Client environment:java.home=/usr/java/jdk-11 (org.apache.zookeeper.ZooKee
per)
[2020-05-27 11:27:27,760] INFO Client environment:java.class.path=.:/usr/java/jdk-11//jre/lib/rt.jar:/usr
/java/jdk-11//lib/dt.jar:/usr/java/jdk-11//lib/tools.jar:/home/quxy/softwares/kafka_2.13-2.5.0/bin/../lib
s/activation-1.1.1.jar:/home/quxy/softwares/kafka_2.13-2.5.0/bin/../libs/aopalliance-repackaged-2.5.0.jar
:/home/quxy/softwares/kafka_2.13-2.5.0/bin/../libs/argparse4j-0.7.0.jar:/home/quxy/softwares/kafka_2.13-2
.5.0/bin/../libs/audience-annotations-0.5.0.jar:/home/quxy/softwares/kafka_2.13-2.5.0/bin/../libs/commons
-cli-1.4.jar:/home/quxy/softwares/kafka_2.13-2.5.0/bin/../libs/commons-lang3-3.8.1.jar:/home/quxy/softwar
es/kafka_2.13-2.5.0/bin/../libs/connect-api-2.5.0.jar:/home/quxy/softwares/kafka_2.13-2.5.0/bin/../libs/c
onnect-basic-auth-extension-2.5.0.jar:/home/quxy/softwares/kafka_2.13-2.5.0/bin/../libs/connect-file-2.5.
```

图 21-6 启动 Kafka（1）

在图 21-7 中可以看到：最后一行的输出中包含有“started”字样，说明已经启动成功了。

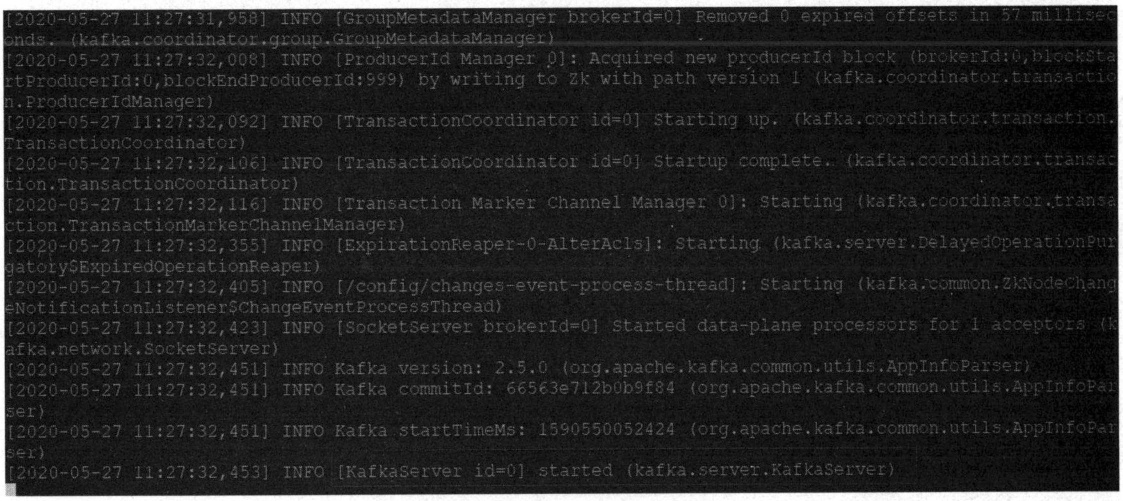

```
[2020-05-27 11:27:31,958] INFO [GroupMetadataManager brokerId=0] Removed 0 expired offsets in 57 millisec
onds. (kafka.coordinator.group.GroupMetadataManager)
[2020-05-27 11:27:32,008] INFO [ProducerId Manager 0]: Acquired new producerId block (brokerId:0,blockSta
rtProducerId:0,blockEndProducerId:999) by writing to Zk with path version 1 (kafka.coordinator.transactio
n.ProducerIdManager)
[2020-05-27 11:27:32,092] INFO [TransactionCoordinator id=0] Starting up. (kafka.coordinator.transaction.
TransactionCoordinator)
[2020-05-27 11:27:32,106] INFO [TransactionCoordinator id=0] Startup complete. (kafka.coordinator.transac
tion.TransactionCoordinator)
[2020-05-27 11:27:32,116] INFO [Transaction Marker Channel Manager 0]: Starting (kafka.coordinator.transa
ction.TransactionMarkerChannelManager)
[2020-05-27 11:27:32,355] INFO [ExpirationReaper-0-AlterAcls]: Starting (kafka.server.DelayedOperationPur
gatory$ExpiredOperationReaper)
[2020-05-27 11:27:32,405] INFO [/config/changes-event-process-thread]: Starting (kafka.common.ZkNodeChang
eNotificationListener$ChangeEventProcessThread)
[2020-05-27 11:27:32,423] INFO [SocketServer brokerId=0] Started data-plane processors for 1 acceptors (k
afka.network.SocketServer)
[2020-05-27 11:27:32,451] INFO Kafka version: 2.5.0 (org.apache.kafka.common.utils.AppInfoParser)
[2020-05-27 11:27:32,451] INFO Kafka commitId: 66563e712b0b9f84 (org.apache.kafka.common.utils.AppInfoPar
ser)
[2020-05-27 11:27:32,451] INFO Kafka startTimeMs: 1590550052424 (org.apache.kafka.common.utils.AppInfoPar
ser)
[2020-05-27 11:27:32,453] INFO [KafkaServer id=0] started (kafka.server.KafkaServer)
```

图 21-7 启动 Kafka（2）

在本机上连接 Kafka 服务器，执行命令：$ bin/kafka-topics.sh --create --bootstrap-server \localhost:9092 --replication-factor 1 --partitions 1 --topic gf_test，创建一个主题 gf_test，如图 21-8 所示。

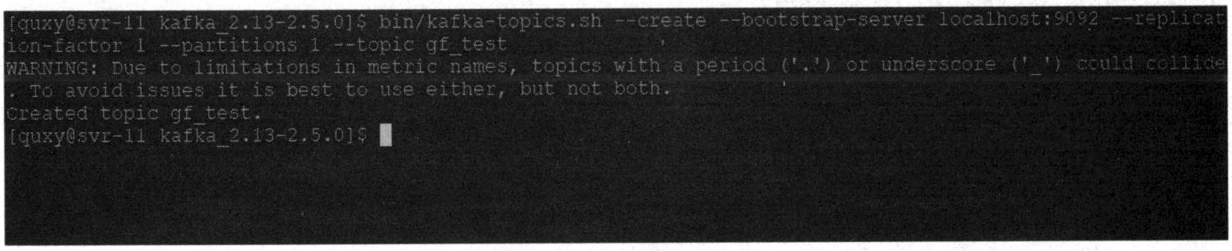

```
[quxy@svr-11 kafka_2.13-2.5.0]$ bin/kafka-topics.sh --create --bootstrap-server localhost:9092 --replicat
ion-factor 1 --partitions 1 --topic gf_test
WARNING: Due to limitations in metric names, topics with a period ('.') or underscore ('_') could collide
. To avoid issues it is best to use either, but not both.
Created topic gf_test.
[quxy@svr-11 kafka_2.13-2.5.0]$
```

图 21-8 创建主题

警告：由于度量名称的限制，带有句点（.）或下划线（_）的主题可能会发生冲突。为了避免出现问题，最好使用其中之一，但不能同时使用两者。执行命令：$ bin/kafka-topics.sh --list --bootstrap-server localhost:9092，要查看所有主题，如图 21-9 所示。

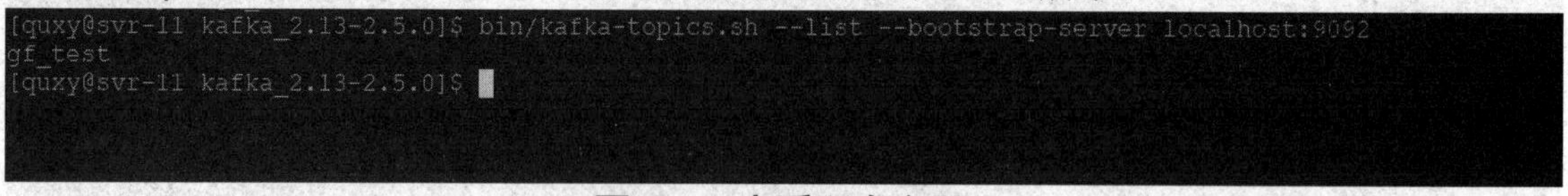
```
[quxy@svr-11 kafka_2.13-2.5.0]$ bin/kafka-topics.sh --list --bootstrap-server localhost:9092
gf_test
[quxy@svr-11 kafka_2.13-2.5.0]$
```

图 21-9 查看所有主题

创建主题后，向该主题发送一些信息，执行命令$ bin/kafka-console-producer.sh --bootstrap-server localhost:9092 --topic gf_test，对主题进行测试，如图 21-10 所示。

```
[quxy@svr-11 kafka_2.13-2.5.0]$ bin/kafka-console-producer.sh --bootstrap-server localhost:9092 --topic g
f_test
>北京欢迎你！
>
```

图 21-10 测试主题

执行命令：$ bin/kafka-console-consumer.sh --bootstrap-server \localhost:9092 --topic gf_test --from-beginning，获取指定主题的信息，如图 21-11 所示。

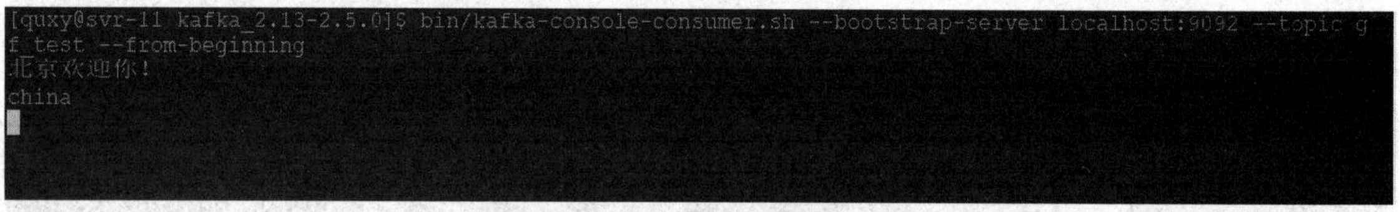
```
[quxy@svr-11 kafka_2.13-2.5.0]$ bin/kafka-console-consumer.sh --bootstrap-server localhost:9092 --topic g
f_test --from-beginning
北京欢迎你！
china
```

图 21-11 获取主题信息

21.3 集群搭建

在这里，我们创建一个拥有 3 个节点的 Kafka 集群。首先，进入到 Kafka 的主目录，打开 zookeeper.properties 配置文件，在文件的底部添加配置项，如下所示：

$ vim config/zookeeper.properties

initLimit=10

syncLimit=5

server.1=192.168.226.130:2888:3888

server.2=192.168.226.131:2888:3888

server.3=192.168.226.132:2888:3888

Kafka 的 3 个节点都进行上面的配置。其中，dataDir 配置项，使用默认值即可，如下所示：

dataDir=/tmp/zookeeper

在 3 个节点上分别新建 myid 文件，并为其添加内容，执行命令，如下所示：

$ echo 1 >/tmp/zookeeper/myid

$ echo 2 >/tmp/zookeeper/myid

$ echo 3 >/tmp/zookeeper/myid

接下来，打开 server.properties 配置文件，在文件的底部添加配置项，如下所示：
$ vim config/server.properties
broker.id=1
#listeners=PLAINTEXT://192.168.226.130:9092
listeners=PLAINTEXT://0.0.0.0:9092
advertised.listeners=PLAINTEXT://192.168.226.130:9092
zookeeper.connect=192.168.226.130:2181,192.168.226.131:2181,192.168.226.132:2181

还可以使用第二种方式的配置项，如下所示：
broker.id=1
listeners=PLAINTEXT://192.168.226.130:9092
#listeners=PLAINTEXT://0.0.0.0:9092
#advertised.listeners=PLAINTEXT://192.168.226.130:9092
zookeeper.connect=192.168.226.130:2181,192.168.226.131:2181,192.168.226.132:2181

Kafka 的 3 个节点的 broker.id 配置项分别设置为 1、2 和 3。listeners 和 advertised.listeners 配置项也分别设为不同的值。其中，log.dirs 配置项，使用默认值即可，如下所示：
log.dirs=/tmp/kafka-logs

Kafka 的 3 个节点都完成上述配置后，就可以启动相关的服务器了，先执行命令依次后台启动 ZooKeeper，如下所示：
$ bin/zookeeper-server-start.sh -daemon config/zookeeper.properties

再执行命令依次后台启动 Kafka，如下所示：
$ bin/kafka-server-start.sh -daemon config/server.properties

主题

Kafka 集群正常启动后，我们通过命令来进行主题相关的操作。列出所有的主题，如下所示：
$ bin/kafka-topics.sh --list \
--bootstrap-server 192.168.226.130:9092,192.168.226.131:9092,192.168.226.132:9092

首先，我们执行命令新建 gf-topic 主题，如下所示：
$ bin/kafka-topics.sh --create --topic gf-topic --partitions 3 --replication-factor 3 \
--bootstrap-server 192.168.226.130:9092,192.168.226.131:9092,192.168.226.132:9092

接着，查看所创建的主题，如下所示：
$ bin/kafka-topics.sh --describe --topic gf-topic \
--bootstrap-server 192.168.226.130:9092,192.168.226.131:9092,192.168.226.132:9092

图 21-12 查看所创建主题

创建完成主题后，在终端启动一个生产者来发送消息，如下所示：

$ bin/kafka-console-producer.sh --topic gf-topic \
--bootstrap-server 192.168.226.130:9092,192.168.226.131:9092,192.168.226.132:9092

还需要在终端启动一个消费者来接收消息，如下所示：

$ bin/kafka-console-consumer.sh --topic gf-topic --from-beginning \
--bootstrap-server 192.168.226.130:9092,192.168.226.131:9092,192.168.226.132:9092

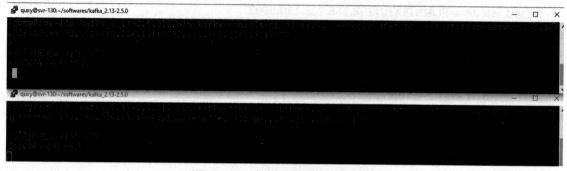

图 21-13 启动消费者接收消息

在终端中，可以通过执行命令来删除不再需要的主题，如下所示：

$ bin/kafka-topics.sh --delete --topic gf-topic \--bootstrap-server 192.168.226.130: 9092,
192.168.226.131:9092,192.168.226.132:9092

21.4 Kafka 安全

在 0.9.0.0 版中，Kafka 社区添加了许多功能，这些功能可以单独使用，也可以一起使用，从而提高 Kafka 集群的安全性。目前支持以下安全措施：

1) 使用 SSL 或 SASL 验证来自客户端（生产者和消费者）、其他代理和工具到代理的连接；
2) 从代理到 ZooKeeper 的连接的身份验证；
3) 使用 SSL 在代理和客户端之间、代理之间或代理和工具之间为传输的数据加密（请注意，启用 SSL 时会出现性能下降，其程度取决于 CPU 类型和 JVM 实现）；
4) 客户端对读/写操作的授权；
5) 授权是可插拔的，并且支持与外部授权服务的集成。

使用 SSL 进行加密和身份验证

Apache Kafka 允许客户端使用 SSL 进行加密通信和身份验证。使用 keytool 和 openssl 命令来生成秘钥、证书、CA 和签名证书等，并配置 Kafka 代理和客户端，从而来实现加密和身份验证。

默认情况下，SSL 是禁用的。请注意，启用 SSL 时会出现性能下降，其程度取决于 CPU 类型和 JVM 实现。有关详细的命令使用和配置信息，请访问以下网址：

http://kafka.apache.org/documentation/#security_ssl

授权和 ACL

Kafka 附带一个可插拔 Authorizer 和一个使用 ZooKeeper 存储所有 ACL 的开箱即用授权者实现。通过在 server.properties 配置文件中设置 authorizer.class.name 来配置 Authorizer。要启

用开箱即用的实现，请添加相应的配置项，如下所示：

authorizer.class.name=kafka.security.authorizer.AclAuthorizer

Kafka 的 bin 目录中的脚本命令 kafka-acls.sh 可以完成与 ACL 相关的各种操作。

ZooKeeper 身份验证

ZooKeeper 从 3.5.x 版本开始支持相互 TLS（mTLS）身份验证。从 2.5 开始，Kafka 支持使用 SASL 和 mTLS 向 ZooKeeper 进行身份验证（可以是单独的，也可以是同时使用两者）。

对于 Kafka 的新集群和现有集群，在处理上是有些不同的，特别是对现有集群的迁移。

ZooKeeper 加密

使用相互 TLS 的 ZooKeeper 连接被加密。从 ZooKeeper 版本 3.5.7（Kafka 2.5 附带的版本）开始，ZooKeeper 支持服务器端配置 ssl.clientAuth，在 ZooKeeper 中将此值设置为 none 以允许客户端通过 TLS 加密的连接进行连接，而无需显示自己的连接证书。

有关 Kafka 安全的更多信息，请访问以下网址：

http://kafka.apache.org/documentation/#security

21.5 简单示例

在 Spring Boot 应用程序中引入 Kafka，添加相应的依赖项，如下所示：

```xml
<dependency>
    <groupId>org.springframework.kafka</groupId>
    <artifactId>spring-kafka</artifactId>
</dependency>
```

接下来，我们添加 Kafka 相关的配置项，以便连接 Kafka 集群，如图 21-14 所示。

```yaml
spring:
  kafka:
    template:
      default-topic: gf-topic-01
    producer:
      key-serializer: org.springframework.kafka.support.serializer.JsonSerializer
      value-serializer: org.springframework.kafka.support.serializer.JsonSerializer
    consumer:
      group-id: gf-group-01
      key-deserializer: org.springframework.kafka.support.serializer.JsonDeserializer
      value-deserializer: org.springframework.kafka.support.serializer.JsonDeserializer
      properties:
        spring:
          json:
            trusted:
              packages: com.gf.kafka.entity
    bootstrap-servers:
      - 192.168.226.130:9092
      - 192.168.226.131:9092
      - 192.168.226.132:9092
```

图 21-14 添加 Kafka 相关配置项

在上图所示的配置中，我们定义一个默认主题 gf-topic-01，并指定生产者和消费者的编解码类以便处理数据实体类的对象，最后，指定了 Kafka 集群的地址。有关 Kafka 可配置的更多配置项，请访问以下网址：

https://docs.spring.io/spring-boot/docs/2.3.1.RELEASE/reference/html/appendix-application-properties.html#integration-properties

现在，我们可以编写具体的业务代码了。先定义一个 controller 类，并基于 KafkaTemplate 实现 Kafka 生产者的逻辑，具体代码，如下所示：

```java
package com.gf.kafka.controller;
import com.gf.kafka.entity.MenuKafka;
import org.springframework.http.ResponseEntity;
import org.springframework.kafka.core.KafkaTemplate;
import org.springframework.web.bind.annotation.PostMapping;
import org.springframework.web.bind.annotation.RequestBody;
import org.springframework.web.bind.annotation.RequestMapping;
import org.springframework.web.bind.annotation.RestController;
import java.util.Collections;
@RestController
@RequestMapping("/kafka")
public class KafkaController {
    private final KafkaTemplate<Long, MenuKafka> kafkaTemplate;
    public KafkaController(KafkaTemplate<Long, MenuKafka> kafkaTemplate) {
        this.kafkaTemplate = kafkaTemplate;
    }
    @PostMapping("/producer")
    public ResponseEntity producer(@RequestBody MenuKafka params) {
        // 发送到配置中指定的默认话题。
        kafkaTemplate.sendDefault(params);
//      kafkaTemplate.sendDefault(params.getId(), params);
        return ResponseEntity.ok(Collections.singletonMap("msg", "发送成功！"));
    }
}
```

使用 @KafkaListener 注解来实现 Kafka 消费者的代码，如下所示：

```java
package com.gf.kafka.consumer;
import com.fasterxml.jackson.core.JsonProcessingException;
import com.fasterxml.jackson.databind.ObjectMapper;
import com.gf.kafka.entity.MenuKafka;
import lombok.extern.slf4j.Slf4j;
import org.springframework.kafka.annotation.KafkaListener;
import org.springframework.stereotype.Component;
@Slf4j
```

```java
@Component
public class KafkaConsumer {
    private final ObjectMapper objectMapper;
    public KafkaConsumer(ObjectMapper objectMapper) {
        this.objectMapper = objectMapper;
    }
    /**
     * 处理接收到消息
     */
    @KafkaListener(topics =
{"${spring.kafka.template.default-topic}"})
    public void processMsg(MenuKafka data) throws JsonProcessingException {
        log.info("接收到的消息：" + objectMapper.writeValueAsString(data));
    }
}
```

定义一个数据实体类 MenuKafka，如下所示：

```java
package com.gf.kafka.entity;
import lombok.Data;
import java.io.Serializable;
@Data
public class MenuKafka implements Serializable {
    private Long id;
    private String name;
    private Long parentId;
}
```

完成以上配置和代码后，打开链接：http://localhost:8080/kafka/producer，启动应用程序，调用 API 以发布消息，如图 21-15 所示。

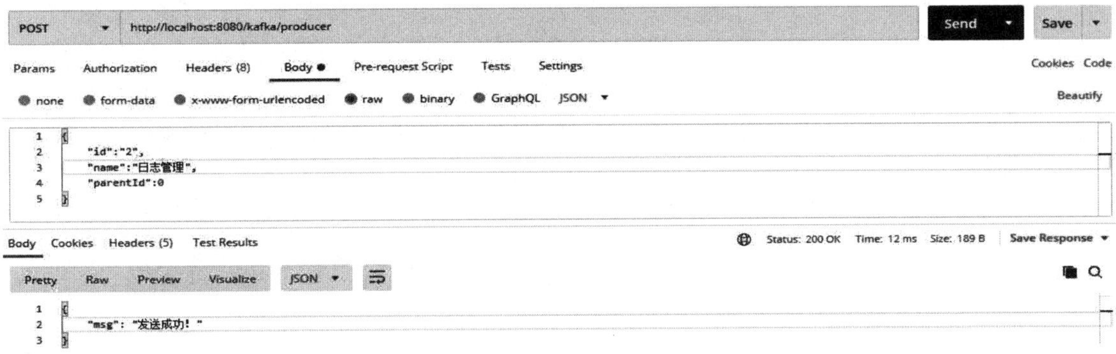

图 21-15 启动应用

收到消息后，把消息输出到控制台，如图 21-16 所示。

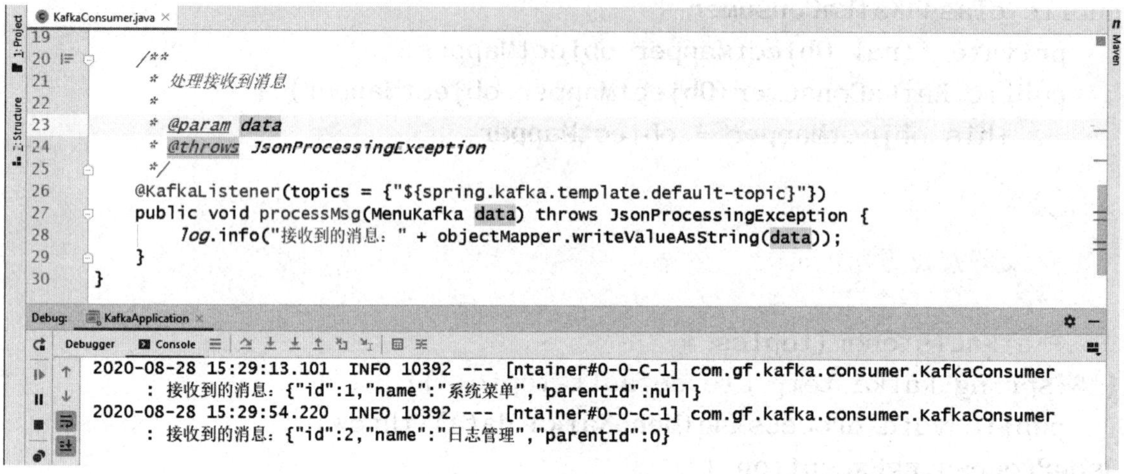

图 21-16 控制台信息

21.6 本章小结

本章讲解了 Kafka 的下载、单节点安装和安全，最后讲解了在 Spring Boot 应用程序中使用 Kafka 的简单示例。希望通过这些知识点的讲解，能把大家带入到 Kafka 的愉快使用之旅当中。

第二十二章 应用监控：Actuator

Spring Boot 包含许多附加功能特性，可以帮助你在将应用程序推送到生产环境时监视和管理它。可以选择使用 HTTP 端点或 JMX 来管理和监视应用程序。审计、运行状况和度量收集也可以自动应用到应用程序中。在 Spring Boot 应用程序中添加应用监控，只需添加相关依赖项，如下所示：

```xml
<dependency>
    <groupId>org.springframework.boot</groupId>
    <artifactId>spring-boot-starter-actuator</artifactId>
</dependency>
```

spring-boot-starter-actuator 的依赖关系，如图 22-1 所示。

```
[INFO] com.gf:gf-actuator:jar:0.0.1-SNAPSHOT
[INFO] \- org.springframework.boot:spring-boot-starter-actuator:jar:2.3.1.RELEASE:compile
[INFO]    +- org.springframework.boot:spring-boot-starter:jar:2.3.1.RELEASE:compile
[INFO]    |  +- org.springframework.boot:spring-boot:jar:2.3.1.RELEASE:compile
[INFO]    |  |  \- org.springframework:spring-context:jar:5.2.7.RELEASE:compile
[INFO]    |  |     +- org.springframework:spring-aop:jar:5.2.7.RELEASE:compile
[INFO]    |  |     +- org.springframework:spring-beans:jar:5.2.7.RELEASE:compile
[INFO]    |  |     \- org.springframework:spring-expression:jar:5.2.7.RELEASE:compile
[INFO]    |  +- org.springframework.boot:spring-boot-autoconfigure:jar:2.3.1.RELEASE:compile
[INFO]    |  +- org.springframework.boot:spring-boot-starter-logging:jar:2.3.1.RELEASE:compile
[INFO]    |  |  +- ch.qos.logback:logback-classic:jar:1.2.3:compile
[INFO]    |  |  |  +- ch.qos.logback:logback-core:jar:1.2.3:compile
[INFO]    |  |  |  \- org.slf4j:slf4j-api:jar:1.7.30:compile
[INFO]    |  |  +- org.apache.logging.log4j:log4j-to-slf4j:jar:2.13.3:compile
[INFO]    |  |  |  \- org.apache.logging.log4j:log4j-api:jar:2.13.3:compile
[INFO]    |  |  \- org.slf4j:jul-to-slf4j:jar:1.7.30:compile
[INFO]    |  +- jakarta.annotation:jakarta.annotation-api:jar:1.3.5:compile
[INFO]    |  +- org.springframework:spring-core:jar:5.2.7.RELEASE:compile
[INFO]    |  |  \- org.springframework:spring-jcl:jar:5.2.7.RELEASE:compile
[INFO]    |  \- org.yaml:snakeyaml:jar:1.26:compile
[INFO]    +- org.springframework.boot:spring-boot-actuator-autoconfigure:jar:2.3.1.RELEASE:compile
[INFO]    +- org.springframework.boot:spring-boot-actuator:jar:2.3.1.RELEASE:compile
[INFO]    +- com.fasterxml.jackson.core:jackson-databind:jar:2.11.0:runtime
[INFO]    |  +- com.fasterxml.jackson.core:jackson-annotations:jar:2.11.0:runtime
[INFO]    |  \- com.fasterxml.jackson.core:jackson-core:jar:2.11.0:runtime
[INFO]    +- com.fasterxml.jackson.datatype:jackson-datatype-jsr310:jar:2.11.0:runtime
[INFO]    \- io.micrometer:micrometer-core:jar:1.5.1:compile
[INFO]       +- org.hdrhistogram:HdrHistogram:jar:2.1.12:compile
[INFO]       \- org.latencyutils:LatencyUtils:jar:2.0.3:runtime
```

图 22-1 spring-boot-starter-actuator 的依赖关系

22.1 端点

Actuator 端点可以监视应用程序并与之交互。Spring Boot 包括许多内置端点，并允许添加自己的端点。例如，health 端点提供基本的应用程序运行状况信息。

可以启用或禁用每个单独的端点，并通过 HTTP 或 JMX 公开。当端点同时被启用和公开时，它被认为是可用的。内置端点只有在可用时才会自动配置。大多数应用程序选择通过 HTTP 进行公开，其中端点的 ID 和前缀 /actuator 被映射到一个 URL。例如：默认情况下，health 端点映射到 /actuator/health。

默认情况下，除了 shutdown 端点之外，所有其他端点都是启用的。如果希望端点启用是选择加入而不是选择退出，请将相应的属性设置为 false，并使用单个启用端点的属性选择重新加入，如下所示：

management.endpoints.enabled-by-default=false
management.endpoint.info.enabled=true

默认情况下，除了 health 和 info 端点之外，所有其他 HTTP 端点都是不公开的；JMX 的端点都是公开的。端点可能包含敏感信息，应仔细考虑何时公开它们。在表中列出了配置项的默认值，如表 22-1 所示。

表 22-1 JMX 的端点配置项列表

配置项	默认值
management.endpoints.jmx.exposure.exclude	
management.endpoints.jmx.exposure.include	*
management.endpoints.web.exposure.exclude	
management.endpoints.web.exposure.include	info, health

1) include 属性列出公开的端点 ID 列表。exclude 属性列出不应该公开的端点的 ID 列表。exclude 属性优先于 include 属性。可以使用端点 ID 列表同时配置 include 和 exclude 属性。
2) * 在 YAML 中有特殊的含义，所以如果想包含（或排除）所有端点，请确保添加双引号。

如果想实现自己的何时公开端点的策略，可以注册一个 EndpointFilter bean。端点自动缓存不带任何参数的读取操作的响应。若要配置端点缓存响应的时间，请使用其 cache.time-to-live 属性，如下所示：

management.endpoint.beans.cache.time-to-live=10s

默认情况下 CORS 支持是禁用的，并且仅在设置了相应的属性后才启用，如下所示：

management.endpoints.web.cors.allowed-origins=https://example.com
management.endpoints.web.cors.allowed-methods=GET,POST

22.2 自定义端点

如果添加一个用 @Endpoint 注解的 @Bean 类，那么该类中任何用 @ReadOperation、@WriteOperation 或 @DeleteOperation 注解的方法都会自动通过 JMX 公开，在 web 应用程序中，也会通过 HTTP 公开。端点可以使用 Jersey、Spring MVC 或 Spring WebFlux 通过 HTTP 公开。

使用 @Endpoint 注解定义 Actuator 端点，并将其注册为 bean，示例代码，如下所示：

```
package com.gf.actuator.endpoints;
import org.springframework.boot.actuate.endpoint.annotation.DeleteOperation;
import org.springframework.boot.actuate.endpoint.annotation.Endpoint;
```

```java
import org.springframework.boot.actuate.endpoint.annotation.ReadOperation;
import org.springframework.boot.actuate.endpoint.annotation.WriteOperation;
import java.util.Collections;
import java.util.Map;
@Endpoint(id = "hello")
public class HelloEndPoint {
    @ReadOperation
    public Map<String, Object> hello() {
        return Collections.singletonMap("city", "天津");
    }
    @WriteOperation
    public Map<String, Object> sayHello(String name) {
        return Collections.singletonMap("name", name);
    }
    @DeleteOperation
    public Map<String, Object> del(String id) {
        return Collections.singletonMap("msg", "删除了 " + id);
    }
}
```

也可以在 HelloEndPoint 类上添加 @Component 注解，来代替下面的代码片段。

```java
package com.gf.actuator.config;
import com.gf.actuator.endpoints.HelloEndPoint;
import org.springframework.context.annotation.Bean;
import org.springframework.context.annotation.Configuration;
@Configuration
public class EndpointsConfig {
    @Bean
    public HelloEndPoint getHelloEndPoint() {
        return new HelloEndPoint();
    }
}
```

使用 @RestControllerEndpoint 注解定义 Actuator 端点，这种方式类似于普通 controller 层类，示例代码，如下所示：

```java
package com.gf.actuator.endpoints;
```

```
import org.springframework.boot.actuate.endpoint.web.annotation.RestControllerEndpoint;
import org.springframework.web.bind.annotation.GetMapping;
import org.springframework.web.bind.annotation.RestController;
import java.util.Collections;
import java.util.Map;
@RestController
@RestControllerEndpoint(id = "city")
public class CityControllerEndpoint {
    @GetMapping("/name")
    public Map<String, Object> name() {
        return Collections.singletonMap("city", "雄安");
    }
}
```

22.3 应用程序信息

应用程序信息公开了从 ApplicationContext 中定义的所有 InfoContributor bean 收集的各种信息。Spring Boot 包含许多自动配置的 InfoContributor bean，也可以自定义。

自动配置的 InfoContributor

表 22-2 即为 InfoContributor bean 在适当的时候通过 Spring Boot 自动配置的内容：

表 22-2 InfoContributor bean 配置项

名称	说明
EnvironmentInfoContributor	从 info 键下的环境中公开任何键
GitInfoContributor	如果 git.properties 文件可用，则公开 git 信息
BuildInfoContributor	如果 META-INF/build-info.properties 文件可用，则公开构建信息

如果不需要它们，则可以通过设置属性来禁用它们，如下所示：

management.info.default.enabled=false

自定义应用程序信息

可以通过设置 info.* Spring 属性自定义 info 端点公开的数据。info 键下的所有 Environment 属性都将自动公开。可以将设置添加到 application.properties 配置文件，如下所示：

info.app.encoding=UTF-8

info.app.java.source=1.8

info.app.java.target=1.8

还可以在构建时扩展 info 属性，而不是对这些值进行硬编码。假设使用的是 Maven，可以按这样的方式重写上面的配置项，如下所示：

info.app.encoding=@project.build.sourceEncoding@
info.app.java.source=@java.version@
info.app.java.target=@java.version@

配置后，访问 URL：http://localhost:8080/actuator/info，启动应用程序，如图 22-2 所示：

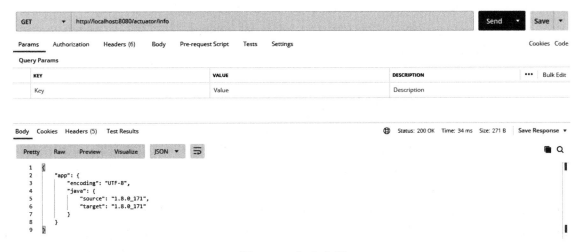

图 22-2 启动应用

更多详细信息，请访问以下网址：
https://docs.spring.io/spring-boot/docs/2.3.1.RELEASE/reference/html/howto.html#howto-automatic-expansion

自定义 InfoContributor

要自定义应用程序信息，请实现 InfoContributor 接口并将其注册为 bean，如下所示：

```
package com.gf.actuator.component;
import org.springframework.boot.actuate.info.Info;
import org.springframework.boot.actuate.info.InfoContributor;
import org.springframework.stereotype.Component;
import java.util.Collections;
@Component
public class MyInfoContributor implements InfoContributor {
    @Override
    public void contribute(Info.Builder builder) {
        // city 作为第一层级的键
        builder.withDetails(Collections.singletonMap("city", "北京"));
    }
}
```

22.4 度量

Spring Boot Actuator 为 Micrometer 提供依赖管理和自动配置，Micrometer 是一种支持众多监控系统的应用度量外观，包括：JMX 等。有关度量的更多信息，请访问以下网址：
https://docs.spring.io/spring-boot/docs/2.3.1.RELEASE/reference/html/production-ready-features.html#production-ready-metrics

22.4.1 metrics 端点

Spring Boot 提供了一个 metrics 端点，可用于诊断检查应用程序收集的度量。该端点在默认情况下不可用，想要公开 metrics 端点，请在 application.yml 配置文件中添加配置项，如下所示：

management.endpoints.web.exposure.include: metrics

要查看可用的 meter 名称的列表，请访问相应的 URL：http://localhost:8080/actuator/metrics，如图 22-3 所示。

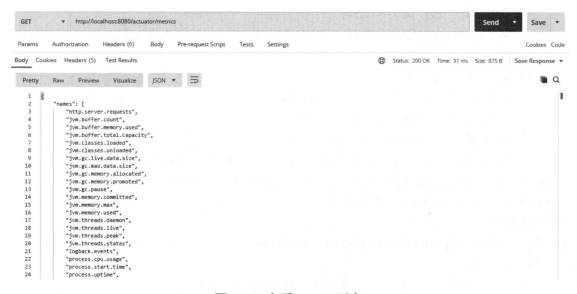

图 22-3 查看 meter 列表

通过提供特定 meter 的名称作为选择器，访问 URL： http://localhost:8080/actuator/metrics/http.server.requests，可以进一步查看该 meter 的信息，如图 22-4 所示。

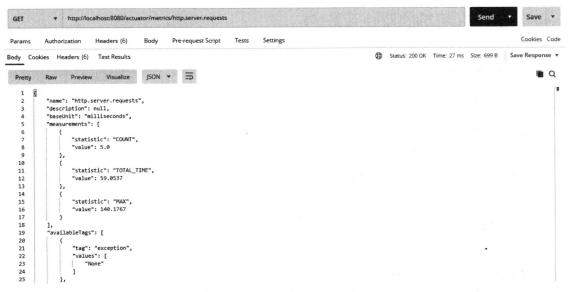

图 22-4 查看 meter 信息

22.4.2 支持的度量

Spring Boot 默认支持的核心度量，如下所示：
1) JVM 度量，报告利用率：
   各种内存和缓冲池
   与垃圾收集相关的统计信息
   线程利用率
   加载/卸载的类数
2) CPU 度量
3) 文件描述符度量
4) Kafka 消费者度量
5) Log4j2 度量：记录每个级别记录到 Log4j2 的事件数；
6) Logback 度量：记录每个级别记录到 Logback 的事件数；
7) 正常运行时间度量：报告正常运行时间的值和表示应用程序绝对启动时间的固定值；
8) Tomcat 度量：server.tomcat.mbeanregistry.enabled 必须设置为 true 才能注册所有 Tomcat 度量。
9) Spring 集成度量
   有关自定义度量等详细信息，请访问以下网址：
   https://docs.spring.io/spring-boot/docs/2.3.1.RELEASE/reference/html/production-ready-features.html#production-ready-metrics-meter

22.4.3 简单的监控系统

Micrometer 附带一个简单的内存后端，如果没有配置其他注册，该后端会自动用作后备。这使你可以查看在 metrics 端点中收集了哪些度量。有关具体信息，请查看上面的小节：metrics 端点。

当使用任何其他可用后端时，内存中的后端会立即禁用自身。也可以显式禁用它，如下所示：

management.metrics.export.simple.enabled=false

22.4.4 Prometheus 监控系统

Prometheus 是一个开源监控解决方案。它具有内存时间序列数据库，具有简单的内置 UI、自定义查询语言和数学运算。Prometheus 被设计为在拉式模型上运行，基于服务发现定期从应用程序实例中获取度量。Prometheus 的官网地址，如下所示：

https://prometheus.io

Spring Boot 提供了一个可用的 /actuator/prometheus 端点，以适当的格式呈现 Prometheus scrape。该端点在默认情况下不可用，必须添加相应的配置项来公开它。要想使用该端点，请添加依赖项，如下所示：

```
<dependency>
    <groupId>io.micrometer</groupId>
    <artifactId>micrometer-registry-prometheus</artifactId>
</dependency>
```

访问 URL：https://prometheus.io/download，进入 Prometheus 官网的下载页面，下载 Prometheus 软件包，如图 22-5 所示。

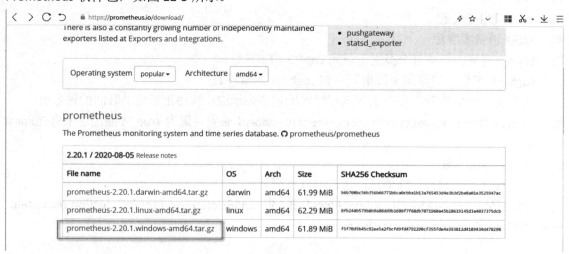

图 22-5 下载 Prometheus

在这里，我们下载的是 Windows 版本的。下载后，解压到指定的目录中，如图 22-6 所示。

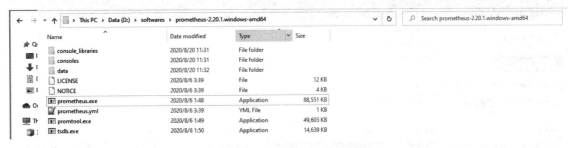

图 22-6 安装 Prometheus

编辑 prometheus.yml 配置文件，添加与 Spring Boot 应用程序相关配置项，如图 22-7 所示。

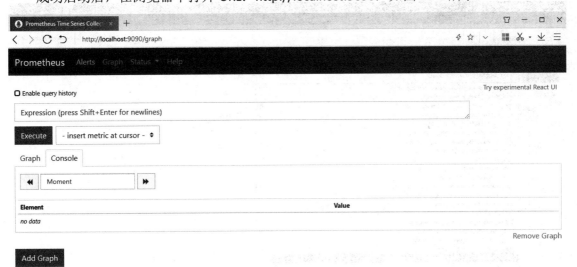

图 22-7 添加相关配置项

执行命令启动 Prometheus，如下所示：

prometheus --config.file=prometheus.yml

成功启动后，在浏览器中打开 URL：http://localhost:9090，如图 22-8 所示。

图 22-8 启动 Prometheus（1）

这时候，8080 端口上的 Spring Boot 应用程序还没有启动，与其相关的页面，如图 22-9 所示。

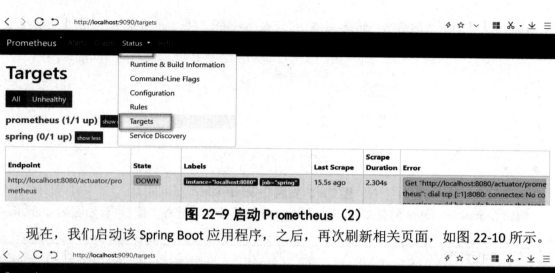

图 22-9 启动 Prometheus（2）

现在，我们启动该 Spring Boot 应用程序，之后，再次刷新相关页面，如图 22-10 所示。

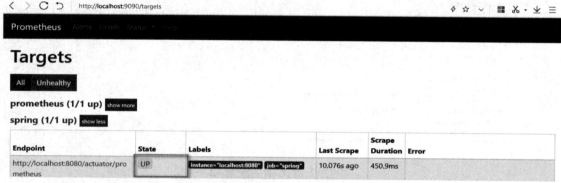

图 22-10 启动 Prometheus（3）

打开菜单[Graph]，从下拉框中选择一个度量，可以查看其图表信息，如图 22-11 所示。

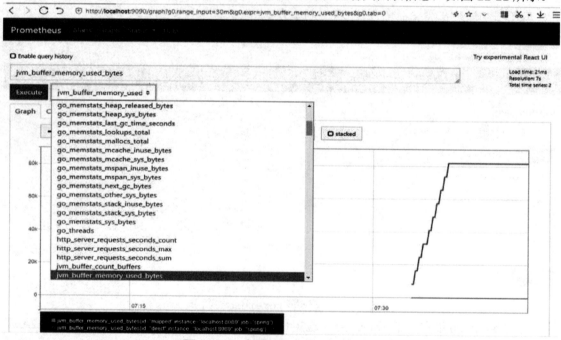

图 22-11 启动 Prometheus（4）

有关 Prometheus 的安装和使用的详细信息，请访问以下网址：
https://prometheus.io/docs/prometheus/2.20/getting_started
有关配置应用程序的更多信息，请访问以下网址：
https://micrometer.io/docs/registry/prometheus

对于可能存在的时间不够长而无法收集的临时或批处理作业，可以使用 Prometheus Pushgateway 支持将其度量公开给 Prometheus。要启用 Prometheus Pushgateway 支持，请添加相应的依赖项，如下所示：

```xml
<dependency>
    <groupId>io.prometheus</groupId>
    <artifactId>simpleclient_pushgateway</artifactId>
</dependency>
```

22.5 HTTP 跟踪

可以通过在应用程序的配置中提供类型为 HttpTraceRepository 的 bean 来启用 HTTP 跟踪。为方便起见，Spring Boot 提供了一个 InMemoryHttpTraceRepository，默认情况下，它存储最近 100 个请求-响应交换的跟踪。与其他跟踪解决方案相比，InMemoryHttpTraceRepository 的能力是有限的，建议仅将其用于开发环境。对于生产环境，建议使用生产就绪跟踪或可观测的解决方案，如 Zipkin 或 Spring Cloud Sleuth。或者，创建自定义的 HttpTraceRepository。

定义一个 InMemoryHttpTraceRepository bean 来启用 HTTP 跟踪，示例代码，如下所示：

```java
package com.gf.actuator.config;
import org.springframework.boot.actuate.trace.http.HttpTraceRepository;
import org.springframework.boot.actuate.trace.http.InMemoryHttpTraceRepository;
import org.springframework.context.annotation.Bean;
import org.springframework.context.annotation.Configuration;
@Configuration
public class MetersConfig {
    @Bean
    public HttpTraceRepository httpTraceRepository() {
        InMemoryHttpTraceRepository httpTraceRepository = new InMemoryHttpTraceRepository();
        // 默认存储最近的 100 条请求记录。
        httpTraceRepository.setCapacity(100);
        // 默认最新的请求排在前面
        httpTraceRepository.setReverse(true);
```

```
        return httpTraceRepository;
    }
}
```

访问 URL：http://localhost:8080/actuator/httptrace，使用 httptrace 端点获取存储在 HttpTraceRepository 中的请求-响应交换的信息，如图 22-12 所示：

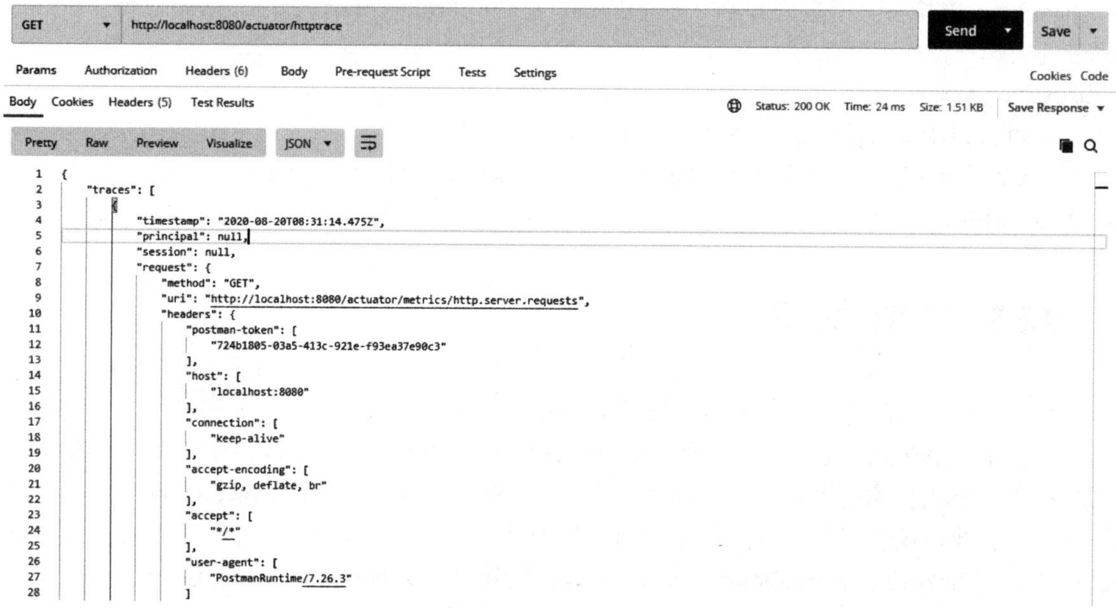

图 22-12 获取请求-响应交换信息

22.6 本章小结

本章讲解了 Actuator 的端点、应用程序信息、度量和 HTTP 跟踪。这些只是应用监控的一部分内容，还有许多相关的知识点等待大家去学习和掌握，一定要详细阅读官方的文档。

第二十三章 网络爬虫

网络爬虫，一般简称为爬虫，它是按照预定的规则自动地抓取互联网信息（如 web 页面）的应用程序或软件工具。在互联网上有各种各样的不同语言开发的爬虫，它们有些使用不同的抓取策略，它们用于不同的应用场景。

按照系统结构和实现技术，网络爬虫可以分为：通用网络爬虫、聚焦网络爬虫、增量式网络爬虫和深层网络爬虫等。

23.1 Apache Solr

Apache Solr 是基于 Apache Lucene 构建的流行、快速和开源的企业搜索平台。Solr 具有高可靠性、可扩展性和容错性，提供分布式索引、复制和负载均衡查询、自动故障转移和恢复、集中配置等功能。Solr 为世界上许多最大的互联网站点提供搜索和导航功能。Solr 的官网，如下所示：

https://lucene.apache.org/solr

Solr 在 Windows 和 Linux 操作系统中都可以运行，在下载页面可以选择所需的版本，如下所示：

https://lucene.apache.org/solr/downloads.html

选择相应的版本，进入镜像下载页面，如下所示：

https://www.apache.org/dyn/closer.lua/lucene/solr/8.6.1/solr-8.6.1.tgz

在镜像下载页面中，可以根据自己当前的网络情况来选择合适的镜像链接，本书选择的镜像链接，如下所示：

$ wget https://mirror.bit.edu.cn/apache/lucene/solr/8.6.1/solr-8.6.1.tgz

下载完成后，把 Solr 软件包解压到指定目录，进入主目录查看，如下所示：

$ tar -xvf solr-8.6.1.tgz

图 23-1 SOLR 主目录

默认后台启动 Solr 服务器，执行命令，如下所示：

bin/solr start

bin/solr stop

默认情况下，访问 URL：http://192.168.226.130:8983，可以本地测试用局域网访问 Solr 服务器，如图 23-2 所示。

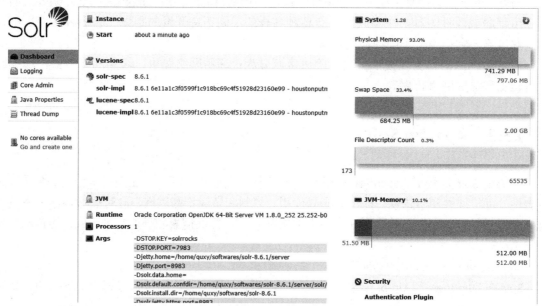

图 23-2 Solr 服务

如果没有使用示例配置启动 Solr，则需要创建一个核心，以便能够进行索引和搜索。执行命令：$ bin/solr create -c gf_index_core，创建名称为 gf_index_core 的核心，如图 23-3 所示。

图 23-3 Solr 服务核心

有关 Solr 的讲解先到这里，本节只是简单地了解一下 Solr 的相关内容，为下一节 Solr 的使用做准备。

23.2 Apache Nutch

Apache Nutch 是用 Java 语言开发的高度可扩展的高度可伸缩的 web 爬虫。它是一个成熟的，可以用于生产环境的的 web 爬虫。Nutch 1.x 支持细粒度配置，依赖于 Apache Hadoop 数据结构，非常适合批处理。Apache Nutch 的官网，如下所示：

http://nutch.apache.org

Apache Nutch 采集到的数据，可以输出到不同的系统中，例如：Solr、RabbitMQ、Dummy、CSV、Elasticsearch、CloudSearch 和 Kafka 等。

23.2.1 Nutch 安装

在 Nutch 的下载页面可以选择所需的版本，如下所示：

http://nutch.apache.org/downloads.html

选择相应的版本，进入镜像下载页面，如下所示：

https://www.apache.org/dyn/closer.lua/nutch/1.17/apache-nutch-1.17-bin.zip

在镜像下载页面中，可以根据自己当前的网络情况来选择合适的镜像链接，本书选择的镜像链接，如下所示：

https://mirrors.bfsu.edu.cn/apache/nutch/1.17/apache-nutch-1.17-bin.zip

下载完成后，把 Nutch 软件包解压到指定目录，进入主目录查看，如图 23-4 所示。

```
[quxy@svr-130 softwares]$ cd apache-nutch-1.17/
[quxy@svr-130 apache-nutch-1.17]$ ll
total 472
drwxr-xr-x.  2 quxy quxy     32 Jun 18 09:12
-rw-rw-r--.  1 quxy quxy 129101 Jun 18 09:12 CHANGES.txt
drwxr-xr-x.  2 quxy quxy   4096 Jun 18 09:12
drwxr-xr-x.  3 quxy quxy     17 Jun 18 09:12
drwxr-xr-x.  8 quxy quxy   8192 Jun 18 09:12
-rw-rw-r--.  1 quxy quxy 326348 Jun 18 09:12 LICENSE.txt
-rw-rw-r--.  1 quxy quxy    427 Jun 18 09:12 NOTICE.txt
drwxr-xr-x. 78 quxy quxy   4096 Jun 18 09:12
[quxy@svr-130 apache-nutch-1.17]$
```

图 23-4 Nutch 主目录

首先，执行命令：$ bin/nutch，验证 Nutch 是否可用，如图 23-5 所示。

```
[quxy@svr-130 apache-nutch-1.17]$ bin/nutch
nutch 1.17
Usage: nutch COMMAND [-Dproperty=value]... [command-specific args]...
where COMMAND is one of:
  readdb            read / dump crawl db
  mergedb           merge crawldb-s, with optional filtering
  readlinkdb        read / dump link db
  inject            inject new urls into the database
  generate          generate new segments to fetch from crawl db
  freegen           generate new segments to fetch from text files
  fetch             fetch a segment's pages
  parse             parse a segment's pages
  readseg           read / dump segment data
```

图 23-5 启动 Nutch（1）

然后，执行命令来启动 Nutch 服务器，如下所示：

$ bin/nutch startserver -host 192.168.226.130 -port 8081

接着，执行命令来启动 Nutch UI，如下所示：

$ bin/nutch webapp

Nutch UI 启动后，在浏览器中访问 URL：http://192.168.226.130:8080，如图 23-6 所示。

图 23-6 启动 Nutch（2）

进入 Nutch UI 界面后，第一步要做的事情就是关联 Nutch 服务器，如图 23-7 所示。

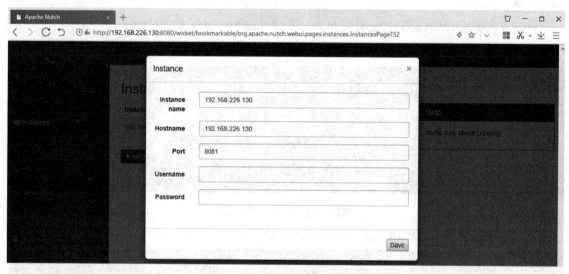

图 23-7 关联 Nutch 服务器

有关 Apache Nutch 的更多信息，请查看教程文档，如下所示：

https://cwiki.apache.org/confluence/display/NUTCH/NutchTutorial

Apache Nutch 的源代码，如下所示：

https://github.com/apache/nutch

23.2.2 简单示例

现在，我们讲解一个抓取数据的简单示例。在对网页进行抓取之前，Nutch 需要添加相应的配置，如下所示：

1) 自定义爬网属性，其中至少要为爬网程序提供名称，以供外部服务器识别；

2) 设置要爬网的 URL 种子列表。

首先，编辑 conf/nutch-site.xml 配置文件，添加爬网程序的名称，如下所示：

```
<property>
  <name>http.agent.name</name>
  <value>Gf Nutch Spider</value>
</property>
```

第二步是执行下列命令创建 URL 种子列表：

```
$ mkdir -p urls && cd urls
$ echo "http://nutch.apache.org" >> seed.txt
```

该种子列表存放于 seed.txt 文件中，如图 23-8 所示。

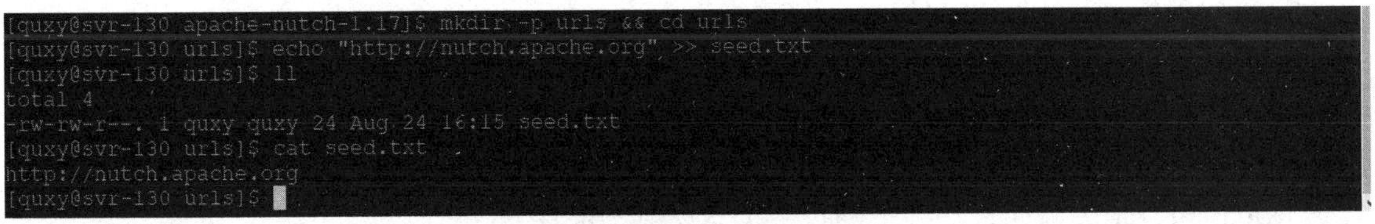

```
[quxy@svr-130 apache-nutch-1.17]$ mkdir -p urls && cd urls
[quxy@svr-130 urls]$ echo "http://nutch.apache.org" >> seed.txt
[quxy@svr-130 urls]$ ll
total 4
-rw-rw-r--. 1 quxy quxy 24 Aug 24 16:15 seed.txt
[quxy@svr-130 urls]$ cat seed.txt
http://nutch.apache.org
[quxy@svr-130 urls]$
```

图 23-8 创建 URL 种子列表

使用 URL 列表为 crawldb 设定种子，执行命令：$ bin/nutch inject crawl/crawldb urls，如图 23-9 所示。

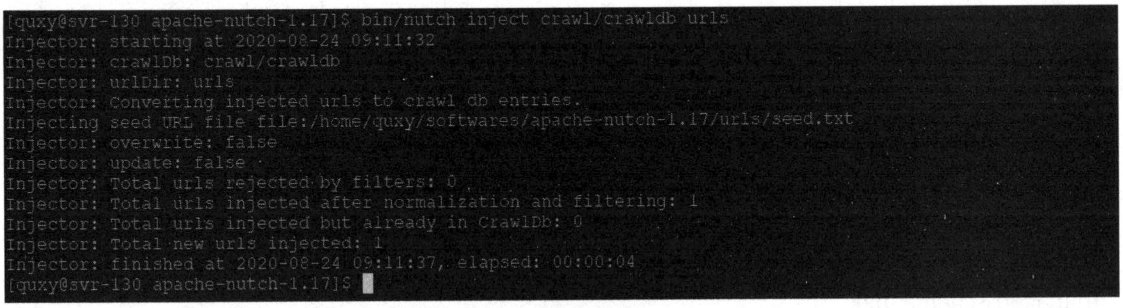

```
[quxy@svr-130 apache-nutch-1.17]$ bin/nutch inject crawl/crawldb urls
Injector: starting at 2020-08-24 09:11:32
Injector: crawlDb: crawl/crawldb
Injector: urlDir: urls
Injector: Converting injected urls to crawl db entries.
Injecting seed URL file file:/home/quxy/softwares/apache-nutch-1.17/urls/seed.txt
Injector: overwrite: false
Injector: update: false
Injector: Total urls rejected by filters: 0
Injector: Total urls injected after normalization and filtering: 1
Injector: Total urls injected but already in CrawlDb: 0
Injector: Total new urls injected: 1
Injector: finished at 2020-08-24 09:11:37, elapsed: 00:00:04
[quxy@svr-130 apache-nutch-1.17]$
```

图 23-9 设定抓取种子

为了抓取数据，先在数据库中生成一个获取列表，执行命令：$ bin/nutch generate crawl/crawldb crawl/segments，如图 23-10 所示。

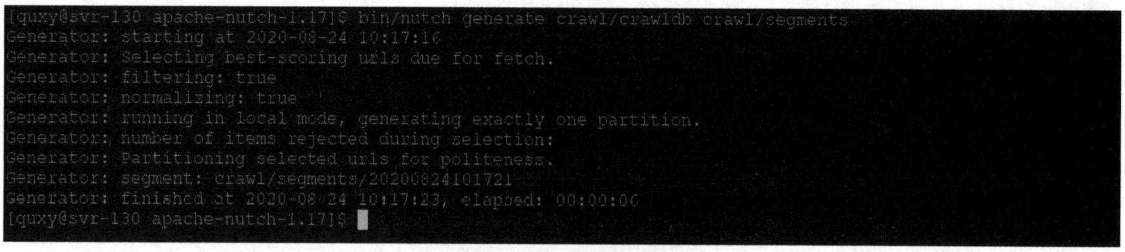

图 23-10 获取抓取列表

上面的操作将为所有要获取的页面生成一个获取列表。获取列表放在新创建的段目录中，该段目录用创建时间命名。我们将此段目录的名称保存在 shell 变量 s1 中：

$ s1=`ls -d crawl/segments/2* | tail -1`
$ echo $s1

执行命令获取网页数据，如下所示：
$ bin/nutch fetch $s1

执行命令解析条目，如下所示：
$ bin/nutch parse $s1

执行命令把得到的结果更新到数据库中，如下所示：
$ bin/nutch updatedb crawl/crawldb $s1

在建立索引之前，反转所有链接，以便可以索引传入的锚文本，执行命令，如下所示：
$ bin/nutch invertlinks crawl/linkdb -dir crawl/segments

经过以上操作后，我们已经获取到了 URL 对应的网页数据。接下来，我们将设置一些配置把 Nutch 和 Solr 关联起来，并把网页数据索引到 Solr 中。

首先，进入到 Solr 的主目录中，新建 nutch 文件夹，如下所示：
$ mkdir -p ./server/solr/configsets/nutch

把相关的配置文件复制到 nutch 目录中，执行命令，如下所示：
$ cp -r ./server/solr/configsets/_default/* ./server/solr/configsets/nutch/

接着，把 Nutch 应用中的 schema.xml 文件复制到 nutch/config 目录中，执行命令，如下所示：
$ cp ../apache-nutch-1.17/plugins/indexer-solr/schema.xml ./server/solr/configsets/nutch/conf/

把 nutch/config 目录中的 managed-schema 文件重命名，执行命令，如下所示：
$ mv ./server/solr/configsets/nutch/conf/managed-schema ./server/solr/configsets/nutch/conf/managed-schema_bak

完成以上操作后，执行命令启动 Solr，如下所示：
$ bin/solr start

创建一个核心 nutch，执行命令，如下所示：
$ bin/solr create -c nutch -d ./server/solr/configsets/nutch/conf/

接下来，编辑 Nutch 服务器中 conf/index-writers.xml 配置文件，以指定 Solr 服务器。默认情况下，连接的是本机的 Solr 服务器，如图 23-11 所示。

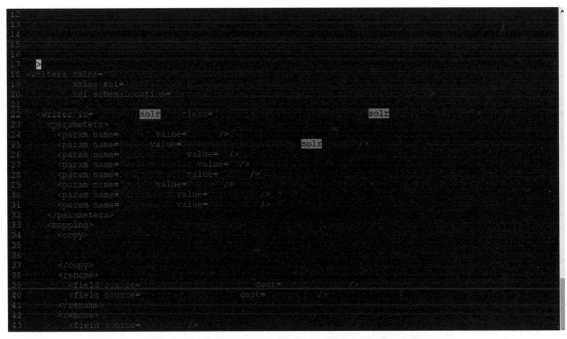

图 23-11 conf/index-writers.xml 配置文件

最后，我们终于可以把网页数据索引到 Solr 服务器了，执行命令，如下所示：

$ bin/nutch index crawl/crawldb/ -linkdb crawl/linkdb/ \
crawl/segments/20200824101721/ -filter -normalize -deleteGone

索引完成之后，访问 URL：http://192.168.226.130:8983/solr/#/nutch/query，查看数据，如图 23-12 所示。

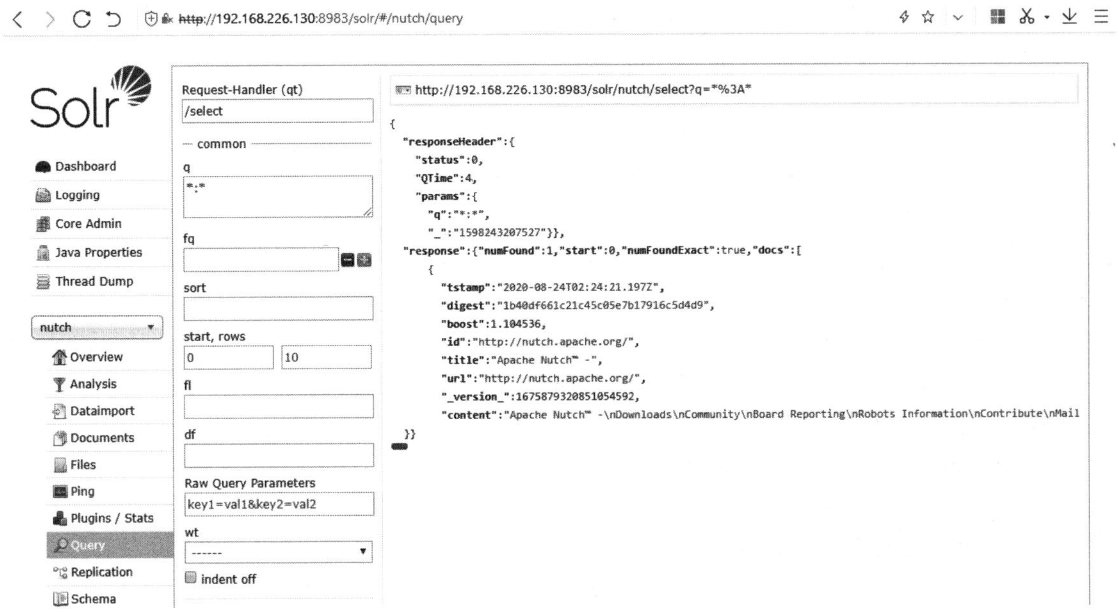

图 23-12 在 Solr 中查看数据

23.3 Spider Flow

Spider Flow 是一个爬虫平台，以图形化方式定义爬虫流程，无需代码即可实现一个爬虫。它有丰富的插件支持，它高度灵活、扩展方便和规则定制灵活，它是基于 Spring Boot 开发的。Spider Flow 的官方网站：https://www.spiderflow.org，如图 23-13 所示。

图 23-13 Spider Flow 的官方网站

功能特性
1) 支持 CSS 选择器和正则提取；
2) 支持 JSON/XML 格式；
3) 支持 Xpath/JsonPath 提取；
4) 支持多数据源和 SQL 操作；
5) 支持爬取用 JavaScript 动态渲染的页面；
6) 支持代理；
7) 支持二进制格式；
8) 支持保存/读取文件（如 csv、xls 和 jpg 等）；
9) 具有常用字符串、日期、文件、加解密、随机等函数；
10) 支持流程嵌套；
11) 支持插件扩展（自定义执行器、自定义函数、自定义 Controller 和类型扩展等）；
12) 支持 HTTP 接口；

在编写本书时，Spider Flow 最新版本为 v0.5.0。查看其最新的版本情况，请访问其源码仓库，在 Gitee 和 GitHub 上都有其源代码，如下所示：

https://gitee.com/jmxd/spider-flow
https://github.com/javamxd/spider-flow

接下来，把 Spider Flow 源代码检出到 IntelliJ IDEA 中，如下所示：

https://gitee.com/jmxd/spider-flow.git

检出后，执行 Maven 的 clean 和 compile 命令，如图 23-14 所示。

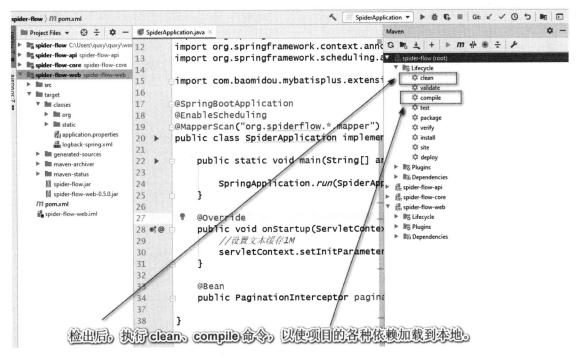

图 23-14 编译 Spider Flow

要想把项目运行起来，我们所要做的第一步就是创建数据表。在这里，我们将数据表导入到 MariaDB 数据库服务器中的 spiderflow 数据库中。数据表定义文件在项目中的位置，如图 23-15 所示。

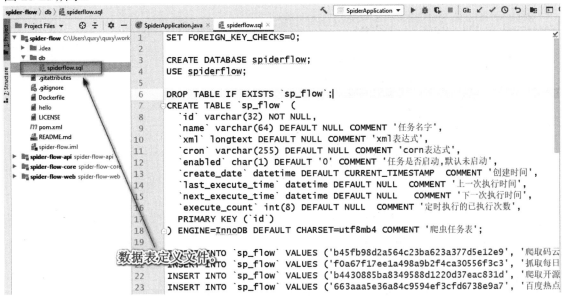

图 23-15 数据表定义文件

我们使用 SQL Front 作为数据库服务器的客户端，将 spiderflow.sql 文件中的 SQL 语句拷贝到该客户端中，并执行所有的 SQL 语句，如图 23-16 所示。

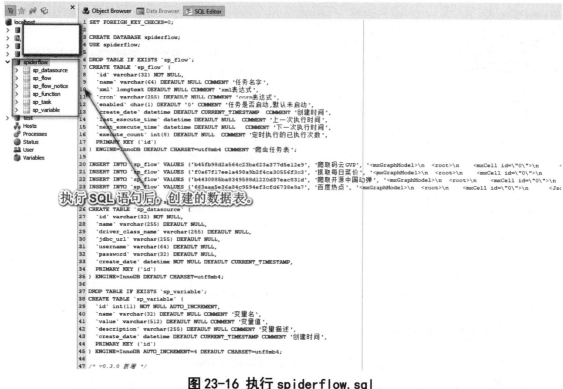

图 23-16 执行 spiderflow.sql

在子模块 spider-flow-web 中，找到主类 SpiderApplication，运行它，控制台的输出，如图 23-17 所示。

```
"C:\Program Files\Java\jdk1.8.0_171\bin\java.exe" ...
Connected to the target VM, address: '127.0.0.1:63241', transport: 'socket'

  .   ____          _            __ _ _
 /\\ / ___'_ __ _ _(_)_ __  __ _ \ \ \ \
( ( )\___ | '_ | '_| | '_ \/ _` | \ \ \ \
 \\/  ___)| |_)| | | | | || (_| |  ) ) ) )
  '  |____| .__|_| |_|_| |_\__, | / / / /
 =========|_|==============|___/=/_/_/_/
 :: Spring Boot ::        (v2.0.7.RELEASE)

2020-05-15 11:44:26.656 [main] INFO  org.spiderflow.SpiderApplication - Starting
 SpiderApplication on DESKTOP-DGFGBUD with PID 7564
 (C:\Users\quxy\quxy\workSpaces\ideaSpaces\IdeaProjects20200212\spider-flow\spider-flow-web
 \target\classes started by quxy in
 C:\Users\quxy\quxy\workSpaces\ideaSpaces\IdeaProjects20200212\spider-flow)
2020-05-15 11:44:26.679 [main] INFO  org.spiderflow.SpiderApplication - No active profile set,
 falling back to default profiles: default
2020-05-15 11:44:26.911 [main] INFO  o.s.b.w.s.c
.AnnotationConfigServletWebServerApplicationContext - Refreshing org.springframework.boot
.web.servlet.context.AnnotationConfigServletWebServerApplicationContext@95e33cc: startup
 date [Fri May 15 11:44:26 CST 2020]; root of context hierarchy
2020-05-15 11:44:32.248 [main] INFO  o.s.boot.web.embedded.tomcat.TomcatWebServer - Tomcat
 initialized with port(s): 8088 (http)
```

图 23-17 控制台输出信息（1）

从图 23-17 中可以看出：Spring Boot 的版本为 v2.0.7.RELEASE，服务的端口为 8088。

```
2020-05-15 11:44:33.579 [main] INFO  c.a.d.s.b.a.DruidDataSourceAutoConfigure - Init
DruidDataSource
2020-05-15 11:44:33.917 [main] INFO  com.alibaba.druid.pool.DruidDataSource - {dataSource-1}
inited
   _ _  |_  _ _|_ ._ __ |
 | | |\/|_) /_\ |_ |_\  |_)|   |_|_\
         /                    |
                           3.1.0
2020-05-15 11:44:35.107 [main] ERROR com.baomidou.mybatisplus.core.MybatisConfiguration -
mapper[org.spiderflow.core.mapper.SpiderFlowMapper.resetNextExecuteTime] is ignored, because
 it exists, maybe from xml file
2020-05-15 11:44:37.152 [main] INFO  org.quartz.impl.StdSchedulerFactory - Using default
implementation for ThreadExecutor
2020-05-15 11:44:37.198 [main] INFO  org.quartz.core.SchedulerSignalerImpl - Initialized
Scheduler Signaller of type: class org.quartz.core.SchedulerSignalerImpl
2020-05-15 11:44:37.199 [main] INFO  org.quartz.core.QuartzScheduler - Quartz Scheduler
v.2.3.0 created.
2020-05-15 11:44:37.201 [main] INFO  org.quartz.simpl.RAMJobStore - RAMJobStore initialized.
2020-05-15 11:44:37.216 [main] INFO  org.quartz.core.QuartzScheduler - Scheduler meta-data:
Quartz Scheduler (v2.3.0) 'quartzScheduler' with instanceId 'NON_CLUSTERED'
  Scheduler class: 'org.quartz.core.QuartzScheduler' - running locally.
  NOT STARTED.
  Currently in standby mode.
  Number of jobs executed: 0
  Using thread pool 'org.quartz.simpl.SimpleThreadPool' - with 10 threads.
  Using job-store 'org.quartz.simpl.RAMJobStore' - which does not support persistence. and is
not clustered.
```

图 23-18 控制台输出信息（2）

从图 23-18 中可以看出：所使用的持久层框架为 MyBatis-Plus 3.1.0。启动后，在浏览器中访问 URL：http://localhost:8088，如图 23-19 所示。

图 23-19 访问 spider-flow

接下来，我们举一个最简单的示例：提取指定的网页文章中的标题、发布日期、概述和正文。在爬虫列表页面，单击按钮[添加爬虫]，并添加开始抓取和输出这两个节点，如图 23-20 所示。

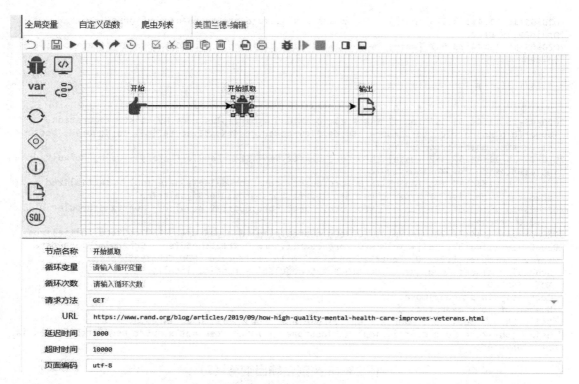

图 23-20 添加抓取节点

在"开始抓取"节点的详情页面中，给 URL 字段添加要抓取的网址：https://www.rand.org/blog/articles/2019/09/how-high-quality-mental-health-care-improves-veterans.html。单击节点[输出]，并添加输出项，如图 23-21 所示。

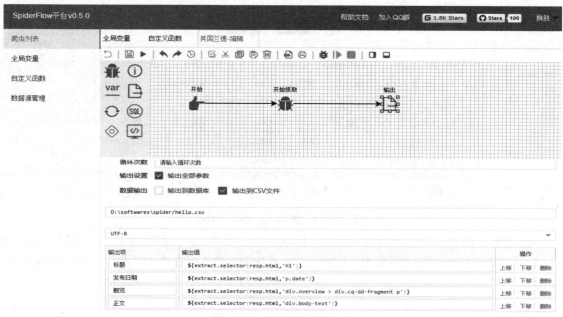

图 23-21 添加输出节点

在这里，我们选择"输出到 CSV 文件"，并指定文件的绝对路径。使用 Spider Flow 的语法格式及 CSS 选择器，从网页中依次提取出：标题、发布日期、概述和正文。配置好后，单击页面左上角的按钮[保存]，再单击按钮[测试]，执行结果，如图 23-22 所示。

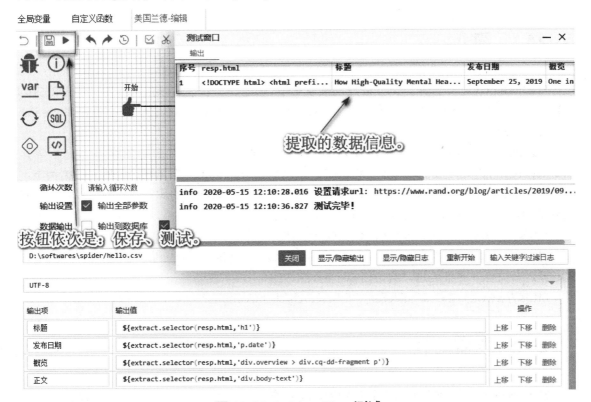

图 23-22 Spider Flow 测试

测试执行完成后，指定的 CSV 文件（若不存在，则会自动生成该文件）中也会生成相应的数据。有关 Spider Flow 的具体用法，这里只是举了一个简单的例子。想要了解更多信息，请访问其官方文档，如下所示：

https://www.spiderflow.org/intro.html

23.4 本章小结

本章先简单地讲解了 Apache Solr 的相关内容，之后讲解了 Apache Nutch，以及二者的结合使用。最后讲解了 Spider Flow，通过其 web 界面的配置即可实现强大的爬虫功能。另外，通过 Python 语言的相关模块也可以实现灵活多变的爬虫功能。

第二十四章 Elasticsearch

24.1 简介

Elasticsearch 是一个基于 Lucene 库的搜索引擎。它是一个分布式、高扩展、高实时的 RESTful 风格的搜索与数据分析引擎。它能够解决不断涌现出的各种用例。本书中所用的 Elasticsearch 版本是：7.7.0。

Elasticsearch 可以用于多个应用场景：应用搜索、安全分析、指标或日志分析等。Elasticsearch 可以处理数字、文本、地理位置、结构化数据和非结构化数据等所有数据类型。

Elasticsearch 作为 Elastic Stack 的核心，它集中存储数据，帮助你发现意料之中以及意料之外的情况。有关更多详细信息，请访问以下网址：

https://www.elastic.co/cn/products/elasticsearch

Elasticsearch 支持多语言客户端：Curl、Java、C#、Python、JavaScript、PHP、Perl、Ruby、SQL 等。有关更多详细信息，请访问以下网址：

https://www.elastic.co/guide/en/elasticsearch/client/community/current/index.html

24.2 发展历程

Shay Banon 在 2004 年创造了 Elasticsearch 的前身 Compass。在考虑 Compass 的第三个版本时，他意识到有必要重写 Compass 的大部分内容，以创建一个可扩展的搜索解决方案。因此，他创建了"一个从头构建的分布式解决方案"，并使用了公共接口，即 HTTP 上的 JSON，它也适用于 Java 以外的编程语言。

Shay Banon 在 2010 年 2 月发布了 Elasticsearch 的第一个版本。

Elasticsearch BV 成立于 2012 年，主要围绕 Elasticsearch 及相关软件提供商业服务和产品。

2014 年 6 月，也就是在成立公司 18 个月后，该公司宣布通过 C 轮融资筹集 7000 万美元。这轮融资由新企业协会（NEA）牵头。

2015 年 3 月，Elasticsearch 公司更名为 Elastic。

在 2018 年 6 月，Elastic 提交了首次公开募股申请，估值在 15 亿到 30 亿美元之间。公司于 2018 年 10 月 5 日在纽约证券交易所挂牌上市。

24.3 基本概念

Elasticsearch 从 7.0 版本开始，在每个发行版中包含 JDK 维护者（GPLv2+CE）提供的 OpenJDK 捆绑版本。当然，还是可以指定自己的 JDK 版本的，有关详情，请参考以下网址：

https://www.elastic.co/guide/en/elasticsearch/reference/7.0/setup.html#jvm-version

OpenJDK 捆绑版本的位置及其版本号，如图 24-1、图 24-2 所示。

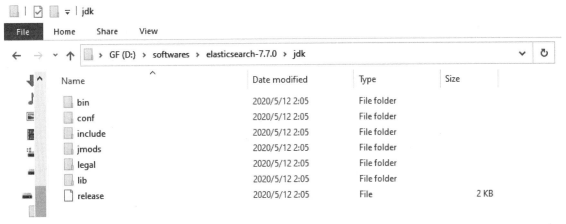

图 24-1 OpenJDK 主目录

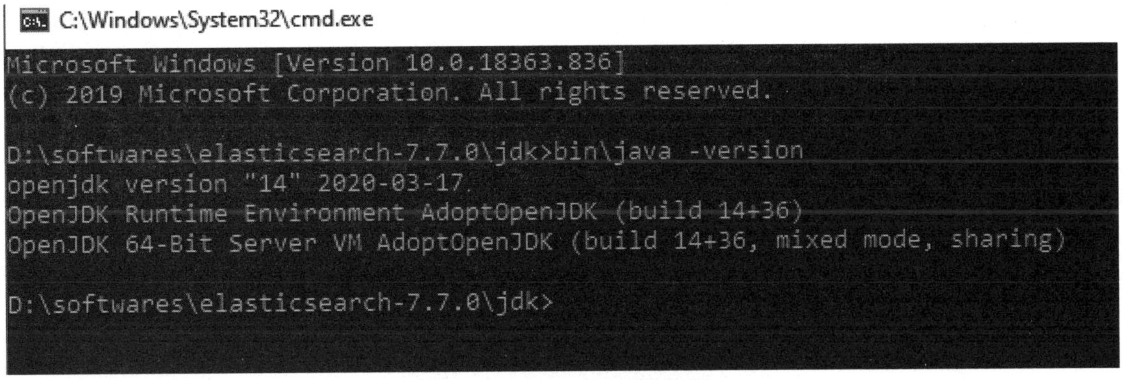

图 24-2 OpenJDK 版本号

24.3.1 正向索引（Forward Index）

在正向索引中，每条数据以文档 ID 作为键，以文档的单词列表作为值。具体形式，如下所示：
1) "文档 1"的 ID --> 单词 1：出现次数、出现位置列表；单词 2：出现次数、出现位置列表；…。
2) "文档 2"的 ID --> 此文档出现的关键词列表。

24.3.2 倒排索引（Inverted Index）

倒排索引是实现"单词-文档矩阵"的一种具体存储形式，通过倒排索引，可以根据单词快速获取包含这个单词的文档列表。倒排索引主要由两个部分组成：单词词典和倒排文件。

假设使用正向索引存储数据，当用户搜索关键词"中国"时，那么就需要扫描索引库中的所有文档，找出所有包含关键词"中国"的文档，并打分，然后排序并呈现。因为从互联网上收录的文档数量非常巨大，这种结构显然无法满足要求。

为了替代正向索引的每个文档的单词列表，倒排索引（反向索引）数据结构开发了出来，它能列出每个单词的所有所在文档的列表。

24.3.3 集群（Cluster）

一个集群是由一或多个集群名称相同的节点组成。推荐 3 个节点作为一个基础集群。在同一台服务器上（主机名称和 IP 地址相同），默认情况下不允许分配多个相同的分片实例，但是可以通过设置参数为 true 来调整，如下所示：

cluster.routing.allocation.same_shard.host=true

在同一台服务器上运行多个节点实例时，可以应用这个设置。想要查看集群健康状况，请在 Kibana 开发控制台中分别输入命令：GET /_cluster/health?pretty、GET /_cat/health?v，系统会显示集群运行状况，如图 24-3、图 24-4 所示。

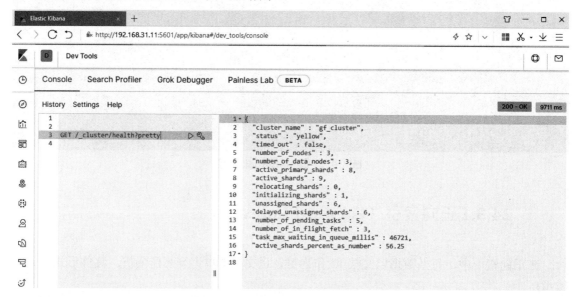

图 24-3 查看集群运行状况（1）

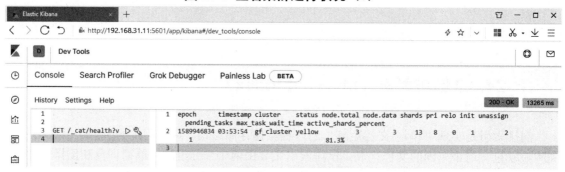

图 24-4 查看集群运行状况（2）

集群的健康状况，用绿、黄和红三种颜色表示，如下所示：

1) 绿（green）：一切正常，集群功能齐全；
2) 黄（yellow）：所有数据可用，但有些副本未分配（不可用），集群功能齐全；
3) 红（red）：有些数据不可用，集群功能有些不可用。

有关集群状态的更详细的信息，请执行命令：GET /_cluster/stats?pretty，如图 24-5 所示。

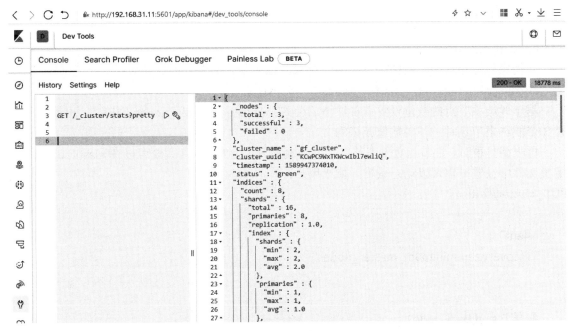

图 24-5 集群状态详细信息（1）

要列出 /_cat 路径下所有的 URI，请执行命令：GET /_cat，如图 24-6 所示。

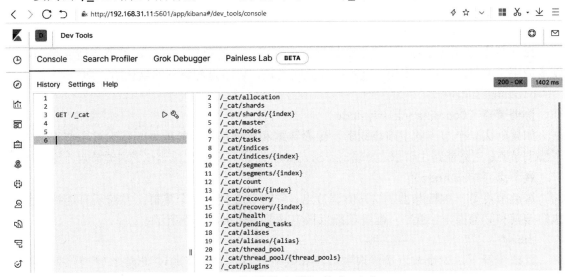

图 24-6 集群状态详细信息（2）

24.3.4 节点（Node）

候选主节点（Master-eligible node）

主节点负责在集群范围内执行轻量级操作，比如创建或删除索引、跟踪哪些节点是集群的一部分，以及决定将哪些分片分配给哪些节点。配置项的默认值，如下所示：

node.master=true

要使集群避免脑裂，请使用下面的配置项，以下是该配置项的默认值：

discovery.zen.minimum_master_nodes=1

基于候选主节点的数量设置该参数值的公式：(master_eligible_nodes / 2) + 1。也就是，当有 3 个候选主节点时，该参数值应该为 2。

再平衡集群时，主节点负责在节点之间移动分片。当然，该参数既可以在 elasticsearch.yml 配置文件中设置，也可以动态设置，如下所示：

```
PUT _cluster/settings
{
 "transient": {
    "discovery.zen.minimum_master_nodes": 2
 }
}
```

数据节点（Data node）

这类节点用于保存数据及执行与数据相关的操作：如 CRUD、搜索和聚合。以下是配置项的默认值：

node.data=true

摄取节点（Ingest node）

摄取节点可以执行由一个或多个摄取处理器组成的预处理管道。这类节点能够将 Ingest 管道应用于文档，以便在索引之前转换和充实文档。以下是配置项的默认值：

node.ingest=true

协调节点（Coordinating only node）

如果除去一个节点的主节点职责、保存数据和预处理文档的能力，那么这个节点就成为了协调节点。它只能路由请求、处理 search reduce 阶段和分发批量索引。这类节点不必太多。

族节点（Tribe node）

这类节点是一种特殊类型的只协调节点。它可以连接到多个集群，并跨所有连接的集群执行搜索和其他操作。族节点配置项都以指定的前缀开头，如下所示：

tribe.*

默认情况下，一个节点是候选主节点和数据节点，并且能通过 Ingest 管道预处理数据（documents）。这对小集群是非常便利的，但当集群变得较大时，考虑把专用的候选主节点与专用的数据节点分离开，这显得比较重要。

机器学习节点（Machine learning node）

想要机器学习节点，请设置如下配置项：

xpack.ml.enabled=true
node.ml=true

有关机器学习节点的详细信息，请访问以下网址：

https://www.elastic.co/guide/en/elastic-stack-overview/7.7/xpack-ml.html

要查看节点列表，请执行命令：GET /_cat/nodes?v，如图 24-7 所示。

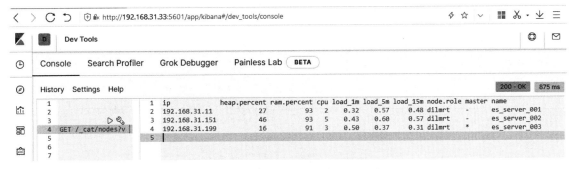

图 24-7 查看所有集群节点列表

要查看所有集群节点的详细信息，请执行命令：GET /_nodes/process?pretty，如图 24-8 所示。

图 24-8 查看所有集群节点详细信息

要查看指定节点的详细信息（es_server_001 为节点名称），请执行命令：GET /_nodes/es_server_001/process?pretty，如图 24-9 所示。

图 24-9 查看指定节点详细信息

有关节点的详细信息，请访问以下网址：
https://www.elastic.co/guide/en/elasticsearch/reference/7.7/modules-node.html

24.3.5 索引库（Index）

一个索引库中包含多个文档。在添加文档（document）时，若对应的索引库不存在，则会自动创建该索引库。索引库名字必须是全部小写，不允许以下划线开头，不能包含逗号。想要查看索引库列表，请执行命令：GET /_cat/indices?v，如图 24-10 所示。

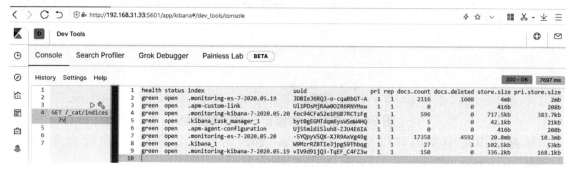

图 24-10 查看索引库列表

24.3.6 类型（Type）

之前，一个 type 表示一个索引库中一组文档的类型。在 7.0 中，mappings 元素在默认情况下不再将 type 作为顶级键。在 Elasticsearch 6.8 中，已经可以选择这一行为，即可以设置相关的请求参数，如下所示：

include_type_name=false，

设置后，可以直接将映射放在索引库创建时调用的 mappings 下，而无需指定类型名称，如下所示：

```
PUT test?include_type_name=false
{
 "mappings": {
    "properties": {
     "foo": {
        "type": "keyword"
     }
    }
 }
}
```

有关详细信息，请访问以下网址：
https://www.elastic.co/guide/en/elasticsearch/reference/7.7/indices-create-index.html#_skipping_types

24.3.7 文档（Document）

一个文档就是一条包含一或多个字段及其值的数据。

字段（Field）

字段是文档下级，一般情况下，多个字段及其值组成一条文档数据。它相当于数据库表中的列。

24.3.8 分片、副本（Shards、Replicas）

一个分片就是一个 Lucene 实例，并且它本身就是一个完整的搜索引擎。索引库中的每个文档属于一个单独的主分片，所以主分片的数量决定了索引库最多能存储多少数据。索引库创建完成时，主分片的数量就固定了，但是副本分片的数量可以随时调整。

注意：可以通过拆分索引库（Split Index）API，基于现有索引库去创建一个新的索引库，以增加主分片的数量。

有关 Elasticsearch 不支持增量分片的详细原因，请访问以下网址：
https://www.elastic.co/guide/en/elasticsearch/reference/7.7/indices-split-index.html#incremental-resharding

一个单一的 Lucene 索引中存储的文档最大值为 2,147,483,519（= Integer.MAX_VALUE - 128）个。一个分片就是一个 Lucene 索引，所以该分片的文档上限也是该值。想要监控分片的大小，请执行命令：GET /_cat/shards，如图 24-11 所示。

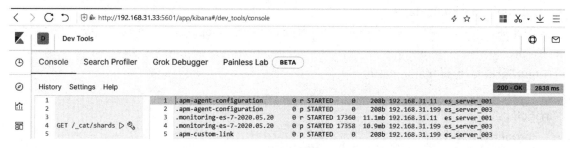

图 24-11 监控分片

注意：分片的平均大小应在几 GB～几十 GB 之间。也要避免分片数量过大的问题，一个节点上分片数量应当与可用的堆栈空间相称。一般情况下，每 GB 堆栈空间对应的分片数量不应当超过 20。

24.4 下载页面

打开 URL：https://www.elastic.co/cn/start，访问最新软件包的下载页面，如图 24-12 所示。

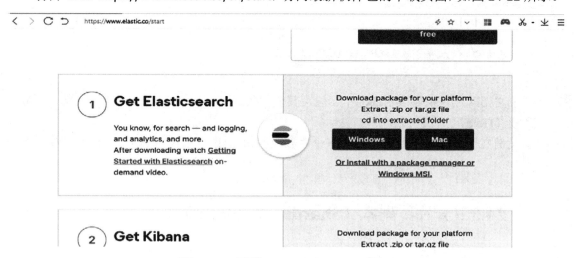

图 24-12 下载 Elasticsearch（1）

更多相关软件包的下载链接为：https://www.elastic.co/cn/downloads，下载页面如图 24-13 所示。

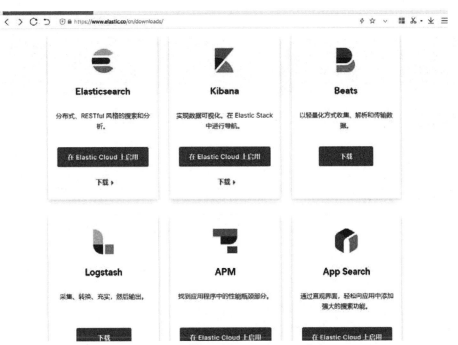

图 24-13 下载 Elasticsearch（2）

在上图中，单击 Elasticsearch 的[下载]按钮，在新打开的页面中会列出最新的 Elasticsearch 的各种格式的下载链接，如图 24-14 所示。

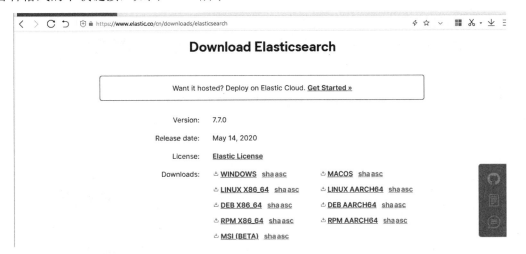

图 24-14 下载 Elasticsearch（3）

GitHub 上的源码包的下载链接 URL 为 https://github.com/elastic/elasticsearch/releases/tag/v7.7.0，下载页面如图 24-15 所示。

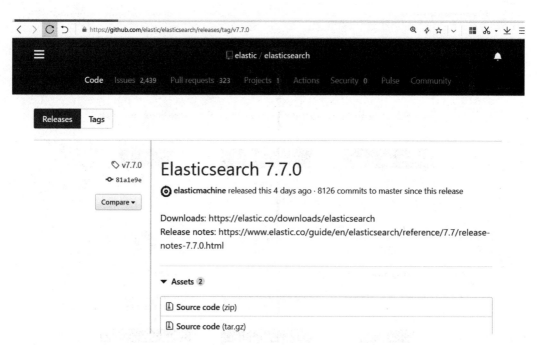

图 24-15 GitHub 上 Elasticsearch 源码包

若想下载 Elasticsearch 和其他软件包的某个历史版本，访问 URL：https://www.elastic.co/cn/downloads/past-releases#elasticsearch，下载页面如图 24-16 所示。

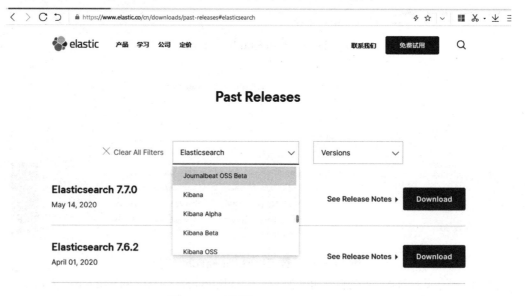

图 24-16 下载 Elasticsearch

24.5 Win 10：Elasticsearch 安装

在这里，使用 JDK 8 作为 Elasticsearch 的运行环境，官方推荐版本为 Java 11。在 Win 10

环境中，本书所有用的下载网址，如下所示：

https://artifacts.elastic.co/downloads/elasticsearch/elasticsearch-7.7.0-windows-x86_64.zip

下载完成后，将其解压到指定目录，然后进入到主目录，如图 24-17 所示。：

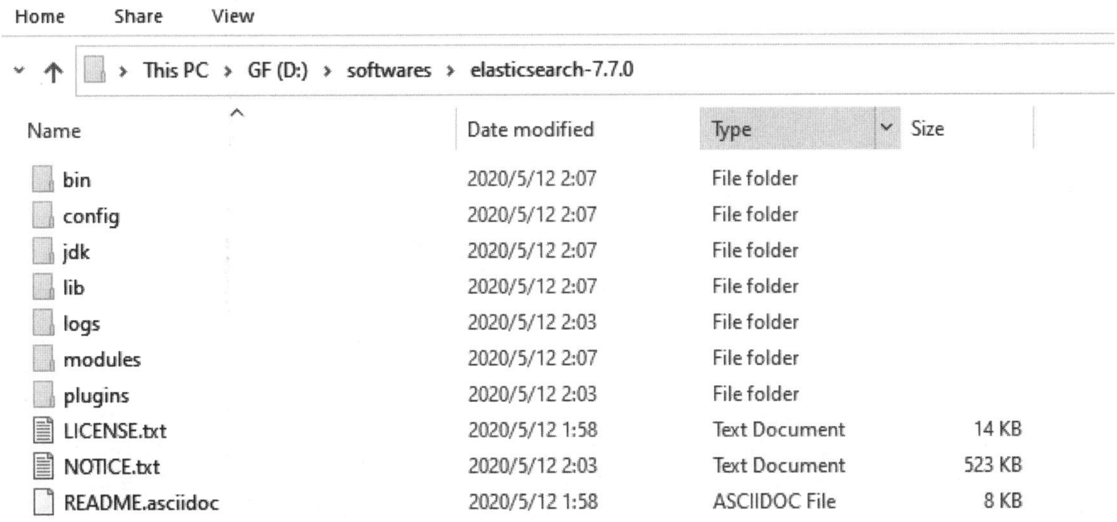

图 24-17 Elasticsearch 主目录

Elasticsearch 运行

单节点 Elasticsearch 服务器，默认情况下无需任何配置。要启动 Elasticsearch 服务器，请打开 CMD 命令行窗口，执行命令：bin\elasticsearch.bat，启动 Elasticsearch 服务，如图 24-18 所示。

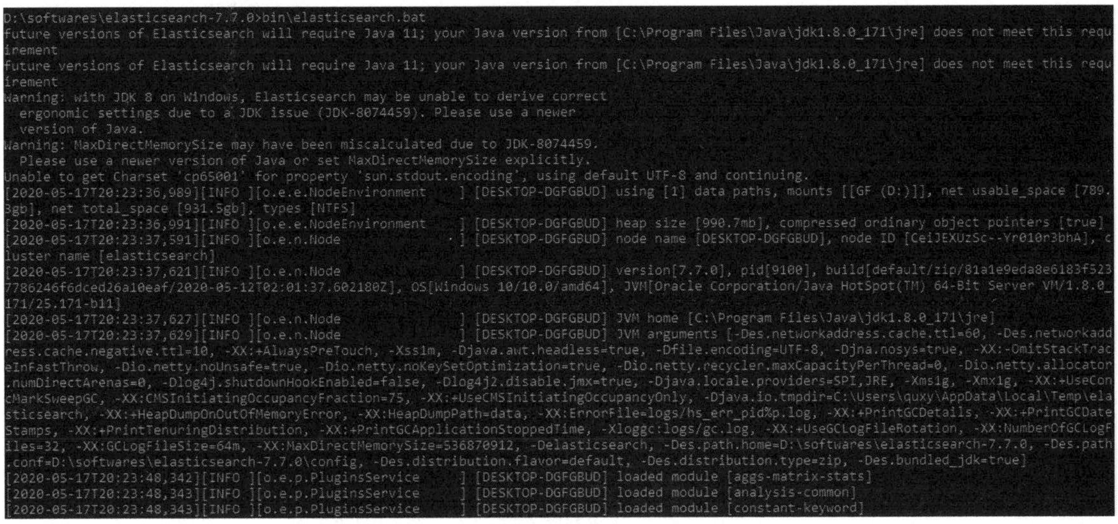

图 24-18 启动 Elasticsearch 服务器

启动成功后，在浏览器中访问 URL：http://localhost:9200，如图 24-19 所示。

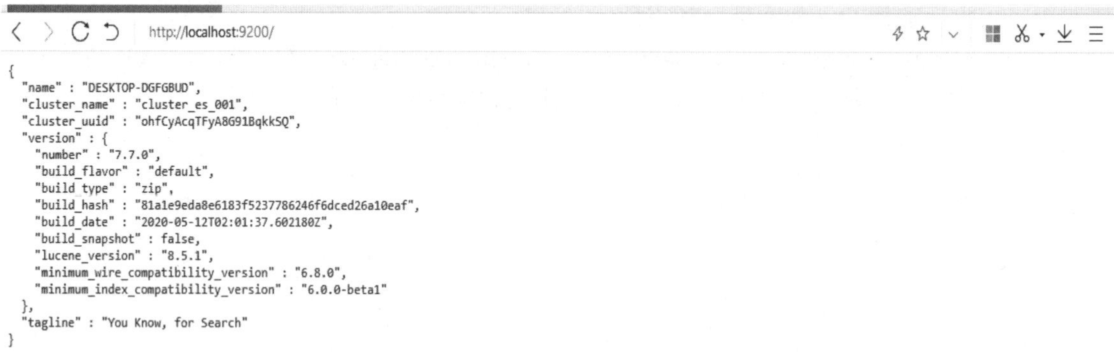

```
{
  "name" : "DESKTOP-DGFGBUD",
  "cluster_name" : "cluster_es_001",
  "cluster_uuid" : "ohfCyAcqTFyA8G91BqkkSQ",
  "version" : {
    "number" : "7.7.0",
    "build_flavor" : "default",
    "build_type" : "zip",
    "build_hash" : "81a1e9eda8e6183f5237786246f6dced26a10eaf",
    "build_date" : "2020-05-12T02:01:37.602180Z",
    "build_snapshot" : false,
    "lucene_version" : "8.5.1",
    "minimum_wire_compatibility_version" : "6.8.0",
    "minimum_index_compatibility_version" : "6.0.0-beta1"
  },
  "tagline" : "You Know, for Search"
}
```

图 24-19 Elasticsearch 服务器基本信息

从上图中可以看出：返回了 Elasticsearch 服务器的基本信息，说明其已经正常启动。要想后台启动，请执行以下命令：bin\elasticsearch.bat –d，后台启动 Elasticsearch 服务，如图 24-20 所示。

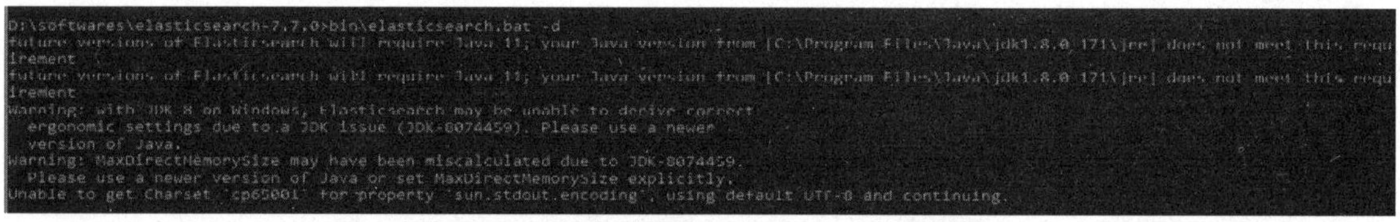

图 24-20 后台启动 Elasticsearch 服务

在 Windows 环境中，这种后台启动只是不输出控制台日志，关闭当前 CMD 窗口 Elasticsearch 服务器会停止的。在 Windows 环境中，可以把 Elasticsearch 服务器注册成为一个后台服务。

24.6　Win 10：Kibana 安装

在 Win 10 环境中，本书所有用的下载网址，如下所示：
https://artifacts.elastic.co/downloads/kibana/kibana-7.7.0-windows-x86_64.zip
下载完成后，将其解压到指定目录，然后进入到主目录，如图 24-21 所示。

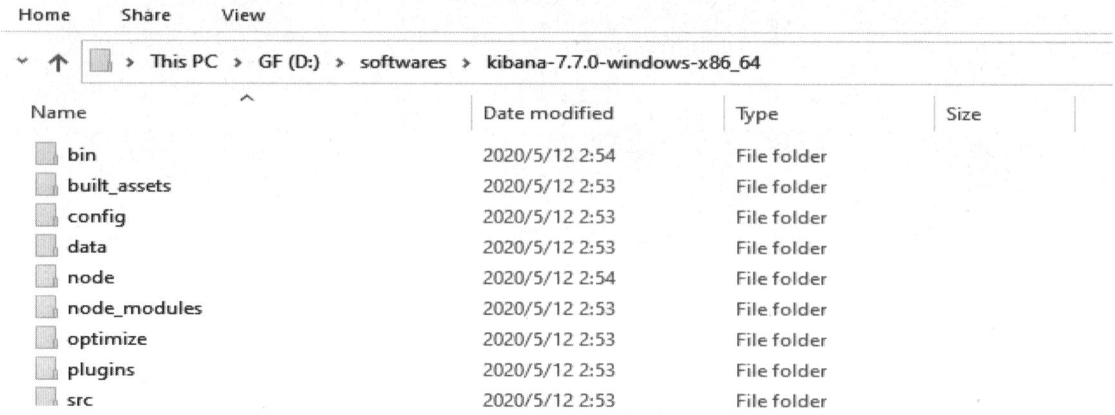

图 24-21 Kibana 主目录

Kibana 运行

Kibana 默认连接的 Elasticsearch 服务器的 URL，如下所示：

http://localhost:9200

Elasticsearch 服务器与 Kibana 在同一台机器上的话，默认情况下无需任何配置，即可实现对前者的访问和管理。要启动 Kibana 服务，请打开 CMD 命令行窗口，执行命令：bin\kibana.bat，启动 Kibana 服务，如图 24-22、图 24-23 所示。

图 24-22 启动 Kibana 服务（1）

图 24-23 启动 Kibana 服务（2）

有时候，Kibana 服务启动比较慢，需要耐心等待一下。启动成功后，在浏览器中访问 URL，如下所示：

http://localhost:5601

立即会跳转到其他页面，并进行初始化操作，请稍等片刻，初始化完成后，所展现的页面，如图 24-24 所示。

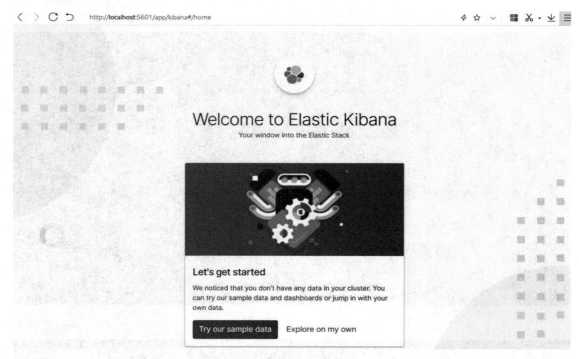

图 24-24 Kibana 服务

如图 24-24 中所示，单击按钮[Try our sample data]，来插入一些示例数据，单击后进入的页面，如图 24-25 所示。

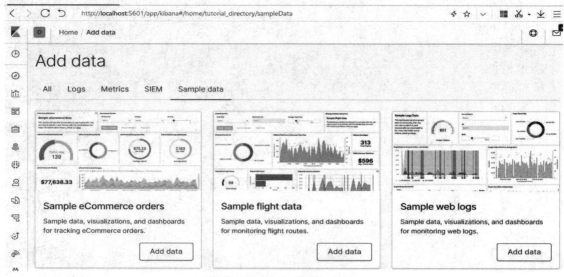

图 24-25 加入数据（1）

如上图中所示，单击按钮[Add data]，即可添加示例数据。数据添加完成后，会显示下拉按钮[View data]，打开它，并从下拉列表中选择一个按钮，如图 24-26 所示。

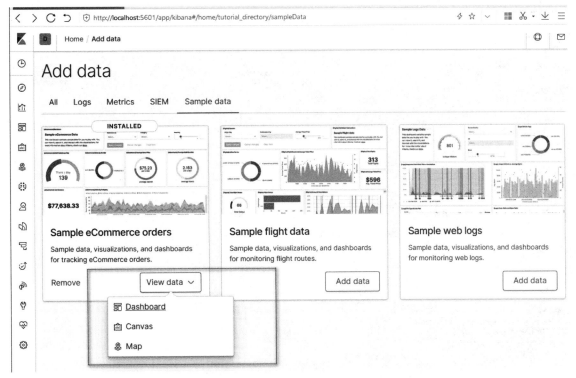

图 24-26 加入数据（2）

在这里，我们单击按钮[Dashboard]，打开后的页面，如图 24-27 所示。

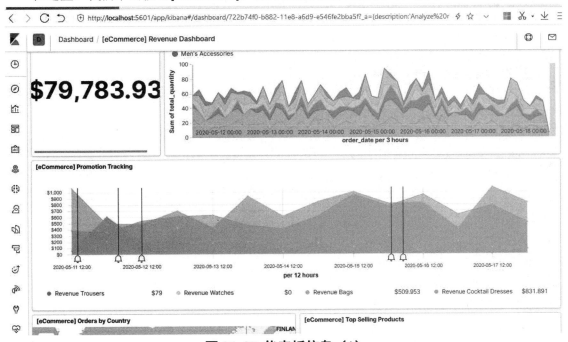

图 24-27 仪表板信息（1）

展开左侧的菜单栏,单击其下方的菜单[Stack Monitoring],并进行几步操作,即可展现相应的页面,如图 24-28 所示。

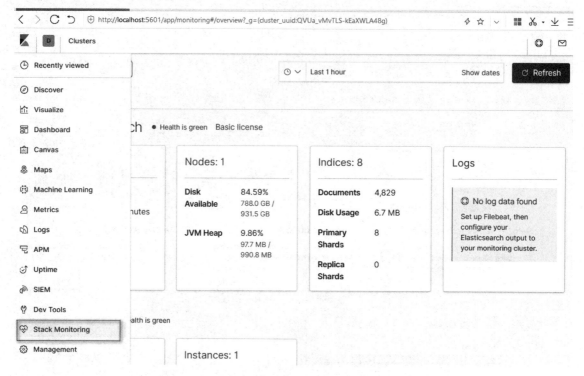

图 24-28 仪表板信息(2)

列出了 Elasticsearch 和 Kibana 的基本运行情况,如图 24-29 所示。

图 24-29 Elasticsearch 和 Kibana 的基本运行

有关菜单栏及其更多页面的情况，请在自己的机器上运行 Elasticsearch 和 Kibana 服务，并浏览查看。

24.7　Win 10：Elasticsearch 集群

在单台机器上，3 个 Elasticsearch 服务器组成的集群，它们使用相同的 Elasticsearch 主目录（$ES_HOME），无需任何配置，使用默认配置即可。启动命令，如下所示：

bin\elasticsearch.bat -E cluster.name=gf_cluster ^
-E node.name=es_server_001 -E http.port=9200 ^
-E path.data=D:\softwares\elasticsearch-7.7.0\data ^
-E path.logs=D:\softwares\elasticsearch-7.7.0\logs
bin\elasticsearch.bat -E cluster.name=gf_cluster ^
-E node.name=es_server_002 -E http.port=9202 ^
-E path.data=D:\softwares\elasticsearch-7.7.0\data2 ^
-E path.logs=D:\softwares\elasticsearch-7.7.0\logs2
bin\elasticsearch.bat -E cluster.name=gf_cluster ^
-E node.name=es_server_003 -E http.port=9203 ^
-E path.data=D:\softwares\elasticsearch-7.7.0\data3 ^
-E path.logs=D:\softwares\elasticsearch-7.7.0\logs3

在浏览器中，访问 URL：http://localhost:9200/_cat/nodes?v，查看这 3 个节点组成的集群情况，如图 24-30 所示。

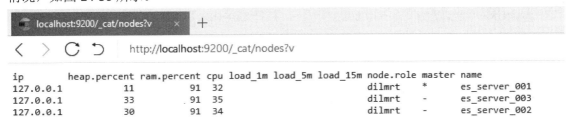

图 24-30 Elasticsearch 集群

在上图中，有 1 个主节点和 2 个从节点，其中带 * 号的表示主节点。这种集群方式不建议在生产环境中使用。对于这个集群，Kibana 也无需任何配置，打开 Kibana 的菜单[Stack Monitoring]，也可以看到这三个实例组成的集群，如图 24-31 所示。

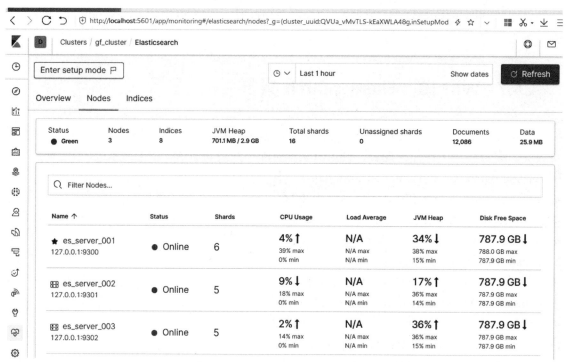

图 24-31 Kibana 中查看 Elasticsearch 集群

24.8 CentOS 8：Java 11 安装

访问 URL：http://jdk.java.net/java-se-ri/11，打开 OpenJDK 的下载页面，如图 24-32 所示。

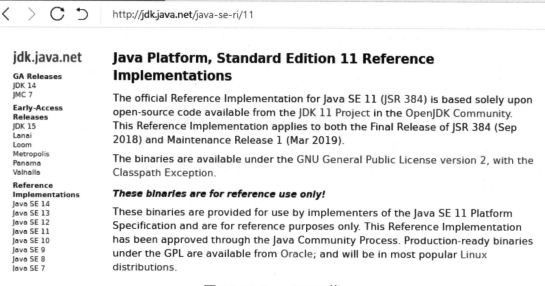

图 24-32 OpenJDK 下载

从上图的页面中，复制下载 URL，然后执行命令，如下所示：

$ wget https://download.java.net/openjdk/jdk11/ri/openjdk-11+28_linux-x64_bin.tar.gz

下载完成后，将其解压到指定目录，解压命令，如下所示：

tar xvf openjdk-11+28_linux-x64_bin.tar.gz -C /usr/java

接下来，还需要将 Java 11 添加到环境变量中，执行命令打开文件：# vim /etc/profile，添加环境变量参数：

export JAVA_HOME=/usr/java/jdk-11/

export CLASSPATH=.:$JAVA_HOME/jre/lib/rt.jar:$JAVA_HOME/lib/dt.jar:$JAVA_HOME/lib/tools.jar

export PATH=$PATH:$JAVA_HOME/bin

在文件中添加后，如图 24-33 所示。

图 24-33 环境变量修改

从上图可以看出，定义了 JAVA_HOME 和 CLASSPATH，并将 Java 的 bin 文件夹添加到了系统变量中。要想使配置生效，还需要执行相应的命令，如下所示：

source /etc/profile

然后，执行 Java 的版本查看命令：# java -version，如图 24-34 所示。

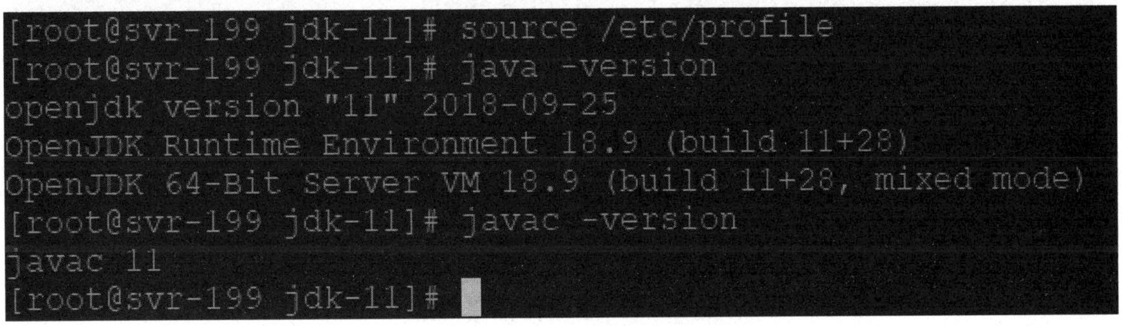

图 24-34 JAVA 版本信息

从上图可以看出，查询出了 Java 的版本信息，说明 Java 11 已经安装完成并且生效了。

24.9 CentOS 8：Elasticsearch 集群

在这里，我们创建一个拥有 3 个节点的集群，规划信息表，如表 24-1 所示：

表 24-1 Elasticsearch 集群节点列表

主机名称	IP	node.name	集群名称
svr-11	192.168.110.11	es_server_001	gf_cluster
svr-151	192.168.110.151	es_server_002	gf_cluster
svr-199	192.168.110.199	es_server_003	gf_cluster

把单节点的 Kibana 部署在 svr-11 机器上。先登录到 svr-11 机器，下载 Elasticsearch，如下所示：

```
$ wget \
https://artifacts.elastic.co/downloads/elasticsearch/elasticsearch-7.7.0-linux-x86_64.tar.gz
```

若上面的方式下载比较慢，推荐使用迅雷等专业下载工具下载，然后通过工具（WinSCP 等）把软件包传到 Linux 机器中。下载完成后，将其解压到指定目录，如下所示：

```
$ tar -xzf elasticsearch-7.7.0-linux-x86_64.tar.gz
```

请注意，不能使用 root 账号启动 Elasticsearch 服务器，否则会报错，如图 24-35 所示。

```
[root@svr-11 elasticsearch-7.7.0]# bin/elasticsearch
[2020-05-18T23:13:49,727][ERROR][o.e.b.ElasticsearchUncaughtExceptionHandler] [svr-11] uncaught excepti
on in thread [main]
org.elasticsearch.bootstrap.StartupException: java.lang.RuntimeException: can not run elasticsearch as
root
        at org.elasticsearch.bootstrap.Elasticsearch.init(Elasticsearch.java:174) ~[elasticsearch-7.7.0
.jar:7.7.0]
        at org.elasticsearch.bootstrap.Elasticsearch.execute(Elasticsearch.java:161) ~[elasticsearch-7.
7.0.jar:7.7.0]
        at org.elasticsearch.cli.EnvironmentAwareCommand.execute(EnvironmentAwareCommand.java:86) ~[ela
sticsearch-7.7.0.jar:7.7.0]
        at org.elasticsearch.cli.Command.mainWithoutErrorHandling(Command.java:127) ~[elasticsearch-cli
-7.7.0.jar:7.7.0]
        at org.elasticsearch.cli.Command.main(Command.java:90) ~[elasticsearch-cli-7.7.0.jar:7.7.0]
        at org.elasticsearch.bootstrap.Elasticsearch.main(Elasticsearch.java:126) ~[elasticsearch-7.7.0
.jar:7.7.0]
        at org.elasticsearch.bootstrap.Elasticsearch.main(Elasticsearch.java:92) ~[elasticsearch-7.7.0.
jar:7.7.0]
Caused by: java.lang.RuntimeException: can not run elasticsearch as root
        at org.elasticsearch.bootstrap.Bootstrap.initializeNatives(Bootstrap.java:111) ~[elasticsearch-
7.7.0.jar:7.7.0]
        at org.elasticsearch.bootstrap.Bootstrap.setup(Bootstrap.java:178) ~[elasticsearch-7.7.0.jar:7.
7.0]
        at org.elasticsearch.bootstrap.Bootstrap.init(Bootstrap.java:393) ~[elasticsearch-7.7.0.jar:7.7
.0]
        at org.elasticsearch.bootstrap.Elasticsearch.init(Elasticsearch.java:170) ~[elasticsearch-7.7.0
.jar:7.7.0]
        ... 6 more
uncaught exception in thread [main]
java.lang.RuntimeException: can not run elasticsearch as root
        at org.elasticsearch.bootstrap.Bootstrap.initializeNatives(Bootstrap.java:111)
        at org.elasticsearch.bootstrap.Bootstrap.setup(Bootstrap.java:178)
        at org.elasticsearch.bootstrap.Bootstrap.init(Bootstrap.java:393)
        at org.elasticsearch.bootstrap.Elasticsearch.init(Elasticsearch.java:170)
        at org.elasticsearch.bootstrap.Elasticsearch.execute(Elasticsearch.java:161)
```

图 24-35 控制台报错信息

要想使 Elasticsearch 服务器能正常启动，需要配置“文件描述符的最大数”和“最大线程数”。编辑配置文件，如下所示：

```
# vi /etc/security/limits.conf
```

在文件结尾添加配置项,如下所示:

quxy - nofile 65535

quxy - nproc 4096

在实际文件中添加后,如图 24-36 所示。

```
49 #
50 #<domain>      <type>   <item>         <value>
51 #
52
53 #*             soft     core           0
54 #*             hard     rss            10000
55 #@student      hard     nproc          20
56 #@faculty      soft     nproc          20
57 #@faculty      hard     nproc          50
58 #ftp           hard     nproc          0
59 #@student      -        maxlogins      4
60
61 quxy - nofile 65535
62 quxy - nproc 4096
63
64 # End of file
```

图 24-36 配置"文件描述符的最大数"和"最大线程数"

退出 Linux 账号重新登录后,以上配置即可生效。另外,还需要配置"最大虚拟内存"。编辑配置文件,如下所示:

vi /etc/sysctl.conf

在文件结尾添加配置项,如下所示:

vm.max_map_count = 262144

在实际文件中添加后,如图 24-37 所示。

```
1 # sysctl settings are defined through files in
2 # /usr/lib/sysctl.d/, /run/sysctl.d/, and /etc/sysctl.d/.
3 #
4 # Vendors settings live in /usr/lib/sysctl.d/.
5 # To override a whole file, create a new file with the same in
6 # /etc/sysctl.d/ and put new settings there. To override
7 # only specific settings, add a file with a lexically later
8 # name in /etc/sysctl.d/ and put new settings there.
9 #
10 # For more information, see sysctl.conf(5) and sysctl.d(5).
11
12 vm.max_map_count = 262144
```

图 24-37 配置"最大虚拟内存"

要想使配置生效,请执行命令,如下所示:

sysctl –p

若只想临时修改上面的配置项,请执行命令,如下所示:

$ ulimit -n 65535
/etc/security/limits.conf
$ ulimit -u 4096
/etc/security/limits.conf
$ sudo sysctl -w vm.max_map_count=262144
/etc/sysctl.conf

有关更多详细信息，请访问以下网址：

https://www.elastic.co/guide/en/elasticsearch/reference/7.7/file-descriptors.html
https://www.elastic.co/guide/en/elasticsearch/reference/7.7/max-number-of-threads.html
https://www.elastic.co/guide/en/elasticsearch/reference/7.7/vm-max-map-count.html

在 elasticsearch.yml 配置文件中，把 network.host 设置为非默认值，且上面的参数没有配置的话，启动时控制台会报错，如图 24-38 所示。

```
[2020-05-19T00:31:30,752][INFO ][o.e.b.BootstrapChecks    ] [es_server_001] bound or publishing to a
non-loopback address, enforcing bootstrap checks
ERROR: [3] bootstrap checks failed
[1]: max file descriptors [4096] for elasticsearch process is too low, increase to at least [65535]
[2]: max number of threads [3046] for user [quxy] is too low, increase to at least [4096]
[3]: max virtual memory areas vm.max_map_count [65530] is too low, increase to at least [262144]
ERROR: Elasticsearch did not exit normally - check the logs at /home/quxy/softwares/elasticsearch-7.7
.0/logs/gf_cluster.log
[2020-05-19T00:31:30,904][INFO ][o.e.n.Node               ] [es_server_001] stopping ...
[2020-05-19T00:31:30,995][INFO ][o.e.n.Node               ] [es_server_001] stopped
[2020-05-19T00:31:30,996][INFO ][o.e.n.Node               ] [es_server_001] closing ...
[2020-05-19T00:31:31,044][INFO ][o.e.n.Node               ] [es_server_001] closed
[2020-05-19T00:31:31,121][INFO ][o.e.x.m.p.NativeController] [es_server_001] Native controller proces
s has stopped - no new native processes can be started
```

图 24-38 控制台报错信息

要想使多个节点之间相互通信，有两个选择：一是关闭防火墙，二是配置相应端口以穿过防火墙。在这里，我们选择前者。先来查看一下防火墙的状态，请执行命令：$ systemctl status firewalld.service，防火墙状态如图 24-39 所示。

```
[quxy@svr-11 ~]$ systemctl status firewalld.service
● firewalld.service - firewalld - dynamic firewall daemon
   Loaded: loaded (/usr/lib/systemd/system/firewalld.service; enabled; vendor preset: enabled)
   Active: active (running) since Tue 2020-05-19 16:55:12 CST; 12s ago
     Docs: man:firewalld(1)
 Main PID: 6409 (firewalld)
    Tasks: 2 (limit: 4874)
   Memory: 28.7M
   CGroup: /system.slice/firewalld.service
           └─6409 /usr/libexec/platform-python -s /usr/sbin/firewalld --nofork --nopid

May 19 16:55:15 svr-11 firewalld[6409]: WARNING: COMMAND_FAILED: '/usr/sbin/iptables -w10 -w --table f>
May 19 16:55:15 svr-11 firewalld[6409]: WARNING: COMMAND_FAILED: '/usr/sbin/iptables -w10 -w --table f>
May 19 16:55:15 svr-11 firewalld[6409]: WARNING: COMMAND_FAILED: '/usr/sbin/iptables -w10 -w --table f>
May 19 16:55:15 svr-11 firewalld[6409]: WARNING: COMMAND_FAILED: '/usr/sbin/iptables -w10 -w --table f>
May 19 16:55:15 svr-11 firewalld[6409]: WARNING: COMMAND_FAILED: '/usr/sbin/iptables -w10 -w --table f>
May 19 16:55:15 svr-11 firewalld[6409]: WARNING: COMMAND_FAILED: '/usr/sbin/iptables -w10 -w --table f>
```

图 24-39 防火墙状态

关闭服务，请执行命令，如下所示：

$ sudo systemctl stop firewalld.service

上面的命令在机器重启后会失效，若想永久生效，请执行命令禁用防火墙，如下所示：

$ sudo systemctl disable firewalld.service

以上两个命令执行完成后，重启机器，再次执行防火墙状态的查看命令，防火墙已经关闭，如图 24-40 所示。

图 24-40

接下来，我们进行最重要的一步，向配置文件中添加配置项。请先进入到 Elasticsearch 主目录中，打开配置文件，如下所示：

$ vim config/elasticsearch.yml

配置文件打开后，在文件结尾添加相应的内容，如下所示：

cluster.name: gf_cluster

node.name: es_server_001

network.host: _site_

#network.bind_host: _site_

#network.publish_host: _site_

#http.port: 9200

discovery.zen.ping.unicast.hosts: ["192.168.31.11","192.168.31.151","192.168.31.199"]

#discovery.zen.minimum_master_nodes: 1

在实际文件中添加后，如图 24-41 所示。

图 24-41 elasticsearch.yml 配置文件

一般情况下，_site_ 表示本地测试用局域网 IP。network.host 可以同时设置 network.bind_host 和 network.publish_host。但有时候需要分别设置，例如：Elasticsearch 服务器运行在一个代理服务器背后时，需要把这两个参数设置为不同的值。

在 Elasticsearch 集群中，可以静态地指定节点的 IP 列表，请在 elasticsearch.yml 配置文件中进行设置，如下所示：

discovery.zen.ping.unicast.hosts: ["192.168.31.11","192.168.31.151","192.168.31.199"]

在 Elasticsearch 集群中，也可以动态地指定节点的 IP 列表，请在 elasticsearch.yml 配置文件中进行设置，如下所示：

discovery.zen.hosts_provider: unicast_hosts.txt

其中，unicast_hosts.txt 是外置的配置文件。当这个文件发生改变时，Elasticsearch 会重新加载这个文件。文件格式，如下所示：

10.10.10.5

10.10.10.6:9305

10.10.10.5:10005

an IPv6 address

#[2001:0db8:85a3:0000:0000:8a2e:0370:7334]:9301

HTTP 接口是异步非阻塞的，要完全禁用 HTTP 模块，请在 elasticsearch.yml 配置文件中设置以下项：

http.enabled: false

path.data 和 path.logs 应当指向 Elasticsearch 主目录之外的位置，这样便于 Elasticsearch 应用复制到其他机器，以形成集群。

至此，svr-11 机器上的 Elasticsearch 服务器已经配置完毕，其他两台机器按照上面的操作步骤执行即可。使用命令把配置好的 Elasticsearch 复制到另外两台机器上，如下所示：

$ scp -r ./elasticsearch-7.7.0 192.168.31.151:/home/quxy/softwares/

$ scp -r ./elasticsearch-7.7.0 192.168.31.199:/home/quxy/softwares/

并把 elasticsearch.yml 配置文件中的 node.name 分别进行修改，如下所示：

es_server_002、es_server_003

然后，修改系统配置文件和防火墙。Elasticsearch 复制到其他机器后，应当删除 data 和 log 目录下的数据，进入到 Elasticsearch 主目录，执行命令，如下所示：

$ rm -rf data/*

$ rm -rf log/*

通过上面的操作，三台机器上的 Elasticsearch 已全部配置完成，分别进入其 Elasticsearch 主目录，使用命令依次后台启动各个服务，如下所示：

$ bin/elasticsearch -d

要查看 3 个节点组成的集群情况，请通过浏览器访问：http://192.168.31.11:9200/_cat/nodes?v，Elasticsearch 集群情况如图 24-42 所示。

```
ip              heap.percent ram.percent cpu load_1m load_5m load_15m node.role master name
192.168.31.199             8          91  37    1.27    0.85     0.59 dilmrt    -      es_server_003
192.168.31.151            25          93   3    0.30    2.07     1.91 dilmrt    -      es_server_002
192.168.31.11             14          92   3    0.00    0.05     0.18 dilmrt    *      es_server_001
```

图 24-42 Elasticsearch 服务器集群

停止命令，其中结尾的数字为 Elasticsearch 服务器的进程 ID，如下所示：

$ kill -SIGTERM 17416

要查看 Elasticsearch 服务器的进程 ID，请执行命令：$ ps -aux|grep elasticsearch，Elasticsearch 集群服务器进程 ID 如图 24-43 所示。

```
[quxy@svr-199 elasticsearch-7.7.0]$ ps -aux|grep elasticsearch
quxy      22261  1.6 45.9 3291496 374932 pts/1  Sl   19:45   2:11 /usr/java/jdk-11//bin/java -Xshare:
auto -Des.networkaddress.cache.ttl=60 -Des.networkaddress.cache.negative.ttl=10 -XX:+AlwaysPreTouch -
Xss1m -Djava.awt.headless=true -Dfile.encoding=UTF-8 -Djna.nosys=true -XX:-OmitStackTraceInFastThrow
-Dio.netty.noUnsafe=true -Dio.netty.noKeySetOptimization=true -Dio.netty.recycler.maxCapacityPerThrea
d=0 -Dio.netty.allocator.numDirectArenas=0 -Dlog4j.shutdownHookEnabled=false -Dlog4j2.disable.jmx=tru
e -Djava.locale.providers=SPI,COMPAT -Xms1g -Xmx1g -XX:+UseConcMarkSweepGC -XX:CMSInitiatingOccupancy
Fraction=75 -XX:+UseCMSInitiatingOccupancyOnly -Djava.io.tmpdir=/tmp/elasticsearch-170967108343926856
80 -XX:+HeapDumpOnOutOfMemoryError -XX:HeapDumpPath=data -XX:ErrorFile=logs/hs_err_pid%p.log -Xlog:gc
*,gc+age=trace,safepoint:file=logs/gc.log:utctime,pid,tags:filecount=32,filesize=64m -XX:MaxDirectMem
orySize=536870912 -Des.path.home=/home/quxy/softwares/elasticsearch-7.7.0 -Des.path.conf=/home/quxy/s
oftwares/elasticsearch-7.7.0/config -Des.distribution.flavor=default -Des.distribution.type=tar -Des.
bundled_jdk=true -cp /home/quxy/softwares/elasticsearch-7.7.0/lib/* org.elasticsearch.bootstrap.Elast
icsearch -d
quxy      22276  0.0  0.0 112096   208 pts/1    Sl   19:45   0:00 /home/quxy/softwares/elasticsearch-
7.7.0/modules/x-pack-ml/platform/linux-x86_64/bin/controller
quxy      23439  1.0  0.1  12108  1072 pts/1    R+   21:55   0:00 grep --color=auto elasticsearch
```

图 24-43 Elasticsearch 服务器的进程 ID

24.10 CentOS 8：Kibana 安装

登录到 svr-11 机器，下载 Kibana，如下所示：

$ wget \

https://artifacts.elastic.co/downloads/kibana/kibana-6.8.0-linux-x86_64.tar.gz

下载完成后，将其解压到指定目录，如下所示：

$ tar -xzf kibana-7.7.0-linux-x86_64.tar.gz

解压完成后，进入到 Kibana 主目录中，用命令打开配置文件，如下所示：

$ vim config/kibana.yml

配置文件打开后，在文件结尾添加内容：

server.host: "192.168.31.11"

elasticsearch.hosts:

["http://192.168.31.11:9200","http://192.168.31.151:9200","http://192.168.31.199:9200"]

在实际文件中添加后，如图 24-44 所示。

图 24-44 kibana.yml 配置文件

有关更多详细信息，请访问以下网址：

https://www.elastic.co/guide/en/kibana/6.8/settings.html

配置完成后，要启动 Kibana 服务，请执行命令，如下所示：

$ bin/kibana

若希望后台启动 Kibana 服务，请执行命令，如下所示：

$ nohup bin/kibana &

在浏览器中访问 Kibana 页面的 URL：http://192.168.31.11:5601

打开 Kibana 的菜单[Stack Monitoring]，就可以看到这三个 Elasticsearch 服务器组成的集群。一般情况下，第一次打开本菜单会出现的页面，如图 24-45 所示。

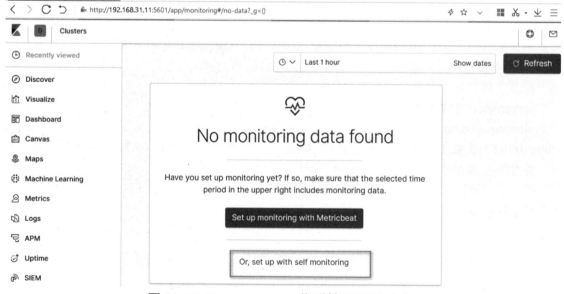

图 24-45 Elasticsearch 集群管理页面（1）

这时，请单击[Or,set up with self monitoring]（有时需要反复操作多次），就会出现接下来的页面，如图 24-46 所示。

图 24-46 Elasticsearch 集群管理页面（2）

在上图的页面中，单击[Nodes:3]链接，即可进入到第 3 个节点的详情页面中，如图 24-47 所示。

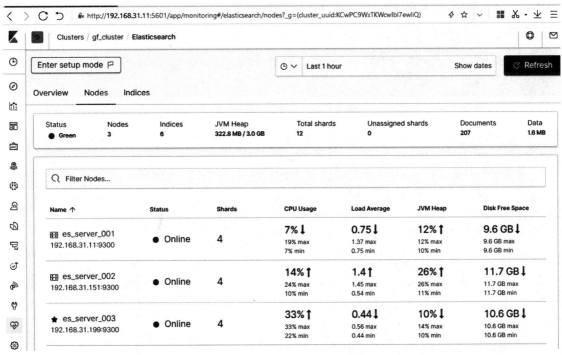

图 24-47 Nodes:3 详情页面

24.11 CentOS 8：Elasticsearch 安全

从 Elastic Stack 6.8 开始，在默认分发包中免费提供多项安全功能，例如：TLS 加密通信、基于角色的访问控制（RBAC），等等。使用 TLS 加密通信，首先是生成证书，它们能保证节点之间安全地通信。登录到 svr-11 机器，进入到 Elasticsearch 主目录，执行命令：$ bin/elasticsearch-certutil cert -out config/elastic-certificates.p12 -pass ""，生成证书如图 24-48、图 24-49 所示。

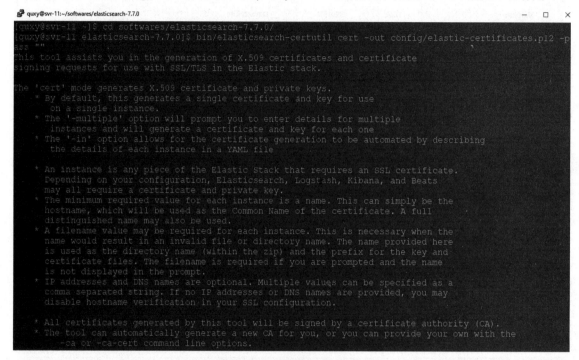

图 24-48 生成证书（1）

图 24-49 生成证书（2）

从上图中可以看出，证书已经被写到 config 文件夹中的 elastic-certificates.p12 文件中。要查看文件，请执行命令：$ cat config/elastic-certificates.p12，如图 24-50 所示。

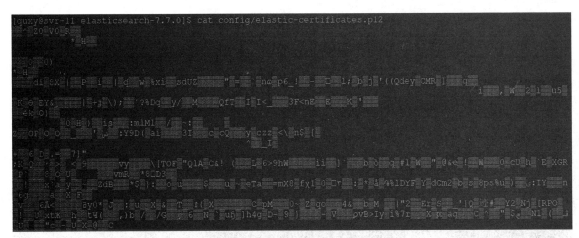

图 24-50 查看证书文件

打开 elasticsearch.yml 配置文件，将相应的内容添加到文件末尾，如下所示：

xpack.security.enabled: true
xpack.security.transport.ssl.enabled: true
xpack.security.transport.ssl.verification_mode: certificate
xpack.security.transport.ssl.keystore.path: elastic-certificates.p12
xpack.security.transport.ssl.truststore.path: elastic-certificates.p12

在实际文件中添加后，如图 24-51 所示。

图 24-51 修改 elasticsearch.yml 配置文件

将证书文件 elastic-certificates.p12 复制到 svr-151 机器的相应目录，请执行命令，如下所示：

$ scp config/elastic-certificates.p12 \
192.168.31.151:/home/quxy/softwares/elasticsearch-7.7.0/config

在 svr-151 机器上，elasticsearch.yml 配置文件的末尾也要添加与上面相同的内容。然后，依次启动这两台机器上的 Elasticsearch 服务器（我们假定 svr-11 机器上的 es_server_001 是 master 节点）。

要生成 Elasticsearch 服务器和 Kibana 服务等的一系列用户名及其密码，请登录到 svr-11 机器，进入到 Elasticsearch 主目录，执行命令：$ bin/elasticsearch-setup-passwords auto，用户名及其密码的生成如图 24-52 所示。

```
[quxy@svr-11 elasticsearch-7.7.0]$ bin/elasticsearch-setup-passwords auto
Initiating the setup of passwords for reserved users elastic,apm_system,kibana,logstash_system,beats_syst
em,remote_monitoring_user.
The passwords will be randomly generated and printed to the console.
Please confirm that you would like to continue [y/N]y


Changed password for user apm_system
PASSWORD apm_system = MiNo4UKcynJYlmRzk6lx

Changed password for user kibana
PASSWORD kibana = aUT5QabBDogedIzdY4Ve

Changed password for user logstash_system
PASSWORD logstash_system = ilXgNFGum8S0ZSMs9cw4

Changed password for user beats_system
PASSWORD beats_system = OnjlIJ6iNyfB4sYFjJWj

Changed password for user remote_monitoring_user
PASSWORD remote_monitoring_user = ZB0M02ukf4s85iKlTSK0

Changed password for user elastic
PASSWORD elastic = XEYeMFtE67Icecben6Tv
```

图 24-52 生成用户名、密码

从上图中可以看出，生成了多个用户及其密码，这里的密码是随机生成的，还可以自定义相应的密码。相关说明，如表 24-2 所示。

表 24-2 用户列表

用户名	说明
apm_system	应用性能管理的系统用户名
kibana	Kibana 服务的用户名
logstash_system	Logstash 服务的系统用户名
beats_system	Beats 服务的系统用户名
remote_monitoring_user	远程监控的用户名
elastic	Elasticsearch 服务器的用户名

有了安全保障后，我们再访问 Elasticsearch 相关的 URL，就需要输入用户名和密码了，如图 24-53 所示。

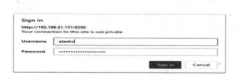

图 24-53 用户登录

至此，我们就为 Elasticsearch 集群添加上安全保障，svr-199 机器上的 es_server_003 节点也需要相同的配置，才能进入集群。

24.12 CentOS 8：Kibana 安全

打开 kibana.yml 配置文件，将配置项放置在末尾，如图 24-54 所示。
elasticsearch.username: "elastic"
elasticsearch.password: "XEYeMFtE67Icecben6Tv"

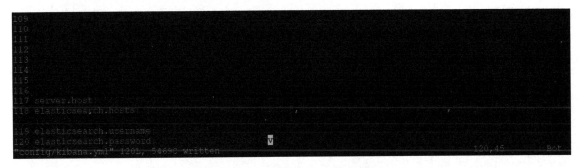

图 24-54 kibana.yml 配置文件

启动 Kibana 服务后，在浏览器中访问 Kibana 服务，会展现登录页面，如图 24-55 所示。

图 24-55

登录成功后，单击菜单栏最下面的菜单[Management]，展现的页面相比于未加认证配置前多了 Security 模块，如图 24-56 所示。

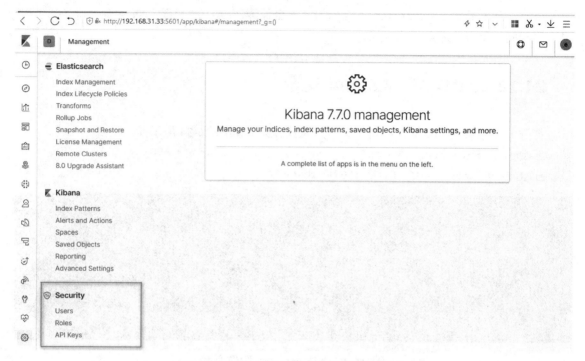

图 24-56 添加认证配置后的页面信息

单击 Security 模块中的链接[Users]，页面中展现了之前使用命令创建的 6 个用户，可以对用户进行编辑操作，如图 24-57 所示。

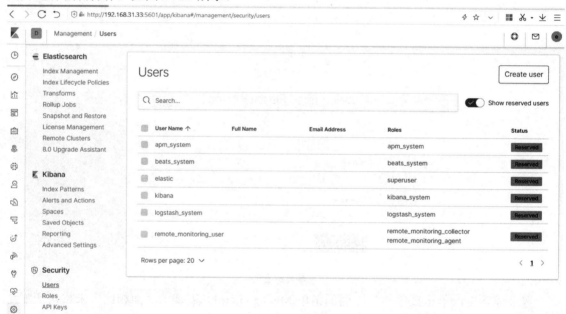

图 24-57 编辑用户

24.13 中文分词

24.13.1 IK 分析插件

IK 分析插件将 Lucene IK 分析器集成到 Elasticsearch 中，支持自定义字典和远程扩展字典。在 IK 分析插件中，分析器的类型有：ik_smart 和 ik_max_word，分词器的类型有：ik_smart 和 ik_max_word。

ik_max_word

它会将文本做最细粒度的拆分，例如：会将"杂粮玉米牛奶发糕"拆分为"杂粮玉米、牛奶发糕、杂粮、玉米、牛奶、发糕、杂、粮、玉、米、牛、奶、发、糕、杂粮玉米牛奶、杂粮玉、粮玉、米牛、奶发"，它会穷尽各种可能的组合，比较适合 TermQuery 方式的查询。

ik_smart

它会做最粗粒度的拆分，比如会将"杂粮玉米牛奶发糕"拆分为"杂粮、玉米、牛奶、发糕"，适合 Phrase 查询。

IK 分析插件的源代码及其使用文档，如下所示：

https://github.com/medcl/elasticsearch-analysis-ik

现在，我们开始安装插件，进入到 Elasticsearch 的主目录中，在插件目录中新建 ik 目录，如下所示：

$ cd plugins/ && mkdir ik

把下载的插件包解压到指定的目录中，如下所示：

$ unzip ./elasticsearch-analysis-ik-7.7.0.zip -d ./elasticsearch-7.7.0/plugins/ik/

三个节点都要进行以上操作，然后重启 Elasticsearch 服务器。启动后，在 Kibana 的 Dev Tools 页面中执行相应的命令，以便查看分词效果，如图 24-58 所示。

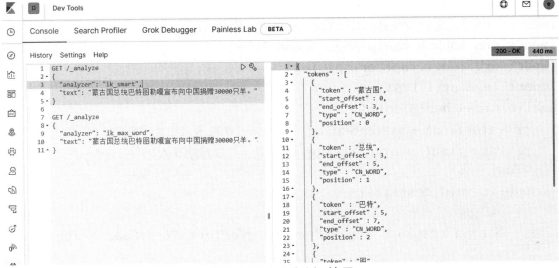

图 24-58 查看分词效果

另外，还可以在线安装 IK 分析插件，如下所示：
$ ./bin/elasticsearch-plugin install \
https://github.com/medcl/elasticsearch-analysis-ik/releases/download/v7.7.0/elasticsearch-analysis-ik-7.7.0.zip

24.13.2　NLP 分词器

HanLP（Han Language Processing）是用 Python 语言开发的国产开源的自然语言（NLP）处理工具包。它是面向生产环境的，基于 TensorFlow 2.0 的，目标是普及落地最前沿的 NLP 技术。HanLP 的官网，如下所示：

https://hanlp.hankcs.com

HanLP 具有的功能有：中文分词、词性标注、命名实体识别、依存句法分析、语义依存分析、新词发现、关键词短语提取、自动摘要、文本分类聚类、拼音简繁转换和自然语言处理等。HanLP 的源代码及其使用文档，如下所示：

https://github.com/hankcs/HanLP

HanLP 的版本发布，可以直接把它的 Maven 依赖项添加到 Maven 项目中，如下所示：

https://github.com/hankcs/HanLP/releases

```xml
<dependency>
    <groupId>com.hankcs</groupId>
    <artifactId>hanlp</artifactId>
    <version>portable-1.7.8</version>
</dependency>
```

添加好依赖项后，我们编写一些测试代码来展现一下 HanLP 的文本处理能力，如下所示：

```java
package com.jijicai.other;
import com.hankcs.hanlp.HanLP;
import com.hankcs.hanlp.dictionary.py.Pinyin;
import com.hankcs.hanlp.seg.common.Term;
import org.junit.jupiter.api.Test;
import java.util.List;
public class HanlpTest {
    private final static String text = "杂粮玉米牛奶发糕";
    private final static String text2 = "我是地球人！";
    @Test
    public void test1() {
        // 提取首字母
        String resulStr = HanLP.convertToPinyinFirstCharString(text, ",", true);
        System.out.println("首字母：" + resulStr);
```

```java
        // 中文转拼音，最后一个数字是声调
        List<Pinyin> list = HanLP.convertToPinyinList(text);
        System.out.println("拼音列表：" + list);
        // 简体中文转变为繁体
        String traditionalChinese = HanLP.convertToTraditionalChinese(text);
        System.out.println("繁体中文：" + traditionalChinese);
    }
    @Test
    public void test2() {
        // 提取关键词
        List<String> list = HanLP.extractKeyword(text2, 2);
        System.out.println("关键词：" + list);
        // 分词和词性标注
        List<Term> terms = HanLP.segment(text2);
        System.out.println(terms);
    }
}
```

想要把 HanLP 把作为 Elasticsearch 插件安装，请访问以下网址：
https://github.com/KennFalcon/elasticsearch-analysis-hanlp

24.14 简单示例

想要在 Spring Boot 应用程序中使用 Elasticsearch，只需添加相应的依赖项，如下所示：

```xml
<dependency>
    <groupId>org.springframework.boot</groupId>
    <artifactId>spring-boot-starter-data-elasticsearch</artifactId>
</dependency>
```

然后，在 application.yml 配置文件中，添加配置项，如图 24-59 所示。

```yaml
spring:
  elasticsearch:
    rest:
      uris:
       - http://192.168.31.11:9200
       - http://192.168.31.151:9200
       - http://192.168.31.199:9200
      password: XEYeMFtE67Icecben6Tv   #若Elasticsearch服务无密码，则可以不填写。
      username: elastic   #若Elasticsearch服务无用户名，则可以不填写。
      connection-timeout: 10s
#      read-timeout: 30s
```

图 24-59

经过以上配置后，使用 ElasticsearchRestTemplate 接口操作 Elasticsearch 服务器，就可以开始编写 controller 层的代码了，如下所示：

```
import com.fasterxml.jackson.databind.ObjectMapper;
import com.gf.common.GfResponseEntity;
import com.gf.mvc.dto.ESMenuCreateDTO;
import com.gf.mvc.entity.ESMenu;
import com.gf.mvc.util.StringUtil;
import lombok.extern.slf4j.Slf4j;
import org.springframework.data.elasticsearch.core.ElasticsearchRestTemplate;
import org.springframework.http.ResponseEntity;
import org.springframework.web.bind.annotation.*;
@Slf4j
@RestController
@RequestMapping("/mvc/es")
public class ElasticsearchController {
    private final ElasticsearchRestTemplate elasticsearchRestTemplate;
    private final ObjectMapper objectMapper;
    public ElasticsearchController(ElasticsearchRestTemplate elasticsearchRestTemplate, ObjectMapper objectMapper) {
        this.elasticsearchRestTemplate = elasticsearchRestTemplate;
        this.objectMapper = objectMapper;
    }
    /**
     * 向索引库中插入一条数据
     */
    @PostMapping("/create")
    public ResponseEntity create(@RequestBody ESMenuCreateDTO params) {
        ESMenu entity = new ESMenu();
        entity.setId(params.getId());
        entity.setName(params.getName());
        entity.setParentId(params.getParentId());
        entity.setSorted(params.getSorted());
        ESMenu entity2 = elasticsearchRestTemplate.save(entity);
        log.info(StringUtil.toString(objectMapper, entity2));
        return GfResponseEntity.self().ok();
```

```java
    }
    /**
     * 根据数据 ID, 查询相应的数据。
     */
    @GetMapping("/view/{id}")
    public ResponseEntity view(@PathVariable String id) {
        ESMenu entity = elasticsearchRestTemplate.get(id, ESMenu.class);
        return GfResponseEntity.self().ok(entity);
    }
}
```

定义 ESMenuCreateDTO 类,用于接收请求参数,如下所示:

```java
import lombok.Data;
import javax.validation.constraints.Min;
import javax.validation.constraints.NotEmpty;
@Data
public class ESMenuCreateDTO {
    private String id;
    @NotEmpty // 不能为空,空字符串也不可以。
    private String name;
    private Long parentId;
    @Min(0) // 不能小于 0
    private int sorted;
}
```

实体类 ESMenu 的定义,如下所示:

```java
package com.gf.mvc.entity;
import lombok.Data;
import org.springframework.data.annotation.Id;
import org.springframework.data.elasticsearch.annotations.Document;
@Data
@Document(indexName = "es_menu")
public class ESMenu {
    @Id
    private String id;
    private String name;
    private Long parentId;
    private int status;
    private int sorted;
    private String remark;
```

}

在上面的代码中，@Document 注解的 indexName 属性的值就是索引库的名称。执行新建数据请求后，默认生成的 mappings 结构，类似于表结构，要查看 es_menu 索引库的结构，请执行命令：GET /es_menu，如图 24-60 所示。

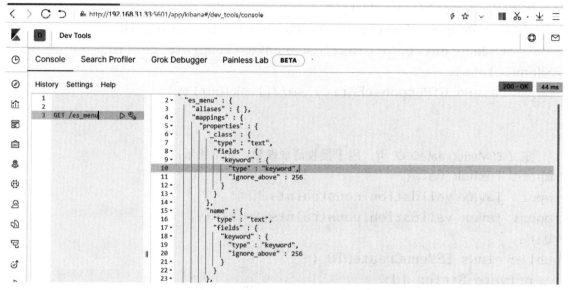

图 24-60 查看 es_menu 索引库结构

要查看 es_menu 索引库中所有的数据，请执行命令，如图 24-61 所示。

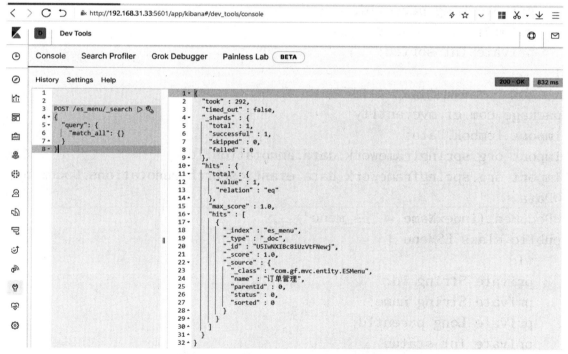

图 24-61 查看 es_menu 索引库数据

删除数据，其中 U5IwNXIBc8iUzVtFNewj 为数据 ID，请执行命令，如图 24-62 所示。

— 322 —

DELETE /es_menu/_doc/ U5IwNXIBc8iUzVtFNewj

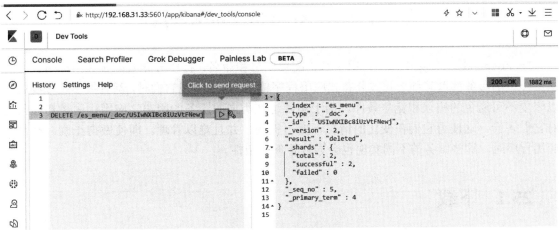

图 24-62 删除数据

24.15 本章小结

本章讲解了 Elasticsearch 和 Kibana 的相关知识点。讲解了 Elasticsearch 的简介、发展历程、基本概念和下载，以及二者在 Win 10 和 CentOS 8 环境中的安装，最后讲解了 Elasticsearch 的安全和中文分词。Kibana 是操作和管理 Elasticsearch 的 UI 界面，当然，Kibana 的功能不止于此，它还有地图和图表绘制等丰富的功能。

第二十五章 Apache ZooKeeper

ZooKeeper 是一种集中服务，用于维护配置信息和命名，提供分布式同步和组服务。所有这些类型的服务都以某种形式被分布式应用程序使用。每次实现它们时，都有很多工作要做，以修复不可避免的错误和竞争条件。由于实现这些类型的服务的困难，应用程序最初通常会对它们吝啬，这使得它们在变化的情况下变得脆弱，并且难以管理。即使做得正确，在部署应用程序时，这些服务的不同实现也会导致管理复杂性。

25.1 下载

访问 ZooKeeper 官网 URL：https://zookeeper.apache.org/releases.html，进入下载页面，如图 25-1 所示。

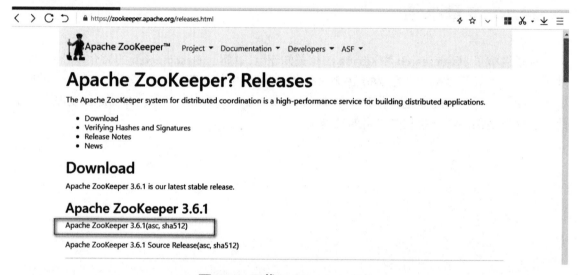

图 25-1 下载 ZooKeeper（1）

在上图的网页中，单击选框中的链接，进入镜像下载页面，如图 25-2 所示。

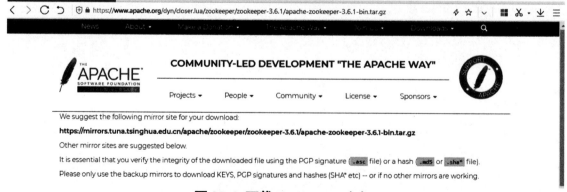

图 25-2 下载 ZooKeeper（1）

在上图的网页中，可以根据自己当前的网络情况来选择合适的镜像链接，本书选择的镜像链接，如下所示：

https://mirrors.tuna.tsinghua.edu.cn/apache/zookeeper/zookeeper-3.6.1/apache-zookeeper-3.6.1-bin.tar.gz

25.2 单节点安装

ZooKeeper 的软件包解压后，在 Linux 和 Windows 环境中都可以使用。在这里，我们举在 Linux 中使用的例子。下载到适当位置后，解压它，执行命令，如下所示：

tar xvf apache-zookeeper-3.6.1-bin.tar.gz

解压后，进入到 ZooKeeper 的主目录中，如图 25-3 所示。

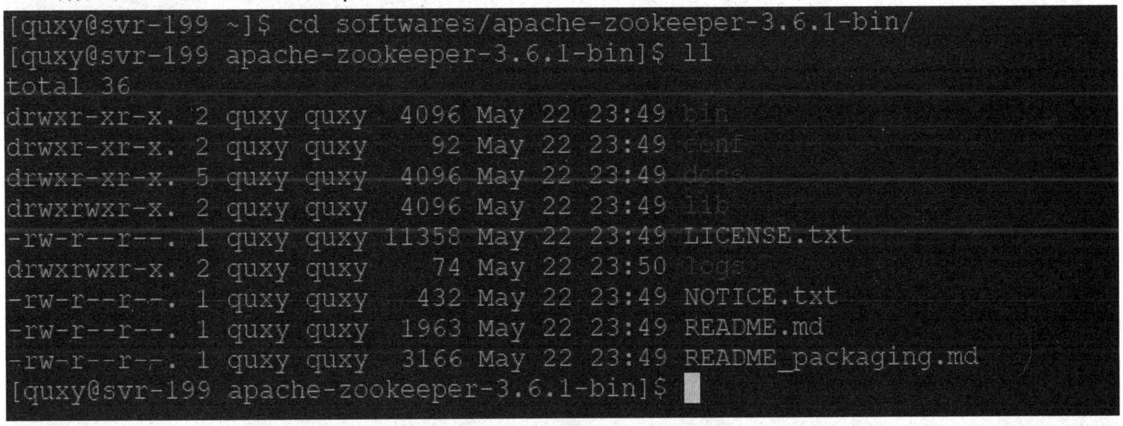

图 25-3 ZooKeeper 主目录

从 ZooKeeper 的样例配置文件复制出一份配置文件，执行命令，如下所示：

$ cp conf/zoo_sample.cfg conf/zoo.cfg

使用 vim 编辑器查看配置文件中的内容，执行命令：$ vim conf/zoo.cfg，zoo.cfg 配置文件内容如图 25-4 所示。

图 25-4 zoo.cfg 配置文件

可以无需修改配置文件中的默认配置项。在 ZooKeeper 主目录中启动 ZooKeeper 服务器，请执行命令，如下所示：

$ bin/zkServer.sh start

ZooKeeper 的启动命令执行完成后，要查看启动状态，请执行命令：$ ps -aux|grep zookeeper，如图 25-5 所示。

图 25-5 启动 ZooKeeper

在本机上连接 ZooKeeper 服务器，请执行命令：$ bin/zkCli.sh，如图 25-6 所示。

图 25-6 连接 ZooKeeper 服务器（1）

在图 25-7 中的最后一行中可以看到"CONNECTED"，说明已经成功连接上 ZooKeeper 服务器。

```
ftwares/apache-zookeeper-3.6.1-bin
2020-05-23 21:04:55,449 [myid:] - INFO  [main:Environment@98] - Client environment:os.memory.free=8MB
2020-05-23 21:04:55,451 [myid:] - INFO  [main:Environment@98] - Client environment:os.memory.max=247MB
2020-05-23 21:04:55,451 [myid:] - INFO  [main:Environment@98] - Client environment:os.memory.total=13MB
2020-05-23 21:04:55,456 [myid:] - INFO  [main:ZooKeeper@1005] - Initiating client connection, connectStri
ng=localhost:2181 sessionTimeout=30000 watcher=org.apache.zookeeper.ZooKeeperMain$MyWatcher@3d36e4cd
2020-05-23 21:04:55,463 [myid:] - INFO  [main:X509Util@77] - Setting -D jdk.tls.rejectClientInitiatedRene
gotiation=true to disable client-initiated TLS renegotiation
2020-05-23 21:04:55,488 [myid:] - INFO  [main:ClientCnxnSocket@239] - jute.maxbuffer value is 1048575 Byt
es
2020-05-23 21:04:55,496 [myid:] - INFO  [main:ClientCnxn@1703] - zookeeper.request.timeout value is 0. fe
ature enabled=false
Welcome to ZooKeeper!
JLine support is enabled
2020-05-23 21:04:55,601 [myid:localhost:2181] - INFO  [main-SendThread(localhost:2181):ClientCnxn$SendThr
ead@1154] - Opening socket connection to server localhost/127.0.0.1:2181.
2020-05-23 21:04:55,601 [myid:localhost:2181] - INFO  [main-SendThread(localhost:2181):ClientCnxn$SendThr
ead@1156] - SASL config status: Will not attempt to authenticate using SASL (unknown error)
2020-05-23 21:04:55,636 [myid:localhost:2181] - INFO  [main-SendThread(localhost:2181):ClientCnxn$SendThr
ead@986] - Socket connection established, initiating session, client: /127.0.0.1:42388, server: localhost
/127.0.0.1:2181
2020-05-23 21:04:55,723 [myid:localhost:2181] - INFO  [main-SendThread(localhost:2181):ClientCnxn$SendThr
ead@1420] - Session establishment complete on server localhost/127.0.0.1:2181, session id = 0x10000207b71
0000, negotiated timeout = 30000

WATCHER::

WatchedEvent state:SyncConnected type:None path:null
[zk: localhost:2181(CONNECTED) 0]
```

图 25-7 连接 ZooKeeper 服务器（2）

想要在本地测试用局域网中的其他机器上连接 ZooKeeper 服务器，请执行命令，如下所示：

$ bin/zkCli.sh -server 192.168.31.199:2181

登录客户端后，有关单条数据的各类操作，如图 25-8 所示。

```
[zk: localhost:2181(CONNECTED) 1] ls /
[config, zk_test, zookeeper]
[zk: localhost:2181(CONNECTED) 2] create /gf_test 改革开发
Created /gf_test
[zk: localhost:2181(CONNECTED) 3] ls /
[config, gf_test, zk_test, zookeeper]
[zk: localhost:2181(CONNECTED) 4] get /gf_test
改革开发
[zk: localhost:2181(CONNECTED) 5] stat /gf_test
cZxid = 0x46
ctime = Sat May 23 21:36:59 CST 2020
mZxid = 0x46
mtime = Sat May 23 21:36:59 CST 2020
pZxid = 0x46
cversion = 0
dataVersion = 0
aclVersion = 0
ephemeralOwner = 0x0
dataLength = 12
numChildren = 0
[zk: localhost:2181(CONNECTED) 6] delete /gf_test
[zk: localhost:2181(CONNECTED) 7] ls /
[config, zk_test, zookeeper]
[zk: localhost:2181(CONNECTED) 8]
```

图 25-8 操作数据

停止 ZooKeeper 服务器时，请使用提供的停止命令。直接关机等情况导致的 ZooKeeper 服务器的进程终止，可能会造成下次不能正常启动。遇到这种情，请删除 dataDir 配置项（在 zoo.cfg 配置文件中）指定的目录中的 zookeeper_server.pid 文件，然后再次执行启动命令即可。要停止 ZooKeeper 服务器，请执行命令：$ bin/zkServer.sh stop，如图 25-9 所示。

```
[quxy@svr-199 apache-zookeeper-3.6.1-bin]$ bin/zkServer.sh stop
ZooKeeper JMX enabled by default
Using config: /home/quxy/softwares/apache-zookeeper-3.6.1-bin/bin/../conf/zoo.cfg
Stopping zookeeper ... STOPPED
[quxy@svr-199 apache-zookeeper-3.6.1-bin]$
```

图 25-9 停止 ZooKeeper 服务器

25.3 集群搭建

在独立模式下运行 ZooKeeper 便于评估、某些开发和测试。但是在生产中，应该在复制模式下运行 ZooKeeper。从同一应用程序中复制的服务器组称为 quorum，在复制模式下，quorum 中的所有服务器都具有相同配置文件的副本。在复制模式下，至少需要三个 ZooKeeper 节点，并且节点数量应该是奇数。在这里，我们搭建具有 3 个节点的集群。

在配置文件中，每个 ZooKeeper 服务器的 IP 后面有两个端口号：2888 和 3888。一个节点使用前一个端口连接到其他节点。这样的连接是必要的，以便节点之间可以进行通信，例如：商定更新的顺序。更具体地说，ZooKeeper 服务器使用此端口将 follower 连接到 leader。当出现新的 leader 时，一个 follower 使用此端口打开到 leader 的 TCP 连接。因为默认的 leader 选举也使用 TCP，所以目前需要另一个端口进行 leader 选举，这就是第二个端口。

首先，打开 zoo.cfg 配置文件，在末尾添加配置项：

server.1=192.168.31.11:2888:3888

server.2=192.168.31.151:2888:3888

server.3=192.168.31.199:2888:3888

在实际文件中添加后，如图 25-10 所示。

图 25-10 zoo.cfg 配置文件

如上图所示，3 个节点都在配置文件中添加相同的配置项。在这里，只列出了必要的配置项，其他的可根据需要添加。然后，需要为这 3 个节点创建 myid 文件来存储 serverid。依次在 3 台机器上执行命令，如下所示：

$ echo 1 >/tmp/zookeeper/myid

$ echo 2 >/tmp/zookeeper/myid
$ echo 3 >/tmp/zookeeper/myid

依次启动这三个节点（启动命令和前文中的一样），启动成功后查看它们的状态，请执行命令：$ bin/zkServer.sh status，如图 25-11 所示。

图 25-11 启动集群节点

在一个节点上存储数据，在另外一个节点上可以查看和修改数据，如图 25-12 所示。

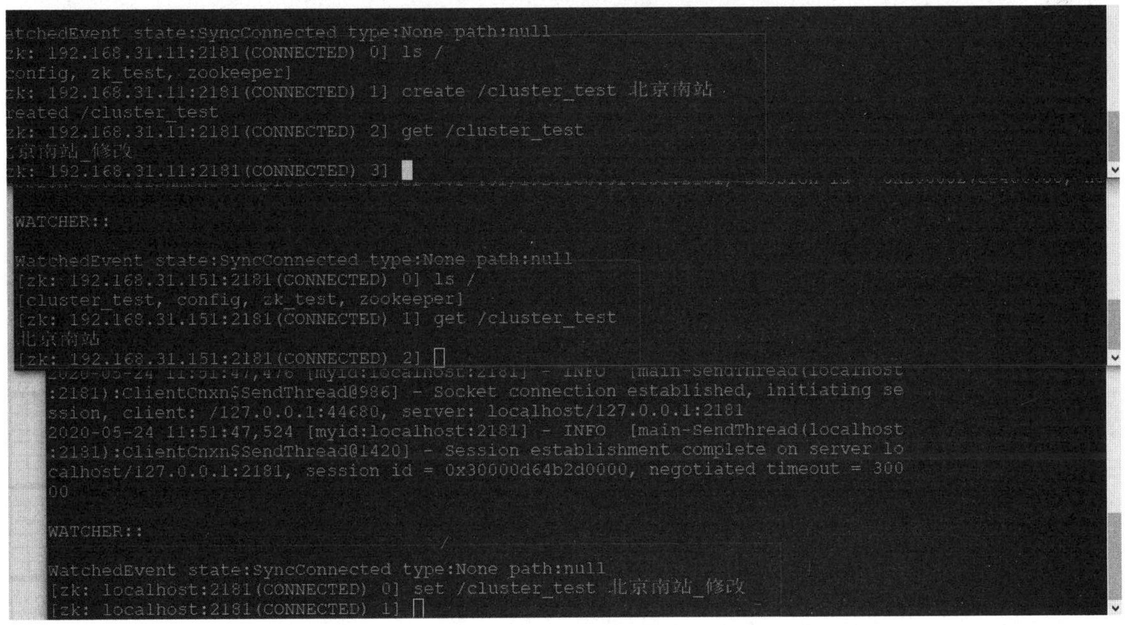

图 25-12 存储节点

AdminServer 是一个嵌入式 Jetty 服务器，它为四字母单词命令提供 HTTP 接口。默认情况下，服务器在端口 8080 上启动，通过转到 URL“/commands/[command name]”发出命令，例如：http://localhost:8080/commands/stat。命令响应以 JSON 形式返回。查看所有可用的命令列表，请在浏览器中打开 URL：http://192.168.31.11:8080/commands，如图 25-13 所示。

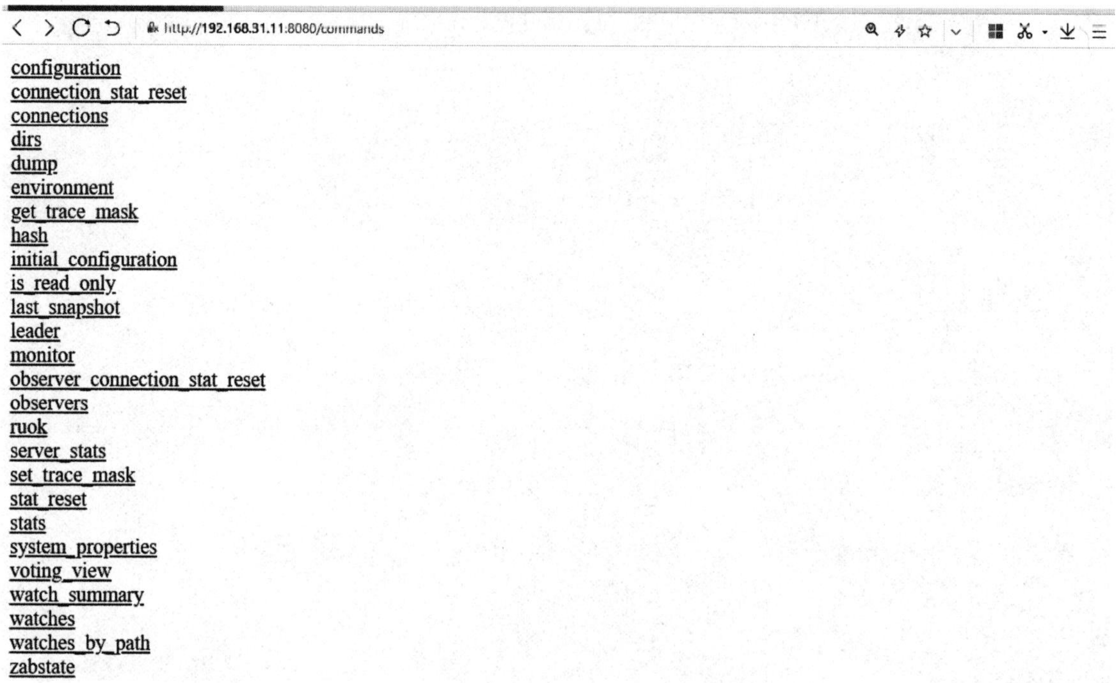
http://192.168.31.11:8080/commands
configuration
connection_stat_reset
connections
dirs
dump
environment
get_trace_mask
hash
initial_configuration
is_read_only
last_snapshot
leader
monitor
observer_connection_stat_reset
observers
ruok
server_stats
set_trace_mask
stat_reset
stats
system_properties
voting_view
watch_summary
watches
watches_by_path
zabstate

图 25-13 AdminServer

默认情况下，AdminServer 是启用的，想用禁用它，请在 zoo.cfg 配置文件中添加配置项，如下所示：

admin.enableServer=false

25.4 ZooKeeper UI

ZooKeeper UI 是对 ZooKeeper 服务器进行 CRUD 操作的 UI 仪表板。它是 GitHub 上的一个开源项目。

功能特性

1) ZooKeeper 属性上的 CRUD 操作
2) 导出属性
3) 通过回调 URL 导入属性
4) 通过文件上传导入属性
5) 变更历史 + 特定路径的变更历史
6) 搜索功能
7) 用于访问 ZooKeeper 属性的 Rest API

8) 基于角色的基本身份验证
9) 支持 LDAP 身份验证
10) 根节点 /zookeeper 为了安全而隐藏
11) ACL 支持全局级别

要查看 ZooKeeper UI 的源码，请访问 URL：https://github.com/DeemOpen/zkui，源码如图 25-14 所示。

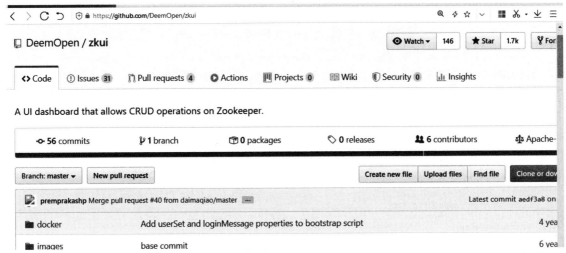

图 25-14 查看 ZooKeeper UI 源码

在 GitHub 网站上，截止到检出本项目前（检出日期：2020 年 5 月 24 日），并未发布版本及版本标签，所以，我们检出最新的项目即可。使用 IntelliJ IDEA 检出 ZooKeeper UI 项目，如下所示：

https://github.com/DeemOpen/zkui.git

项目检出后，执行 clean、compile 和 install 命令，生成了可执行 jar 文件，如图 25-15 所示。

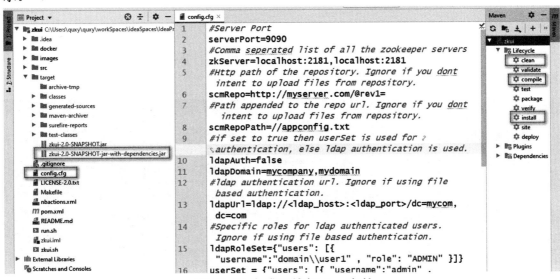

图 25-15 生成可执行 jar 文件

将在 target 目录中生成的 zkui 可执行 jar 文件和配置文件 config.cfg 拷贝到指定目录，如图 25-16 所示。

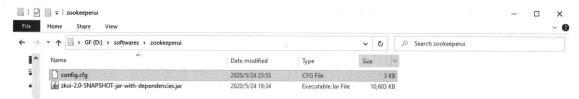

图 25-16 拷贝 jar 文件和 config.cfg

然后，打开配置文件 config.cfg，设置 zkServer，如下所示：
zkServer=192.168.31.11:2181,192.168.31.151:2181,192.168.31.199:2181

在实际文件中设置后，如图 25-17 所示。

```
#Server Port
serverPort=9090
#Comma seperated list of all the zookeeper servers
zkServer=192.168.31.11:2181,192.168.31.151:2181,192.168.31.199:2181
#Http path of the repository. Ignore if you dont intent to upload files from repository.
scmRepo=http://myserver.com/@rev1=
#Path appended to the repo url. Ignore if you dont intent to upload files from repository.
scmRepoPath=//appconfig.txt
#if set to true then userSet is used for authentication, else ldap authentication is used.
ldapAuth=false
ldapDomain=mycompany,mydomain
#ldap authentication url. Ignore if using file based authentication.
ldapUrl=ldap://<ldap_host>:<ldap_port>/dc=mycom,dc=com
#Specific roles for ldap authenticated users. Ignore if using file based authentication.
ldapRoleSet={"users": [{ "username":"domain\\user1" , "role": "ADMIN" }]}
userSet = {"users": [{ "username":"admin" , "password":"manager","role": "ADMIN" },{ "username":"appconfig" , "password":"appconfig","role": "USER" }]}
#Set to prod in production and dev in local. Setting to dev will clear history each time.
env=prod
jdbcClass=org.h2.Driver
jdbcUrl=jdbc:h2:zkui
jdbcUser=root
jdbcPwd=manager
#If you want to use mysql db to store history then comment the h2 db section.
#jdbcClass=com.mysql.jdbc.Driver
#jdbcUrl=jdbc:mysql://localhost:3306/zkui
```

图 25-17 置 zkServer

另外，ZooKeeper 的 zoo.cfg 配置文件中也需要添加配置项：4lw.commands.whitelist=stat,envi，如图 25-18 所示。

图 25-18 zoo.cfg 配置文件

配置完成后，启动 zkui 服务，请在主目录中执行命令：java -jar zkui-2.0-SNAPSHOT-jar-with-dependencies.jar，运行结果如图 25-19 所示。

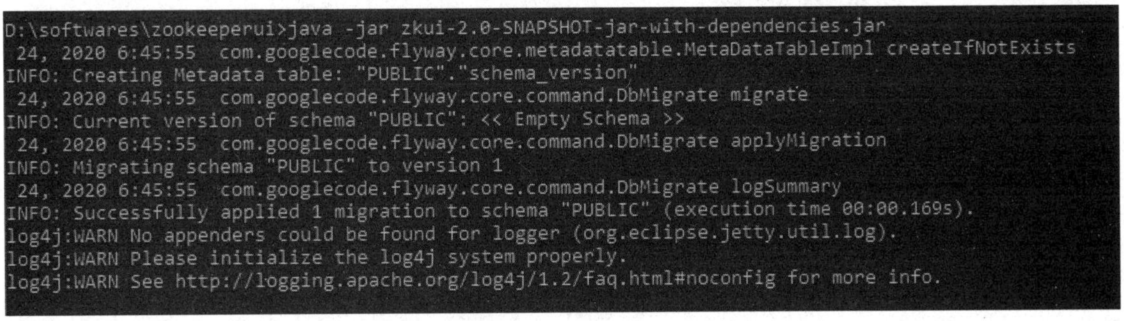

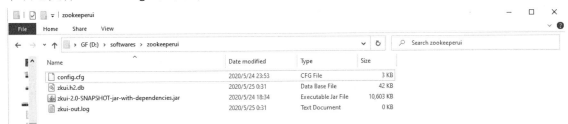

图 25-19 启动 zkui

在 Linux 机器上启动 zkui 服务，请执行命令，如下所示：

nohup java -jar zkui-2.0-SNAPSHOT-jar-with-dependencies.jar &

zkui 服务启动后，再去查看 zkui 的主目录，其中多了 2 个文件，分别是数据库文件 zkui.h2.db 和日志文件 zkui-out.log，如图 25-20 所示。

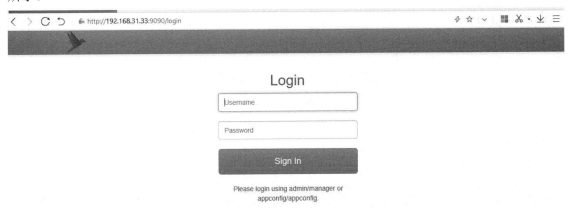

图 25-20 查看 zkui 的主目录

zkui 服务启动后，在浏览器中访问 URL：http://192.168.31.33:9090，登录界面如图 25-21 所示。

图 25-21 登录 zkui（1）

在上图所示的页面中，我们可以看到 2 个登录用户名及其密码，即：admin/manager 和 appconfig/appconfig。这里，我们使用前者。登录成功后，进入到相应的页面，如图 25-22 所示。

— 333 —

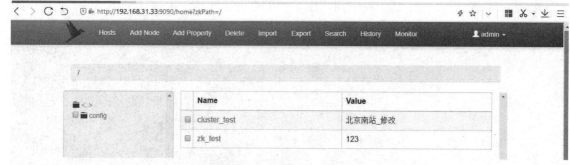

图 25-22 登录 zkui（2）

在上图的页面中，页面顶部是各种操作项，下面是结构树的列表，存储到 ZooKeeper 集群中的所有数据都在其中。单击页面顶部的链接[Monitor]，进入到监控页面，其中展现了所有节点的详细信息，如图 25-23 所示。

图 25-23 ZooKeeper 集群节点

关于上图中的页面，本书中所用的版本还有待改进优化。其中比较明显的问题是：当某个节点掉线时，整个页面都无法正常展现。关闭掉 ZooKeeper 集群的 1 个节点后出现问题，如图 25-24 所示。

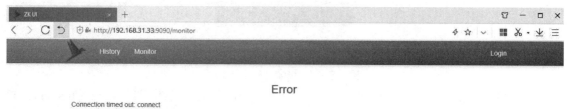

图 25-24 关闭 ZooKeeper 集群 1 个节点（1）

接着，再关闭掉 ZooKeeper 集群的 1 个节点，之后又出现问题，如图 25-25 所示。

图 25-25 关闭 ZooKeeper 集群 1 个节点（2）

25.5 配置中心

ZooKeeper 提供了一个分层命名空间，允许客户端存储任意数据，如配置数据。配置数据在特定的"引导"阶段被加载到 Spring 环境中。默认情况下，配置存储在 /config 命名空间中。根据应用程序的名称和活动配置文件创建多个 PropertySource 实例，以模拟解析属性的 Spring Cloud Config 顺序。例如，将分隔符设置为"::"，名为 gf-zookeeper 且具有 dev 配置文件的应用程序具有为其创建的属性源，如下所示：

/config/gf-zookeeper::dev

/config/gf-zookeeper

/config/app::dev

/config/app

ZooKeeper 集群作为 Spring Boot web 应用程序的配置中心，只需添加相应的依赖项，如下所示：

```xml
<dependency>
    <groupId>org.springframework.cloud</groupId>
    <artifactId>spring-cloud-starter-zookeeper-config</artifactId>
</dependency>
```

完整的 pom.xml 配置文件内容，如图 25-26 所示。

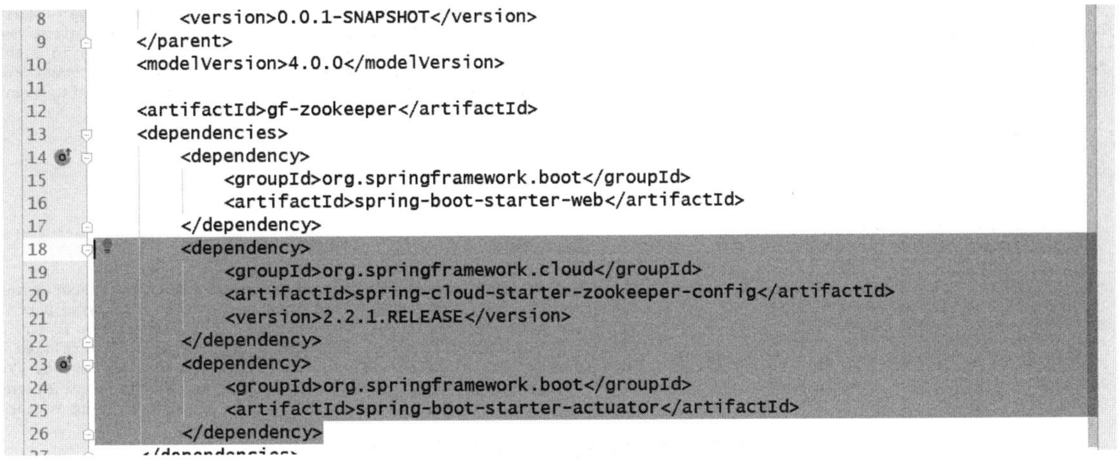

图 25-26 pom.xml 配置文件

然后，在 src\main\resources 目录中新建 bootstrap.yml 配置文件并添加配置项，如图 25-27

所示。

图 25-27 bootstrap.yml 配置文件

接着，添加应用程序主类，示例代码，如下所示：

```
package com.gf.zookeeper;
import org.springframework.boot.SpringApplication;
import org.springframework.boot.autoconfigure.SpringBootApplication;
@SpringBootApplication
public class ZooKeeperApplication {
    public static void main(String[] args) {
        SpringApplication.run(ZooKeeperApplication.class, args);
    }
}
```

经过以上配置及操作后，就可以启动应用程序了，控制台的输出信息如图 25-28 所示。

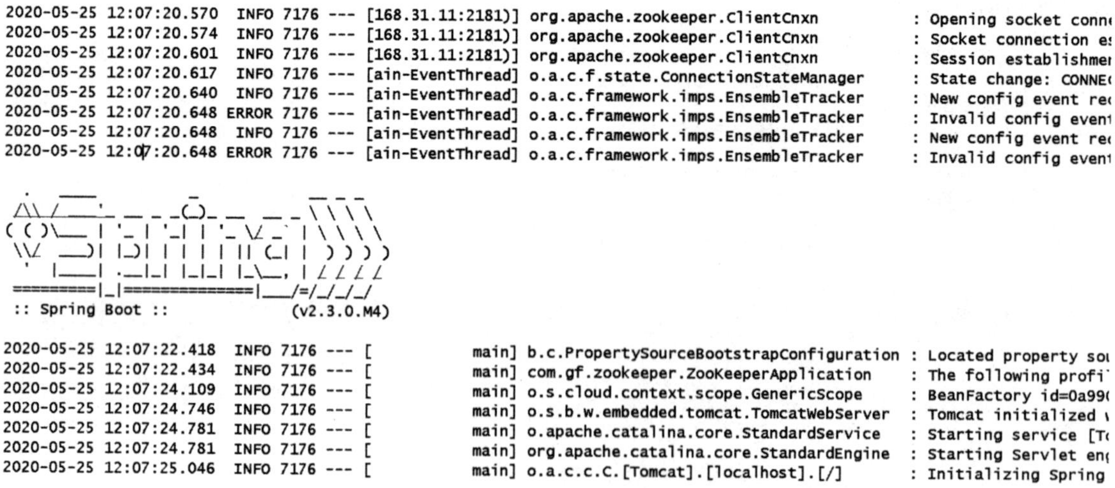

图 25-28 ZooKeeper 控制台信息

从上图中可以看出，与之前 web 应用程序的控制台输出不同的是：在打印 Spring 图标之前，会根据 boostrap.yml 中的配置进行 ZooKeeper 集群连接等相关的处理。之后，根据从

ZooKeeper 集群获取到的配置信息才真正开始启动 Spring Boot 应用程序。从 ZooKeeper 服务器获取配置项的前置路径，如图 25-29 所示。

```
2020-05-23 15:55:42.074  INFO 10972 --- [           main]
b.c.PropertySourceBootstrapConfiguration : Located property source:
[BootstrapPropertySource {name='bootstrapProperties-config/gf-zookeeper::dev'},
 BootstrapPropertySource {name='bootstrapProperties-config/gf-zookeeper'},
 BootstrapPropertySource {name='bootstrapProperties-config/app::dev'},
 BootstrapPropertySource {name='bootstrapProperties-config/app'}]
```

图 25-29 ZooKeeper 服务器路径

控制台打印的应用程序所用的端口为 8084，这是从 ZooKeeper 服务器获取的值，如图 25-30 所示。

```
eBootstrapConfiguration    : Located property source: [BootstrapPropertySource {name='bootstrapProper
.ZooKeeperApplication      : The following profiles are active: dev
t.scope.GenericScope       : BeanFactory id=0a990699-7871-3de4-9a19-324e3e9d384d
.tomcat.TomcatWebServer    : Tomcat initialized with port(s): 8084 (http)
a.core.StandardService     : Starting service [Tomcat]
ina.core.StandardEngine    : Starting Servlet engine: [Apache Tomcat/9.0.33]
t].[localhost].[/]         : Initializing Spring embedded WebApplicationContext
ContextLoader              : Root WebApplicationContext: initialization completed in 2561 ms
.ThreadPoolTaskExecutor    : Initializing ExecutorService 'applicationTaskExecutor'
s.util.InetUtils           : Cannot determine local hostname
dpointLinksResolver        : Exposing 2 endpoint(s) beneath base path '/actuator'
.tomcat.TomcatWebServer    : Tomcat started on port(s): 8084 (http) with context path ''
s.util.InetUtils           : Cannot determine local hostname
.ZooKeeperApplication      : Started ZooKeeperApplication in 33.54 seconds (JVM running for 34.799)
```

图 25-30 ZooKeeper 端口

使用 ZooKeeper 命令行客户端修改应用程序的端口，如图 25-31 所示。

```
[zk: localhost:2181(CONNECTED) 8] get /config/gf-zookeeper::dev/server/port
8084
[zk: localhost:2181(CONNECTED) 9] set /config/gf-zookeeper::dev/server/port 8085
[zk: localhost:2181(CONNECTED) 10] get /config/gf-zookeeper::dev/server/port
8085
[zk: localhost:2181(CONNECTED) 11]
```

图 25-31 修改 ZooKeeper 端口

当然，也可以使用 ZooKeeper UI 和 ZooInspector 等应用修改相应的配置信息。修改端口后，应用程序收到端口变更的信息并在控制台中打印相关信息，如图 25-32 所示。

```
o.s.cloud.commons.util.InetUtils           : Cannot determine local hostname
o.a.c.f.imps.CuratorFrameworkImpl          : Starting
org.apache.zookeeper.ZooKeeper             : Initiating client connection, connectString=192.168.31.1]
org.apache.zookeeper.ClientCnxnSocket      : jute.maxbuffer value is 4194304 Bytes
org.apache.zookeeper.ClientCnxn            : Opening socket connection to server 192.168.31.199/192.16
org.apache.zookeeper.ClientCnxn            : Socket connection established, initiating session, client
o.a.c.f.imps.CuratorFrameworkImpl          : Default schema
org.apache.zookeeper.ClientCnxn            : Session establishment complete on server 192.168.31.199/1
o.a.c.f.state.ConnectionStateManager       : State change: CONNECTED
o.a.c.framework.imps.EnsembleTracker       : New config event received: {server.1=192.168.31.11:2888:3
o.a.c.framework.imps.EnsembleTracker       : Invalid config event received: {server.1=192.168.31.11:28
o.a.c.framework.imps.EnsembleTracker       : New config event received: {server.1=192.168.31.11:2888:3
o.a.c.framework.imps.EnsembleTracker       : Invalid config event received: {server.1=192.168.31.11:28
o.s.cloud.commons.util.InetUtils           : Cannot determine local hostname
b.c.PropertySourceBootstrapConfiguration   : Located property source: [BootstrapPropertySource {name='
o.s.boot.SpringApplication                 : The following profiles are active: dev
o.s.boot.SpringApplication                 : Started application in 3.523 seconds (JVM running for 420
o.a.c.f.imps.CuratorFrameworkImpl          : backgroundOperationsLoop exiting
org.apache.zookeeper.ZooKeeper             : Session: 0x300000156510000 closed
org.apache.zookeeper.ClientCnxn            : EventThread shut down for session: 0x300000156510000
o.s.c.e.event.RefreshEventListener         : Refresh keys changed: [server.port]
```

图 25-32 控制台信息

虽然 Spring Boot 应用程序已经获取到变更的端口，但当前端口仍然是原来的。因为，端口这样的参数只在容器初始化时获取一次。

25.6 本章小结

本章讲解了 ZooKeeper 的下载、安装和集群搭建，还讲解了 ZooKeeper UI 客户端的使用，以及如何把 ZooKeeper 作为配置中心。ZooKeeper 还作为了 Kafka 的依赖应用，来存储 Kafka 的元数据和集群维护。另外，可以作为配置中心的软件还有很多（如 Consul 和 Nacos 等），有兴趣的读者也可以将 ZooKeeper 与它们做一下对比。

附录

附录A 参考

A.1 Spring Boot 官方

Spring Boot 官方网站：
https://spring.io/projects/spring-boot
Spring Boot 官方文档：
https://docs.spring.io/spring-boot/docs/2.3.1.RELEASE/reference/html

A.2 开发环境中的软件

JDK
1) 各个版本的下载地址：
   https://www.oracle.com/java/technologies/javase-downloads.html
2) JDK 8 下载地址：
   https://www.oracle.com/java/technologies/javase/javase-jdk8-downloads.html

OpenJDK
1) 各个版本的下载地址：
   https://jdk.java.net/archive
2) OpenJDK 8 下载地址：
   https://jdk.java.net/java-se-ri/8-MR3
3) OpenJDK 8 下载地址：
   https://download.java.net/openjdk/jdk8u41/ri/openjdk-8u41-b04-windows-i586-14_jan_2020.zip

IntelliJ IDEA
1) 下载地址：
   https://www.jetbrains.com/idea/download/#section=windows
2) 下载地址：
   https://download.jetbrains.8686c.com/idea/ideaIC-2020.1.3.win.zip

Maven
1) 下载地址：
   http://maven.apache.org/download.cgi
2) 下载地址：

https://downloads.apache.org/maven/maven-3/3.6.3/binaries/apache-maven-3.6.3-bin.zip

MariaDB

1) 下载地址：

   https://downloads.mariadb.org/mariadb/+releases

2) 下载地址：

   https://mirrors.tuna.tsinghua.edu.cn/mariadb/mariadb-10.5.4/winx64-packages/mariadb-10.5.4-winx64.zip

CentOS

1) 下载地址：

   https://www.centos.org/centos-linux

2) CentOS 8（x86_64）的下载地址：

   http://isoredirect.centos.org/centos/8/isos/x86_64

3) 7.7GB 文件，下载地址：

   https://mirrors.tuna.tsinghua.edu.cn/centos/8.2.2004/isos/x86_64/CentOS-8.2.2004-x86_64-dvd1.iso

4) 1.6GB 文件，下载地址：

   https://mirrors.tuna.tsinghua.edu.cn/centos/8.2.2004/isos/x86_64/CentOS-8.2.2004-x86_64-minimal.iso

附录B 推荐

B.1 Spring

Spring 是众多项目的一个集合，这些项目共同构成了 Spring 生态。官方网址：
https://spring.io
在 Spring 社区，可以获得更多更及时的支持，网址：
https://spring.io/community
查看 Spring 项目的源代码，可参考 GitHub 上相关内容，网址：
https://github.com/spring-projects

B.2 其他

Spring Boot 参考指南中文文档：
https://github.com/jijicai/Spring/tree/master/spring-boot/reference
Spring Cloud 中文网：
https://www.springcloud.cc
反应式编程 Reactor3 参考指南中文文档：
https://github.com/jijicai/ProjectReactor/tree/master/book/Reactor3